
Engineering Principles for Electrical Technicians

Engineering Principles for Electrical Technicians

BY

K. M. SMITH, C.ENG., M.I.E.E.

AND

P. HOLROYD, C.ENG., M.I.MECH.E., A.M.I.STRUCT.E.

PERGAMON PRESS

OXFORD · LONDON · EDINBURGH · NEW YORK
TORONTO · SYDNEY · PARIS · BRAUNSCHWEIG

Pergamon Press Ltd., Headington Hill Hall, Oxford
4 & 5 Fitzroy Square, London W.1
Pergamon Press (Scotland) Ltd., 2 & 3 Teviot Place, Edinburgh 1
Pergamon Press Inc., Maxwell House, Fairview Park,
Elmsford, New York 10523
Pergamon of Canada Ltd., 207 Queen's Quay West, Toronto 1
Pergamon Press (Aust.) Pty. Ltd., 19a Boundary Street,
Rushcutters Bay, N.S.W. 2011
Pergamon Press S.A.R.L., 24 rue des Écoles, Paris 5e
Vieweg & Sohn GmbH, Burgplatz 1, Braunschweig

First edition 1969

Library of Congress Catalog Card No. 68-26947

Printed in Great Britain by Thomas Nelson (Printers) Ltd., London and Edinburgh

08 012985 4 (flexicover)
08 012986 2 (hard cover)

Contents

Acknowledgements

THE authors wish to acknowledge permission given by Edward Arnold (Publishers) Ltd. for the Abridged Callendar Steam Tables printed at the end of Chapter 8. They also wish to express their gratitude to the following bodies who have given permission for their examination questions to be included:

City and Guilds of London Institute. (C.G.L.I.)
Northern Counties Technical Examinations Council. (N.C.T.E.C.)
Union of Lancashire and Cheshire Institutes. (U.L.C.I.)
Union of Educational Institutions. (U.E.I.)
East Midland Educational Union. (E.M.E.U.)

They also wish to acknowledge the assistance given by colleagues at the Huddersfield College of Technology and Ramsden Technical College, Huddersfield.

Introduction

THE rapid development of courses for technicians has shown the need for a different type of textbook suited to their special requirements. Ideas and principles must be emphasised by demonstrations and laboratory work so that the technician can develop his skills and techniques. This book has been written for the technician to use in lectures and in the laboratory. By using typical results taken from simple laboratory experiments and with worked examples the authors have provided the necessary introduction to fundamental engineering principles.

This book was written specifically for Part I of the Electrical Technicians Course (C.G.L.I. No. 57) and the Telecommunications Technicians Course (C.G.L.I. No. 49). It will also be useful for the Mechanical Technicians Course (C.G.L.I. No. 293), and will cover Part I of the new Electrical Installation Technicians Course.

N. HILLER

Cambridge *Editor*

Symbols, Units and Abbreviations

Quantity	*Symbol*	*Units*	*Abbreviations*
Length	l	inch, foot, metre	in., ft, m
Mass	m	pound, kilogramme	lb, kg
Time	t	second, minute, hour	s or sec, min, h
Force	F, P	pound-force, newton	lbf, N
		kilogramme-force	kgf
Torque	T	pound-force-feet	lbf-ft
		newton-metre	N-m
Power	P	foot-pound-force per second	ft-lbf/s
		newton-metre per second (or Watt)	Nm/s (or W)
		horsepower	hp
Work, energy	W	foot-pound-force	ft-lbf
Displacement, linear	s	foot	ft
Displacement, angular	θ	radian	rad
Velocity, linear	v	foot per second	ft/s
		metre per second	m/s
Velocity, angular	ω	radian per second	rad/s
	N	revolutions per minute	rev/min, or r.p.m.
Acceleration, linear	f	foot per second per second	ft/s^2
Acceleration, angular	α	radian per second per second	rad/s^2
Acceleration, gravitational	g	foot per second per second	ft/s^2
Coefficient of friction	μ	—	—
Stress, tensile and compressive	f	pound-force per square inch	lbf/in^2
Stress, shear	q	pound-force per square inch	lbf/in^2
Strain, tensile and compressive	e	—	—
Strain, shear	ϕ	—	—
Modulus of Elasticity	E	pound-force per square inch	lbf/in^2
Coefficient of linear expansion	α	—	—
Heat, quantity of	Q	joule	J
		British thermal unit	Btu

Quantity	*Symbol*	*Units*	*Abbreviations*
Heat, quantity of		Centigrade heat unit gram-calorie kilocalorie	Chu cal kcal
Specific heat	c	—	—
Heat, total	H_s	British thermal unit per pound	Btu/lb
Heat, latent	L	British thermal unit per pound	Btu/lb
Heat, sensible	h	British thermal unit per pound	Btu/lb
Capacitance	C	farad, microfarad	F, μF
Charge (electric)	Q	coulomb, microcoulomb	C, μC
Current	I	ampere, milliampere	A, mA
Electric flux	ψ	coulomb	C
Electric flux density	D	coulomb per square metre	C/m^2
Electric force or potential gradient	E	volt per metre	V/m
Electromotive force	E	volt, millivolt, kilovolt	V, mV, kV
Electrical energy	W	joule (or volt-coulomb) watt-second, watt-hour, kilowatt-hour	J (or V-C) Ws, Wh, kWh
Frequency	f	hertz	Hz
Impedance	Z	ohm	Ω
Inductance (self)	L	henry, millihenry	H, mH
Inductance (mutual)	M	henry, millihenry	H, mH
Magnetic flux	Φ	weber, milliweber	Wb, mWb
Magnetic flux density	B	weber per square metre	Wb/m^2
Magnetising force	H	ampere-turn per metre	AT/m
Magnetomotive force	F	ampere-turn	AT
Permeability of free space	μ_0	henry per metre	H/m
Permeability (relative)	μ_r	—	—
Permittivity of free space	ε_0	farad per metre	F/m
Permittivity (relative)	ε_r	—	—
Potential difference	V	volt, millivolt, kilovolt	V, mV, kV
Power (electric)	P	joule per second (or watt) kilowatt, megawatt	J/s (or W) kW, MW
Power (apparent)	—	volt-ampere, kilovolt-ampere	VA, kVA
Power factor	—	—	—
Reactance (capacitive)	X_C	ohm	Ω
Reactance (inductive)	X_L	ohm	Ω
Reluctance	S	ampere-turn per weber	AT/Wb
Resistance	R	ohm, microhm, megohm	Ω, μΩ, MΩ
Resistivity	ρ	ohm-inch, ohm-metre, ohm-centimetre	Ω-in, Ω-m, Ω-cm

CHAPTER 1

Forces and Equilibrium

1.1. Types of force

When we think of forces we usually consider the action of lifting a weight, bending a piece of wire or setting a vehicle in motion. Whenever we move, or move other objects around us, then forces are brought into play. In fact, approximately three centuries ago Sir Isaac Newton gave us a definition of a force which includes all the forces that we can imagine. "Force changes or tends to change the state of a body's rest or uniform motion in a straight line." One of the most important forces we meet is the one which causes objects to fall downwards, the *Force of Gravity*.

It has been found that between all bodies there is a force of attraction defined as the gravitational force. The larger the bodies the greater the attraction between them, but as the distance between the bodies increases the attraction decreases. Normally bodies on the surface of the earth do not show this attraction between themselves because the force is very small, but if we consider two masses, one of which is the earth, then the force of attraction between these masses can be considerable. The earth being the larger body will not move perceptibly and therefore the smaller body, if free to do so, will move towards the centre of the earth. This force of gravity acting on the body is called its *Weight*.

It should be understood that not all the forces we meet are gravitational forces. Friction forces are introduced when two surfaces slide over each other; inertia forces occur when a mass has to be accelerated; tensile, compressive and shear forces are considered in dealing with the strength of materials (Fig. 1.1). The idea of the gravitational force can be used to give us a unit of measurement of a force.

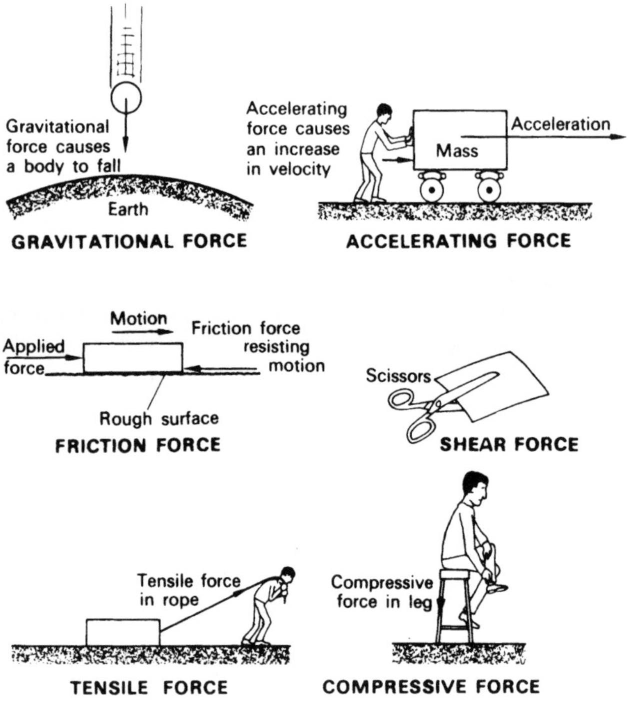

FIG. 1.1. Different applications of forces. The gravitational force causes all bodies to fall towards the centre of the earth. An accelerating force causes a body to change its speed or velocity. When two surfaces slide over each other the resistance to motion is called the friction force. A pair of scissors cuts paper by a shearing force. A tensile force is introduced into a rope when pulling a box along the ground. The stool legs are subject to a compressive force when anyone sits on the stool.

1.2. Unit of force

If we wish to compare forces, then we must have a standard of reference. In England a cylinder of platinum of a certain size is kept at the Board of Trade in London representing a mass of one pound. The gravitational force exerted on this mass at mean sea level is called the pound-force. Other weights are then made to

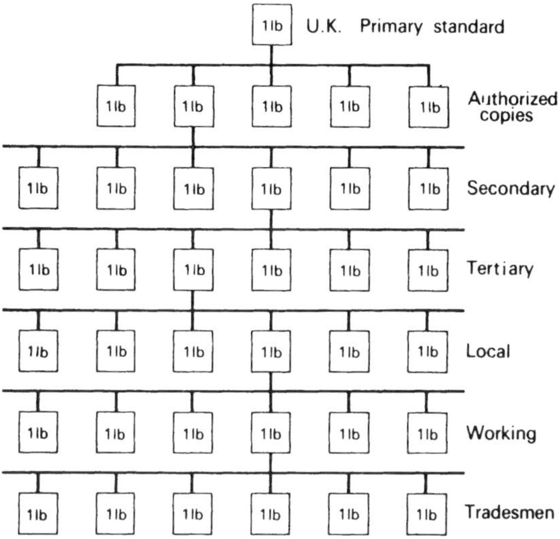

FIG. 1.2. The U.K. Standard One Pound Mass. A cylinder of platinum kept at the Board of Trade represents the United Kingdom Primary Standard one pound mass. Five copies of this are distributed to various other offices in London. Secondary standards are made to compare with the authorised copies and tertiary standards are compared to the secondary standards. In each town and city a set of local standards are kept and periodically checked with the tertiary standards. The Weights and Measures Officer uses a set of working standards for checking tradesmen's weights which he maintains by checking with the local standards. This flow enables a continual check to be made on the accuracy of all weights used in the country.

compare with the standard and distributed to Weights and Measures offices throughout the country from which the weights you see in shops or factories are checked. See Fig. 1.2.

Another standard of mass used on the Continent is the kilogramme, represented by a cylinder of platinum–iridium kept in

France. The gravitational force exerted on this mass is called the kilogramme-force.

The relationship between the pound-force and the kilogramme-force is

1 pound-force (lbf) = 0·4536 kilogrammc-force (kgf)

or 1 kilogramme-force (kgf) = 2·205 pound-force (lbf).

We can use multiples of these forces or weights.

The pound-force is related to the ton-force (tonf) such that

$$2240 \text{ lbf} = 1 \text{ tonf}.$$

Similarly $$1 \text{ gramme force (gf)} = \frac{1}{1000} \text{ kgf}.$$

1.3. Effects of forces

If we consider a mass of 5 lb hanging from a rope (Fig. 1.3) we should say that the force in the rope is 5 lbf. If we now support the rope over a pulley and hang a 5 lb mass on each end (Fig. 1.4) would the force on the rope now be 5 lbf? If you are in doubt, cut the string and insert a spring balance in the system (Fig. 1.5) and take a reading from the balance. This should be 5 lbf. Now ask yourself why this is only 5 lbf and not 10 lbf. The answer was given again by Sir Isaac Newton, who stated: "To every action there is an equal and opposite reaction."

In Fig. 1.3 the support is providing an upward force of 5 lbf to balance the weight hanging downwards, therefore in Fig. 1.4. to keep the system balanced, a mass of 5 lb must be attached at each end of the string. If one mass were taken away then the other would fall to the ground. This situation occurs when anything rests on the ground. If you weigh 145 lbf and you are standing on the ground, then the earth is pressing upwards with an equal force of 145 lbf to keep you at rest. When you stand on the bathroom scales, the scales are sandwiched between your feet and the floor so that pressure is exerted on the top and bottom of the scales to enable a reading of your weight to be taken. If you attempted to jump off a high diving board with the scales strapped to your feet then no reading would be registered until you hit the water, with disastrous results to the scales.

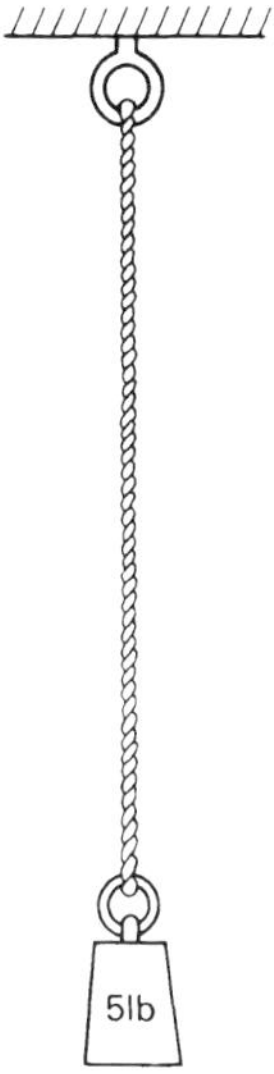

FIG. 1.3

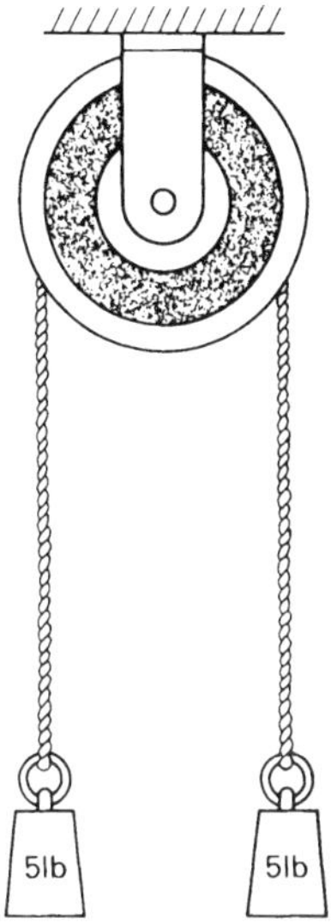

FIG. 1.4

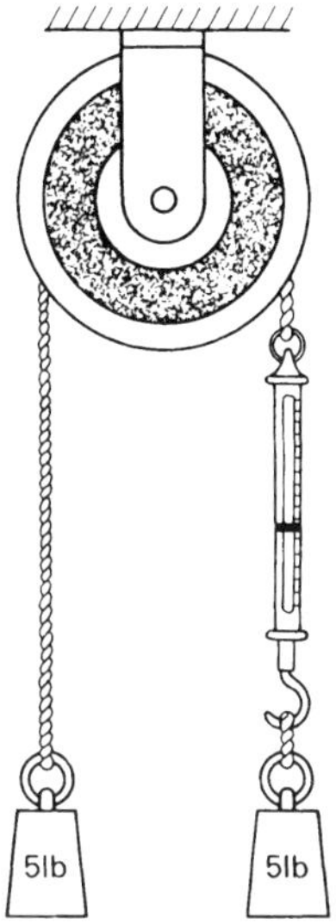

FIG. 1.5

1.4. Equilibrium of forces

In the previous paragraph we considered someone standing on the ground and the ground pressing upwards with a force equal to the weight of the person. Under these conditions no movement occurs. If the person steps off the high diving board, then there is nothing to provide an upward force to stop the person falling and motion occurs. In the first case we say the forces acting are in equilibrium, and in the second case they are not in equilibrium.

Now imagine two tug-of-war teams straining on a rope. Each team pulls away from the centre of the rope and if both teams are as strong as each other then no movement occurs. If, however, one team is stronger than the other then the rope moves towards the stronger team and the system becomes unbalanced and ceases to be in equilibrium.

When you ride a cycle on a level road, the action of pedalling is to overcome the resistance due to friction of the moving parts and any air resistance. See Fig. 1.6.

If you are travelling at a constant speed then the force pushing you forward due to pedalling is equal to the resistance tending to stop the motion and the system is in equilibrium. If you stop

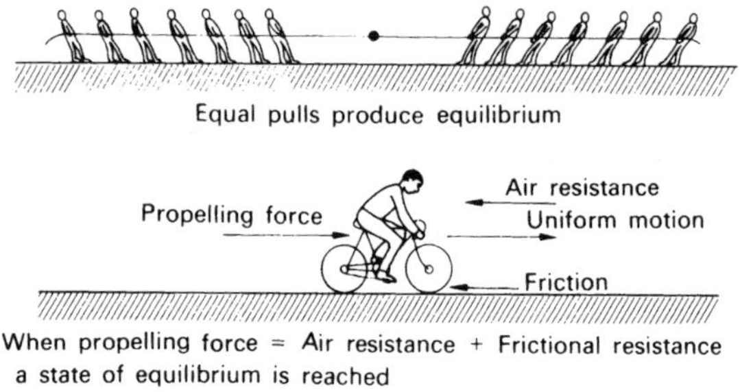

FIG. 1.6. Equilibrium conditions. A state of equilibrium can occur if the forces are stationary such as a tug-of-war when both teams are pulling with equal forces. Alternatively if a cyclist is moving along a road at a constant speed then he is in equilibrium when the propelling force is equal to the friction force plus the force due to air resistance.

pedalling the cycle retards, or if you pedal harder the cycle goes faster. In either case the system is unbalanced and ceases to be in equilibrium. "Equilibrium can occur if a body remains stationary under the action of a number of forces or alternatively if a body is moving at a uniform speed."

1.5. Scalar and vector quantities

Quantities involving magnitude only, e.g. money, time, temperature, resistance, speed, etc., are called SCALAR quantities and can be expressed by a number and the appropriate unit. If, however, direction is involved in addition to magnitude, e.g.

velocity, acceleration, force, etc., then they are called VECTOR quantities.

Scalar quantities can be added together or subtracted arithmetically but vector quantities require special treatment.

1.6. Representation of a force

When a force acts on a body, the effect the force has on this body depends on (a) the magnitude, (b) the direction, (c) the point of application and (d) the sense of the force. Figure 1.7 shows how these quantities may be represented on a simple diagram called a "space diagram".

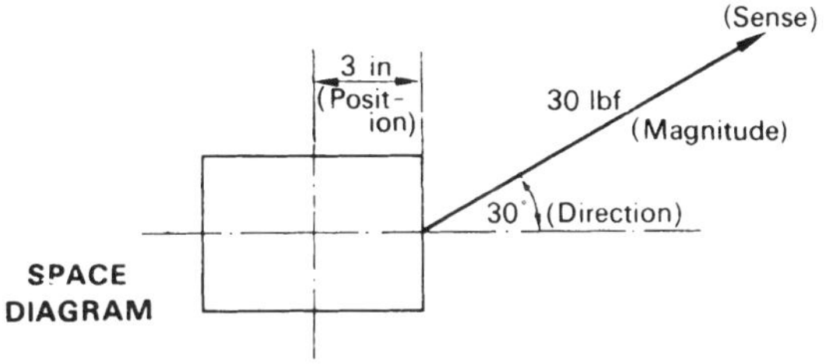

FIG. 1.7

The vector representation of the force will be made by drawing a straight line to a suitable scale of force, say 1 inch to represent 20 lbf, in the same direction as the force is applied to the body and placing an arrow head on the line to indicate the sense (Fig. 1.8).

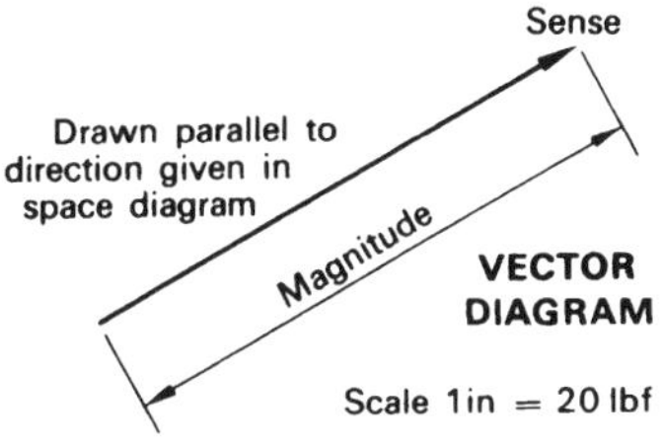

FIG. 1.8

1.7. Addition of forces

Figure 1.9 shows forces of 3 lbf and 4 lbf acting on a mass *M*. Notice that these forces are not acting in the same direction. To find the effect these two forces have on the mass requires a vector addition of the two forces to be carried out. A vector $\overline{ab}$ is drawn in the same direction as the 3 lbf to a suitable scale, say 1 inch to 2 lbf, and an arrow is placed on the vector to indicate the sense. On to the end of this vector another line $\overline{bc}$ is drawn to represent

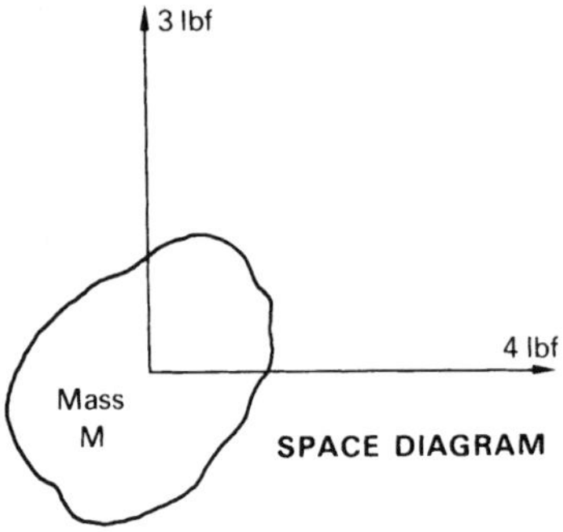

FIG. 1.9

the 4 lbf, 2 inches long in the same direction as the 4 lbf, and the arrow is placed on the vector. Now vectorially:

$$\overline{ab}+\overline{bc}=\overline{ac}$$

The lines over the letters indicate vectorial quantities. The length of the line $\overline{ac}$ to the same scale as $\overline{ab}$ and $\overline{bc}$ gives the result of the addition. If the diagram, (Fig. 1.10) is drawn carefully the answer is found to be 5 lbf. Therefore a force of 5 lbf acting in a direction 37° to the 4 lbf will have the same effect on the mass *M* as the 4 lbf and 3 lbf acting at right angles. This single force is called the RESULTANT of the two forces.

"The RESULTANT of a set of forces is that single force which will replace the system of forces and have the same effect."

For two forces to be in equilibrium they must be (a) equal in magnitude, (b) acting in the same direction, (c) of opposite sense and (d) passing through the same point.

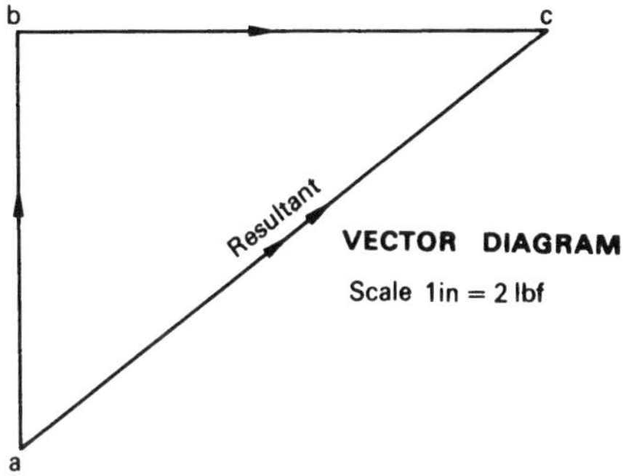

FIG. 1.10

If now we have a number of forces acting through a point and we wish to put the system of forces in equilibrium by the use of a single force, all we need to do is to find the resultant of the system and balance the resultant. This single force which balances a number of forces is called the EQUILIBRANT.

The equilibrant is exactly the same as the resultant except for the sense.

EXAMPLES

Find the resultant of forces *A* and *B* (Fig. 1.11), stating clearly the magnitude and direction relative to force *A*.

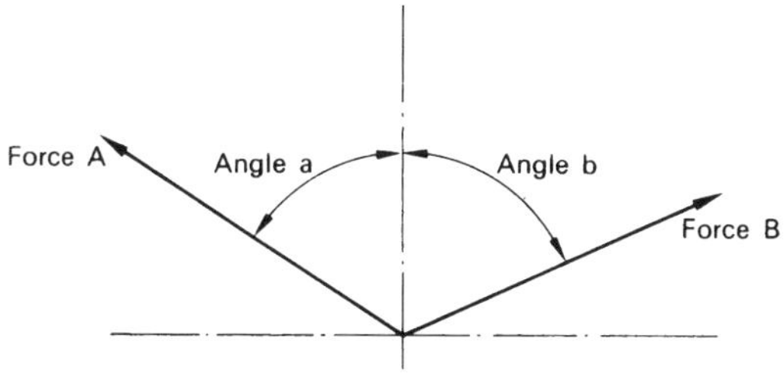

FIG. 1.11

	Force A	angle a	Force B	angle b
(a)	6 lbf	30°	8 lbf	20°
(b)	35 lbf	45°	45 lbf	45°
(c)	10 kgf	25°	15 kgf	50°
(d)	20 kgf	40°	10 kgf	35°
(e)	3 tonf	50°	4 tonf	40°
(f)	6·7 lbf	23°	4·6 lbf	37°

[(a) 12·7 lbf, 29°; (b) 57 lbf, 52°; (c) 20 kgf, 46°; (d) 24·75 kgf, 24°; (e) 5 tonf, 53°; (f) 9·9 lbf, 23·75°.]

Experimental verification of equilibrant and resultant

Equipment required. Wall board, thin string, weights, paper and protractor.

Join three lengths of string at a point. Let them hang over three pulleys attached to a vertical wall board to which a piece of paper has been attached. Make sure the pulleys run freely. From the ends of the strings hang weights of, say, $\frac{1}{2}$ lbf, $\frac{3}{4}$ lbf and 1 lbf and

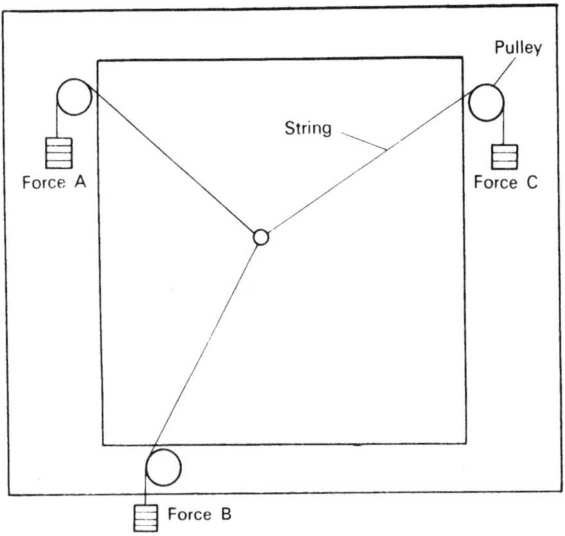

Fig. 1.12

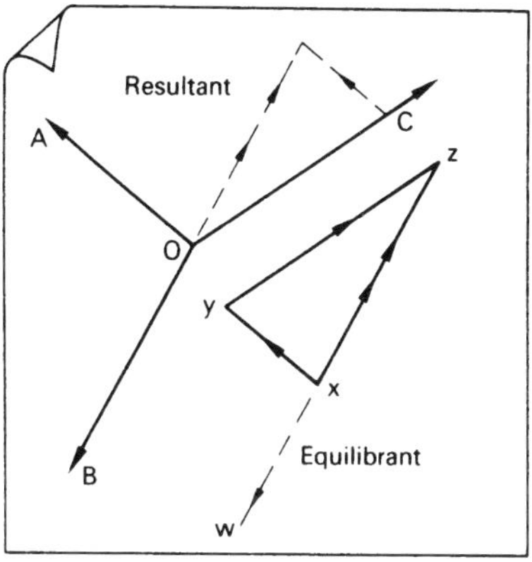

FIG. 1.13

let the system come freely to rest (Fig. 1.12). Carefully mark on the paper the directions of the three strings and the magnitude of the weight attached to each string, then take the paper from the board.

Draw the directions clearly on the paper *OA*, *OB*, and *OC* as shown in Fig. 1.13. Draw $\overline{xy}$ to represent force *A*, to scale, parallel to *OA*. Draw $\overline{yz}$, to scale, parallel to *OC* to represent force *C* and join $\overline{xz}$.

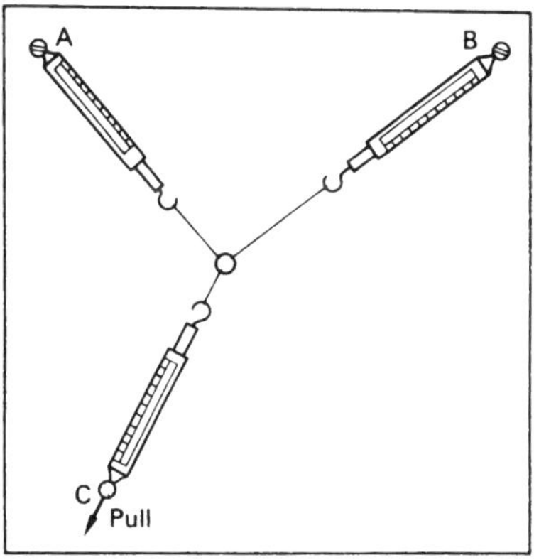

FIG. 1.14

The length of $\overline{xz}$ should be equal to the force B and the angle yxw should be equal to the angle AOB within the limits of experimental error. The experiment can be repeated using different weights at A, B and C and different positions of the pulleys.

An alternative method of finding the resultant is to use three spring balances attached to a board as shown in Fig. 1.14. Great care must be taken that the balances are free to take up their correct directions when they are under tension. Screws are used at A and B to allow the balances at A and B to be supported. Short lengths of string are attached to a small ring and to the hook of each balance. A pull is applied to the balance C in any direction and readings are taken from the balances. The directions of the strings are marked on a piece of paper attached to the board and the same procedure carried out as before to determine the resultant and equilibrant.

1.8. Vector representation of forces in equilibrium

From the results of the previous experiment it is possible to give a condition of equilibrium for three forces acting at a point. If the

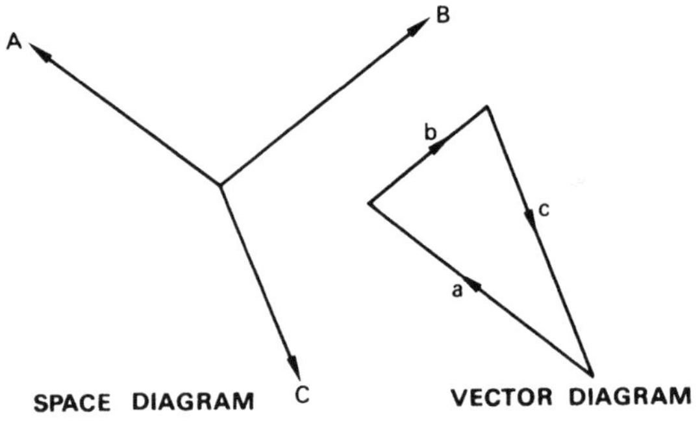

FIG. 1.15. If three forces A, B and C are in equilibrium the vector diagram forms a closed triangle.

vectors representing forces A, B and C are drawn, the resulting figure is a triangle. We can say, therefore: "If three forces acting at a point in the same plane are in equilibrium, the vector diagram forms a closed triangle."

This can be extended to any number of forces in the same plane acting at a point. If the forces are in equilibrium the vector diagram forms a closed polygon.

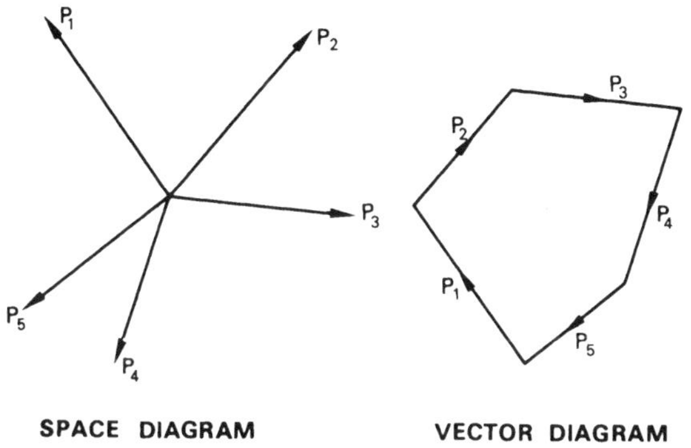

FIG. 1.16. If any number of forces P_1, P_2, P_3, P_4, and P_5 are in equilibrium the vector diagram forms a closed polygon.

When a graphical construction is used to find the solution to a problem, and there are many instances in engineering practice where this is done, the accuracy of the answer will depend on the quality of the drawing. Make sure pencils are sharp and drawing equipment is in good condition.

EXAMPLES

Check the value of the equilibrant E and the angle α (Fig. 1.17) from the experimental results obtained (Table 1.1) and state which experimental results you consider to be incorrect.

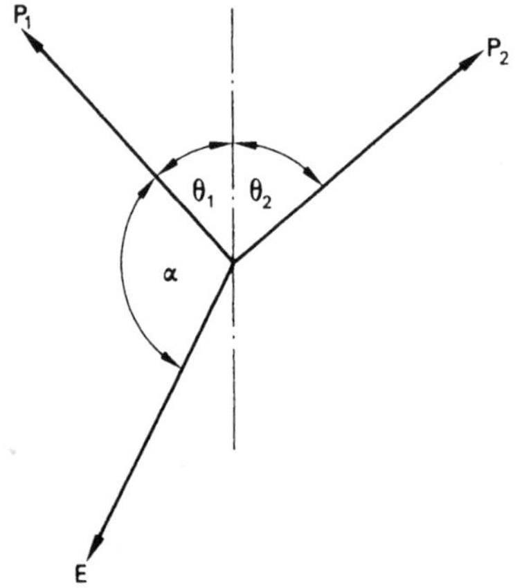

Fig. 1.17

Table 1.1

Exp. no.	P_1	θ_1	P_2	θ_2	E	α
1	4 lbf	45°	3 lbf	30°	5·6 lbf	148·5°
2	4 kgf	20°	2 kgf	90°	3·8 kgf	150·5°
3	3 lbf	0°	5 lbf	120°	4·75 lbf	106°
4	3 kgf	90°	6 kgf	30°	5·2 kgf	70°
5	5 lbf	23°	3·5 lbf	55°	6·66 lbf	149°
6	4 kgf	35°	3·7 kgf	9°	7·15 kgf	159°
7	5·4 lbf	57½°	2·3 lbf	96°	4·5 lbf	153°
8	3·75 kgf	14½°	4·58 kgf	29°	7·71 kgf	156°

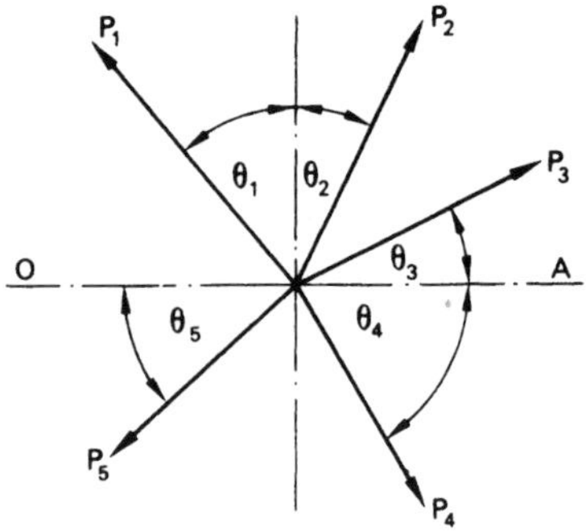

Fig. 1.18

Find the resultant of each set of forces listed below (Table 1.2), stating its magnitude and direction clockwise to the axis *OA* (Fig. 1.18).

TABLE 1.2

Exp. no.	P_1	θ_1	P_2	θ_2	P_3	θ_3	P_4	θ_4	P_5	θ_5
1	2 lbf	20°	3 lbf	10°	1 lbf	30°	2 lbf	20°	3 lbf	45°
2	4 gf	30°	2 gf	25°	3 gf	35°	2 gf	45°	1 gf	20°
3	15 lbf	45°	20 lbf	5°	10 lbf	10°	12 lbf	60°	30 lbf	60°
4	4 kgf	50°	0	0	3 kgf	40°	2·5 kgf	45°	4 kgf	50°

1.9. Resolution of forces

We have seen that a resultant force can replace two forces acting at a point. The opposite is true and a single force can be replaced

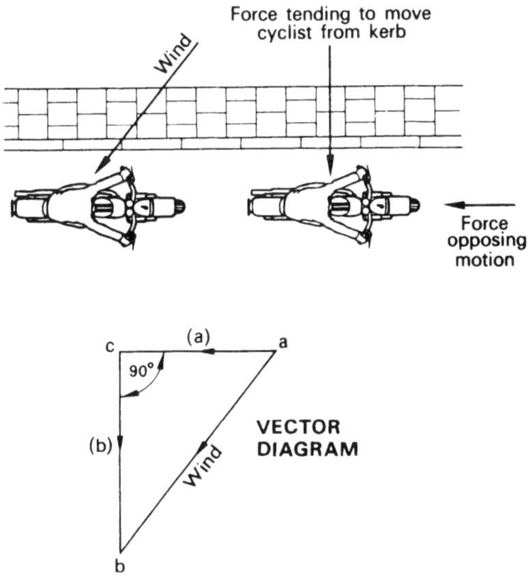

FIG. 1.19

by two other forces in specified directions, which are generally, for convenience, chosen to be at right angles to each other. These are called the "components of the force".

A motor cyclist travelling against wind force, as indicated in Fig. 1.19, has two components to contend with,

(a) the force tending to prevent motion along the road, and
(b) the force tending to move him from the kerb.

The magnitude of these two forces can be obtained from a vector diagram (Fig. 1.19).

$\overline{ab}$ represents to scale the wind force in the given direction, $\overline{ac}$ and $\overline{cb}$ are drawn at right angles to give the components (a) and (b).

EXAMPLE. A fitter holds a chisel at an angle of 55° to the horizontal and applies a blow of 500 lbf to the head of the chisel by means of a hammer. Find (a) the force digging into the metal and (b) the force shearing the metal.

The space diagram (Fig. 1.20) shows the arrangement of the system of forces. Vector $\overline{ab}$ is drawn to scale parallel to the chisel to

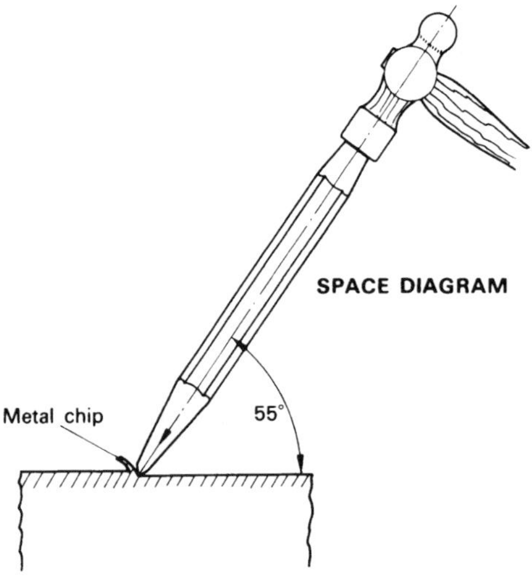

FIG. 1.20

represent the force of 500 lbf. Vectors $\overline{bc}$ and $\overline{ac}$ are then drawn to represent the vertical and horizontal components respectively of the 500 lbf (Fig. 1.21). By measurement of $\overline{cb}$, the force digging into the metal is found to be 410 lbf, and measurement of $\overline{ac}$ indicates that the shearing force is 287 lbf.

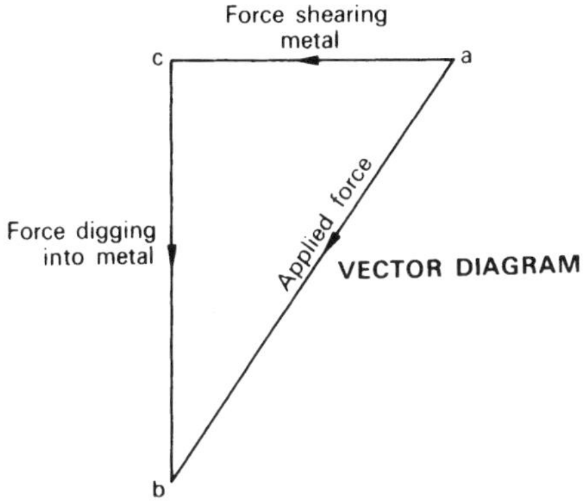

FIG. 1.21

EXAMPLE. A load of 800 lbf is hanging from a crane. It is pulled away from the vertical by a horizontal rope until the angle of the

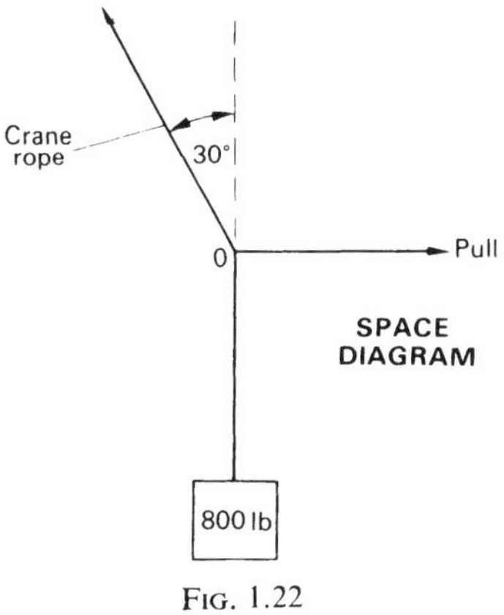

FIG. 1.22

crane rope is 30° to the vertical. Find the pull in the horizontal rope and in the crane hoisting rope.

Figure 1.22 shows the forces acting in the system. At point *O* there are three forces in equilibrium. If, therefore, a vector diagram is drawn it will form a triangle.

Vector $\overline{ab}$ (Fig. 1.23) is drawn to scale to represent the load of

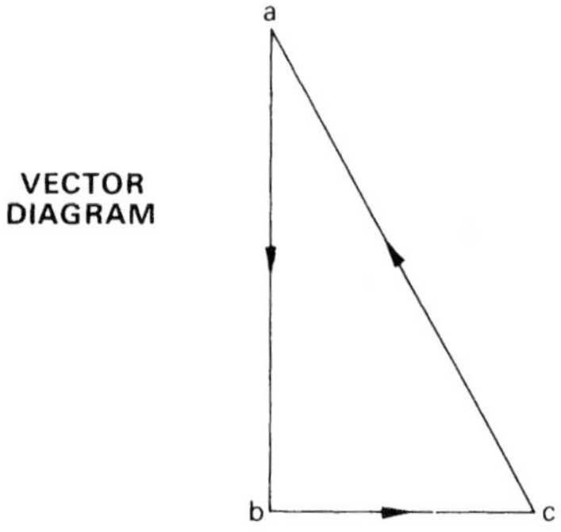

FIG. 1.23

800 lbf, the direction being vertical. From "*a*", vector $\overline{ac}$ is drawn parallel to the hoisting rope *OP*, and from "*b*", vector $\overline{bc}$ parallel to the horizontal rope *OP*. The vector diagram forms a closed triangle.

By measurement of $\overline{bc}$ the horizontal force is found to be 462 lbf and the tension in the hoisting rope from the length of $\overline{ac}$ is 924 lbf.

1.10. Turning effects of forces

The effect of a number of forces acting on a body depends not only on the magnitude of the forces and their directions but also on the points at which the forces are applied to the body.

This action of the force turning about an axis or fulcrum is called the turning moment or moment of the force, or torque.

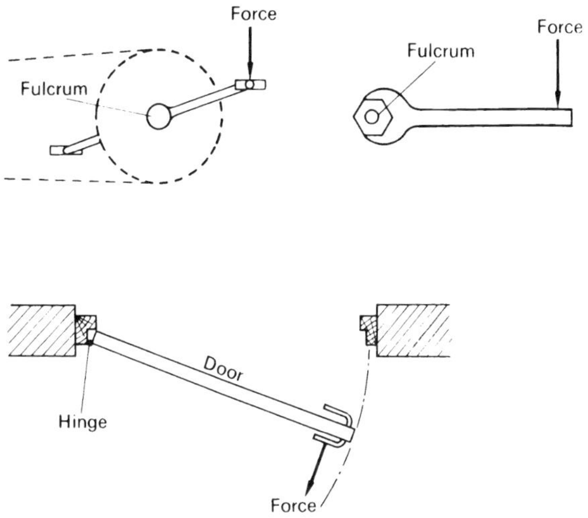

FIG. 1.24. Turning effects of forces. A cyclist presses on the pedal to rotate the crank about the fulcrum. A nut and bolt is tightened by exerting a force at the end of the spanner. To open a door the handle is placed at the opposite end to the hinge so that a force applied to the door handle will cause the door to rotate about the hinge.

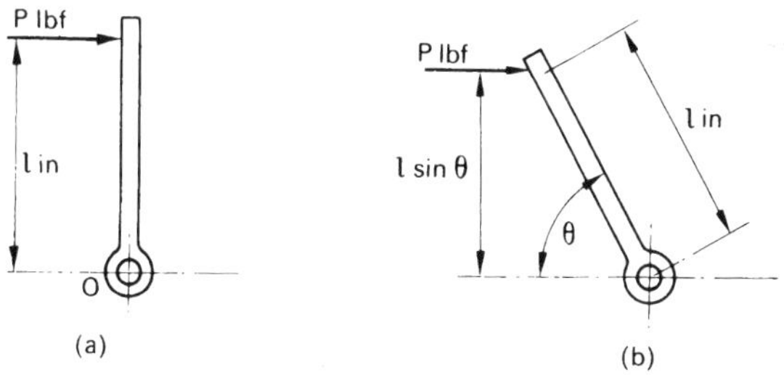

FIG. 1.25. Turning moment = Force × perpendicular distance to the line of action of the force. (*a*) Turning moment $P \times = l$ lbf-in. (*b*) Turning moment $= P \times l \sin \theta$ lbf-in.

"The magnitude of the turning moment of a force about a point is given by the product of the force and the perpendicular distance of the point from the line of action of the force."

This definition is important and particular attention should be paid to the idea of the distance being perpendicular to the line of action of the force (Fig. 1.25).

1.11. Unit of moment

When the moment of a force is calculated the value is found by multiplying a force by a distance. If the force is measured in pound-force and the distance is measured in feet, then the moment is expressed as pound-force-feet (lbf-ft).

If the force is measured in ton-force and the distance is measured in inches, then the moment is ton-force-inches (tonf-in).

If the force is measured in kilogramme-force and the distance is measured in metres, then the moment is kilogramme-force-metres (kgf-m).

When the force tends to cause rotation anticlockwise about a fulcrum or point of rotation the moment is said to be anticlockwise, and if the force tends to rotate clockwise about the point, then it is a clockwise moment (Fig. 1.26).

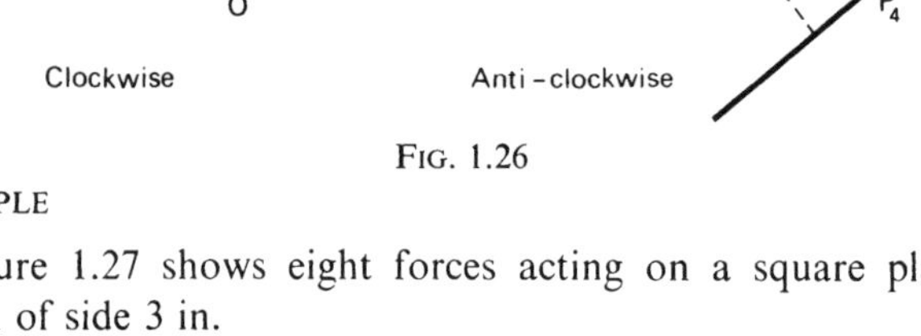

FIG. 1.26

EXAMPLE

Figure 1.27 shows eight forces acting on a square plate of length of side 3 in.

Make a list of

(a) the clockwise moments,
(b) the anticlockwise moments,

(i) assuming corner *A* as the fulcrum,
(ii) ,, ,, *B* ,, ,,
(iii) ,, ,, *C* ,, ,,
(iv) ,, ,, *D* ,, ,,

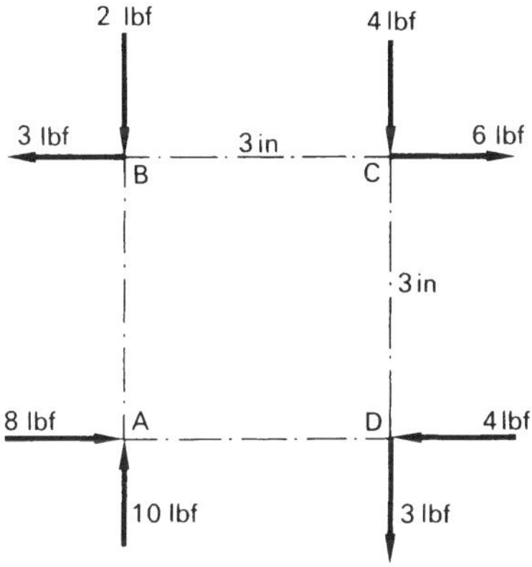

Fig. 1.27

(i) (a) $(4\times 3)+(6\times 3)+(3\times 3)$ lbf-in.,
(i) (b) (3×3) lbf-in.,
(ii) (a) $(4\times 3)+(4\times 3)+(3\times 3)$ lbf-in.,
(ii) (b) (8×3) lbf-in.,
(iii) (a) $(10\times 3)+(4\times 3)$ lbf-in.,
(iii) (b) $(8\times 3)+(2\times 3)$ lbf-in.,
(iv) (a) $(10\times 3)+(6\times 3)$ lbf-in.,
(iv) (b) $(2\times 3)+(3\times 3)$ lbf-in.

1.12. Principle of moments

If a number of forces are acting about a point and the sum of the clockwise moments equals the sum of the anticlockwise moments, then the resultant moment will be zero.

Under these conditions a position of equilibrium will have been reached as clockwise moments = anticlockwise moments.

Experimental verification of principle of moments

Equipment required. Metre rule with hole drilled through centre, various weights, stand, piece of string.

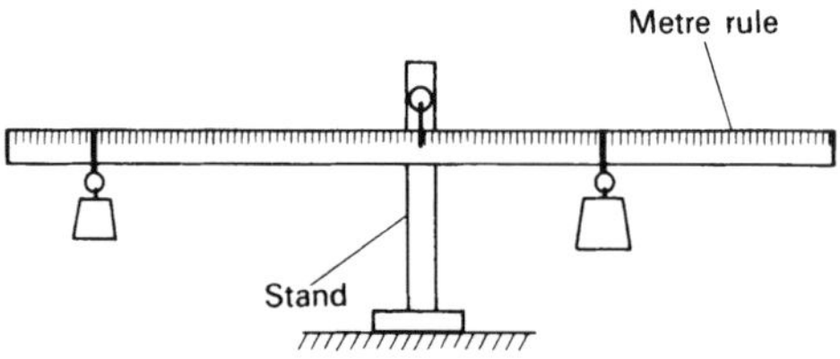

Fig. 1.28

Support the metre rule by a loop of string passed through the hole in the centre. With no weights on, the rule should just balance. Attach a ½-lbf weight, say 30 cm, from the centre and balance this using 1-lbf weight on the opposite side of the fulcrum. Note very carefully the distance of each weight from the fulcrum. Now calculate the clockwise moments and anticlockwise moments about the fulcrum (Fig. 1.28).

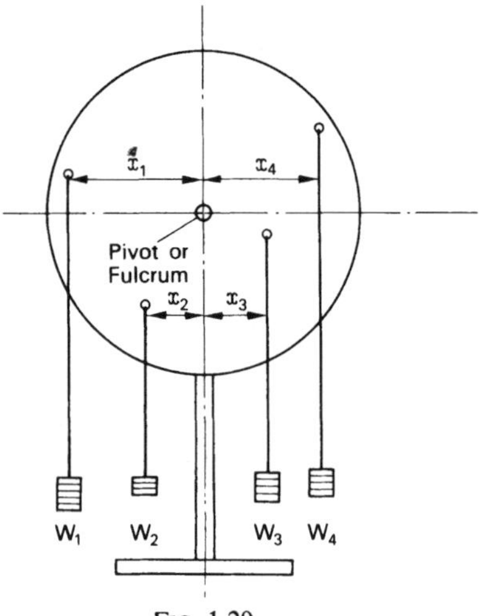

Fig. 1.29

Within the limits of experimental error the two values should be the same.

Repeat the experiment using, say, two or three weights on one side of the fulcrum balanced by two or three weights on the other side and compare the sum of the clockwise moments with that of the anticlockwise moments.

As an alternative experiment a circular board may be used pivoted about the centre and the weights hung on pieces of thin string (Fig. 1.29).

The system is balanced as before and the sum of the clockwise moments compared with the sum of the anticlockwise moments.

The experimental results can be tabulated for easy comparison (Table 1.3).

TABLE 1.3

Clockwise moments			Anticlockwise moments		
Force	Perp. dist. from fulcrum	Moments	Force	Perp. dist. from fulcrum	Moments
2 lbf	6 in.	12 lbf-in.	1 lbf	6 in.	6 lbf-in.
$\frac{1}{2}$ lbf	18 in.	9 lbf-in.	2 lbf	$7\frac{1}{2}$ in.	15 lbf-in.
Total		**21 lbf-in.**	Total		**21 lbf-in.**

1.13. Beam reactions

The principle of moments can be used to find the supporting forces for a beam with a number of loads on it.

Figure 1.30 shows a "weightless" beam *AB* on supports 10 in. apart. If we take moments of the forces acting in the system about any point in the plane of the forces, then the clockwise moments will equal the anticlockwise moments if the beam is in equilibrium.

The simplest way to find the value of one of the support reactions is to take moments about the other support, thus reducing the number of unknowns in the equation to one.

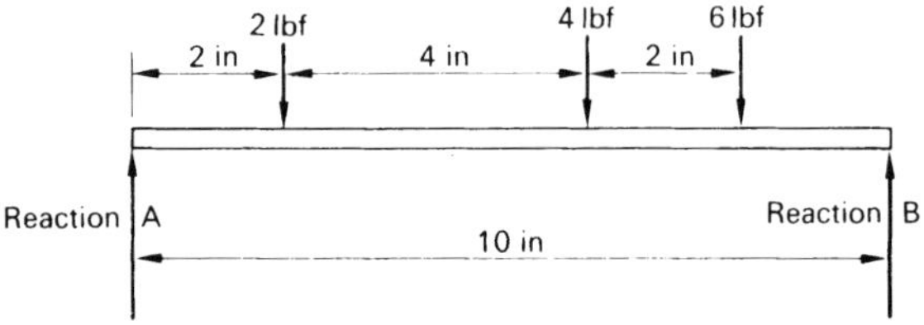

FIG. 1.30

Take moments about reaction A:

$$\text{clockwise moments} = \text{anticlockwise moments}$$
$$2\times 2+4\times 6+6\times 8 = \text{reaction } B\times 10$$
$$4+24+48 = \text{reaction } B\times 10$$
$$\text{reaction } B = \frac{76}{10}\text{ lbf}$$
$$= \mathbf{7{\cdot}6\ lbf.}$$

Having found reaction B, then by considering the vertical forces acting on the beam,

$$\text{upward forces} = \text{downward forces}$$
$$\text{reaction } A + 7{\cdot}6 = 2+4+6$$
$$\text{reaction } A = 12-7{\cdot}6\text{ lbf}$$
$$= \mathbf{4{\cdot}4\ lbf.}$$

The forces indicated in Fig. 1.31 keep the beam in equilibrium.

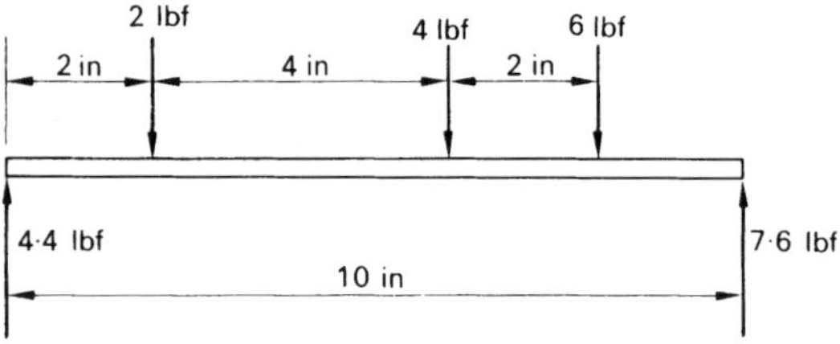

FIG. 1.31

EXAMPLE. Find the reactions to the beam shown loaded as in Fig. 1.32 and supported at A and B.

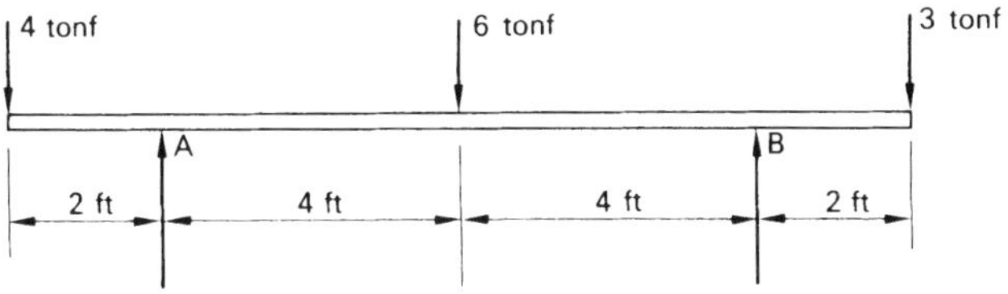

FIG. 1.32

Take moments about A:

clockwise moments = anticlockwise moments.

$$6 \times 4 + 3 \times 10 = 4 \times 2 + R_B \times 8$$

$$24 + 30 = 8 + 8R_B$$

$$R_B = \frac{54-8}{8} \text{ tonf}$$

$$= \mathbf{5{\cdot}75\ tonf.}$$

Now consider the vertical forces:

upward forces = downward forces.

$$R_A + 5{\cdot}75 = 4 + 6 + 3$$

$$R_A = 13 - 5{\cdot}75 \text{ tonf}$$

$$= \mathbf{7{\cdot}25\ tonf.}$$

EXPERIMENTAL VERIFICATION OF BEAM REACTIONS

Equipment required. Metre rule, various weights, two spring balances, two stands.

Mount the spring balances on the stands and hang the rule from the hook by means of a loop of string (Fig. 1.33). With no weights on the ruler take a preliminary reading from each spring balance. Add weights along the ruler at varying distances from balance B_1 and then take a final reading from each spring balance.

The experimental value of the reactions will be the final reading minus the preliminary reading. To check the reactions by calculations take moments about the line through B_1:

$$\text{clockwise moments} = \text{anticlockwise moments.}$$
$$W_1x_1 + W_2x_2 + W_3x_3 = B_2L.$$

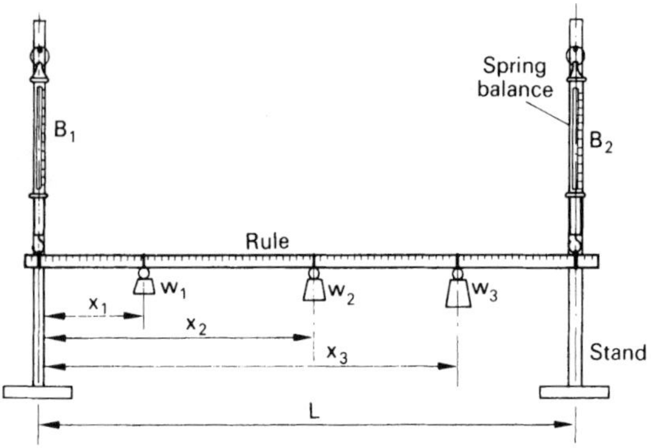

FIG. 1.33

EXAMPLE. Suppose in Fig. 1.33 $W_1 = 2$ lbf, $W_2 = 1$ lbf, $W_3 = 2$ lbf, and $x_1 = 10$ cm, $x_2 = 48$ cm, $x_3 = 74$ cm, and $L = 90$ cm.

Taking moments about B_1:

$$\text{clockwise moments} = \text{anticlockwise moments}$$
$$2 \times 10 + 1 \times 48 + 2 \times 74 = B_2 \times 90$$
$$\therefore\ B_2 = \frac{20 + 48 + 148}{90}\ \text{lbf}$$
$$= \mathbf{2{\cdot}4\ lbf.}$$

Also

$$\text{upward forces} = \text{downward forces}$$
$$B_1 + 2{\cdot}4 = 2 + 1 + 2$$
$$\therefore\ B_1 = 5 - 2{\cdot}4\ \text{lbf}$$
$$= \mathbf{2{\cdot}6\ lbf.}$$

The experiment should be repeated for different weights at various distances along the beam and different positions of reactions (see Fig. 1.34).

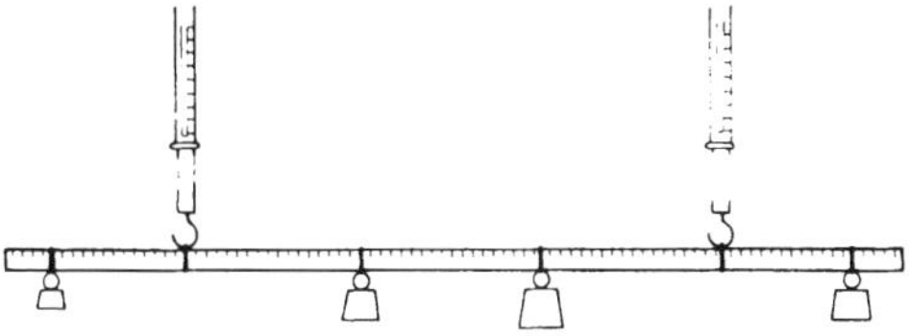

FIG. 1.34

1.14. Friction

When two surfaces slide over one another a resisting force is set up. This force is due to friction between the surfaces and is called the Force of Friction or Frictional Resistance.

It is found that the force required to keep a body moving is slightly less than that necessary to start the body moving. Hence kinetic friction involving movement of the adjacent surfaces is slightly less than static friction where the adjacent surfaces are at rest.

The magnitude of the friction force depends on the two surfaces in contact. If we are designing brakes and clutches, then we require as much resistance as possible between the surfaces, whereas bearings for shafts and electric motors should be as frictionless as possible.

There are certain Laws of Friction which can be verified using suitable experimental apparatus.

1. The force of friction always acts in the direction to oppose motion.
2. It depends on the materials and quality of surfaces in contact.
3. It is independent of the area of contact.
4. It is proportional to the pressure between the surfaces.

Figure 1.35 shows the forces acting on a block of material moving slowly over a horizontal table under the action of a force P.

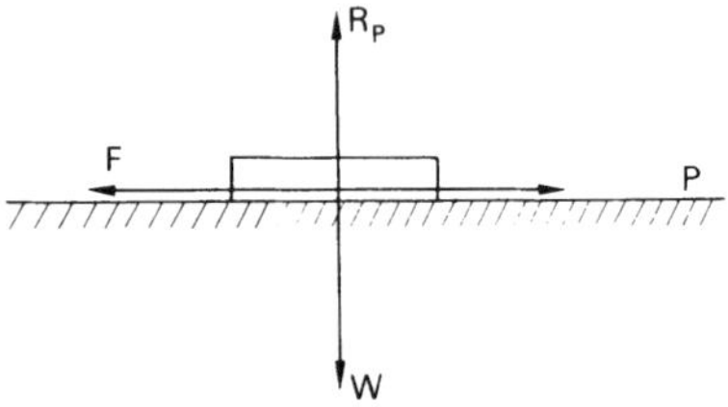

FIG. 1.35

F represents the friction force.
W represents the weight of the body.
R_p represents the reaction of the table on the body.
For equilibrium $P = F$
and $R_p = W$.
From the fourth law $P \propto W$
and $P = \mu W$,

where the Greek letter μ (mu) represents the coefficient of friction.

The coefficient of friction $\mu = P/W$ and is constant for the two given surfaces in contact.

If the two forces R_p and F are combined into a resultant force R then this resultant is always inclined backwards to the direction of motion at an angle ϕ (see Fig. 1.36).

The angle ϕ is called the Angle of Friction.

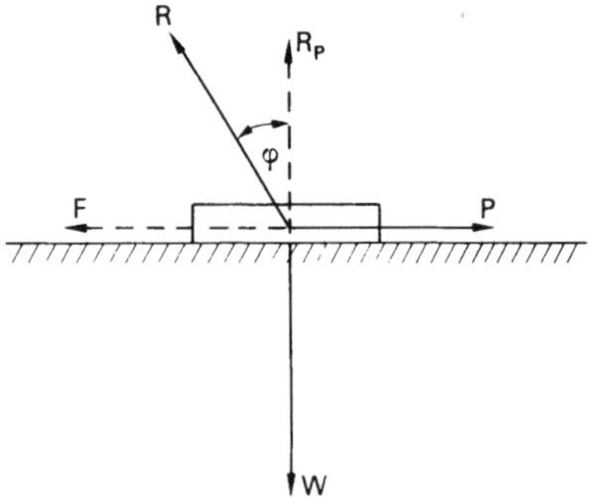

FIG. 1.36

EXPERIMENT. *To find the coefficient of friction between two surfaces in contact.*

Equipment required. Slider with friction material fastened underneath, smooth plane with pulley attached to the end, weights.

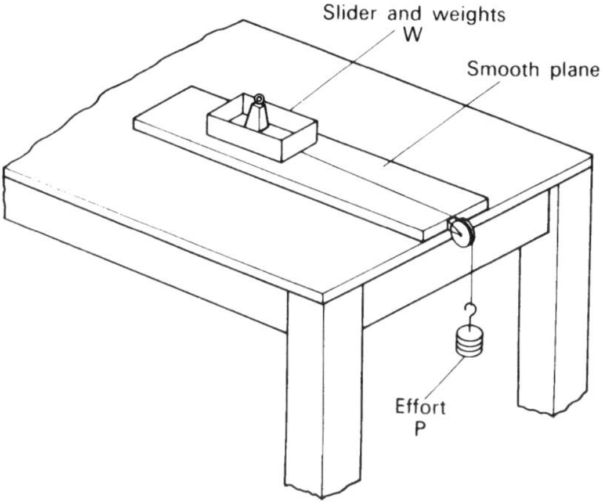

FIG. 1.37

Weigh the slider and then place it on the plane. Attach a piece of string to the slider, pass it over the pulley and fasten a weight hanger to the other end. Add weights to the hanger until the slider moves along the plane at a uniform speed. Place a 1-lbf weight in the slider and increase the weights on the hanger until the slider again moves at a uniform speed along the plane. Repeat the experiment for 1-lbf increments of load until 5 lbf is in the slider. Tabulate the results.

The results in Table 1.4 were obtained from a cast-iron slider on a steel plane.

TABLE 1.4

Weight of slider and additional weights (W lbf)	Effort (P lbf)
2·5	0·5
3·5	0·7
4·5	0·9
5·5	1·1
6·5	1·3

A graph of effort (vertical axis) against force between the surfaces is shown in Fig. 1.38.

The graph verifies the fourth law showing $P \propto W$.

The slope of the graph $P/W = \mu$.

The coefficient of friction for the two given surfaces, cast iron and steel, is found from the slope of the graph.

$$\mu = \frac{0{\cdot}8}{4}$$

$$= \mathbf{0{\cdot}2.}$$

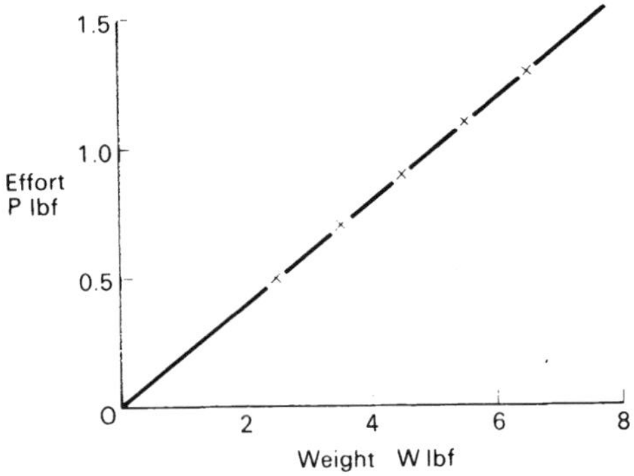

FIG. 1.38

This experiment can be repeated (a) for sliders with different materials fastened to the underside to determine the different values of μ for the different surfaces in contact, and (b) for sliders of the same material but with different areas of contact, to verify Law 3.

TABLE 1.5

Coefficients of friction

Materials in contact	Coefficient of friction
Metal on metal	0·15 to 0·2
Leather on metal	0·3 to 0·6
Metal on wood	0·2 to 0·6

EXAMPLE. A block of wood weighing 12 lbf rests on a horizontal surface. A horizontal force of 5 lbf is just sufficient to cause the block to slide. Find (a) the coefficient of friction and (b) the angle of friction.

$$\text{(a)} \qquad \mu = \frac{P}{W} = \frac{5}{12} = \mathbf{0{\cdot}416.}$$

$$\text{(b)} \quad \tan\phi = \mu = 0{\cdot}416$$

$$\therefore \quad \phi = \mathbf{22^\circ 36'.}$$

EXAMPLE. A block of material weighing 20 lbf rests on a horizontal plane and μ for the materials in contact is 0·2. Find the force required to keep the block moving if the pull is applied (a) horizontally, (b) at 30° to the horizontal. (c) Find the least value of the pull to keep the block moving.

$$\text{(a)} \quad P = \mu W = 0{\cdot}2 \times 20 = \mathbf{4\ lbf.}$$

(b) The pull P must be resolved into a horizontal component and a vertical component as shown in Fig. 1.39.

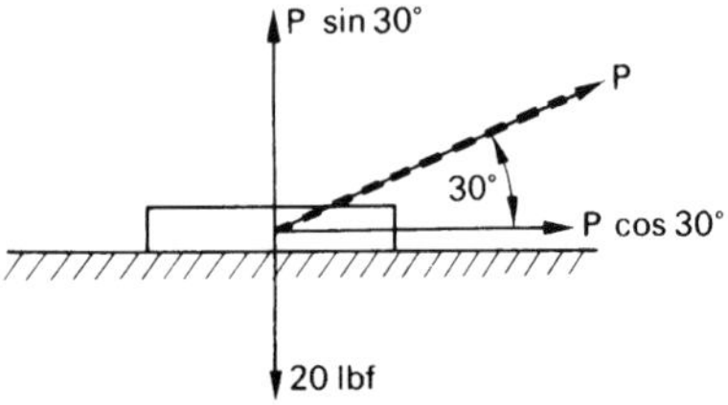

FIG. 1.39

Now the force between the surfaces $= 20 - P \sin 30°$

and the pull $= P \cos 30°$

Hence

$$P \cos 30° = \mu(20 - P \sin 30°),$$
$$P \times 0{\cdot}866 = 0{\cdot}2(20 - 0{\cdot}5P),$$
$$0{\cdot}866P + 0{\cdot}1P = 4,$$
$$P = \frac{4}{0{\cdot}966} \text{ lbf}$$
$$= \mathbf{4{\cdot}14 \text{ lbf.}}$$

(c) From the diagram Fig. 1.40 and the vector diagram Fig. 1.41, the least force required to move the block will be when angle acb is a right angle, i.e. the shortest length of $\overline{bc}$.

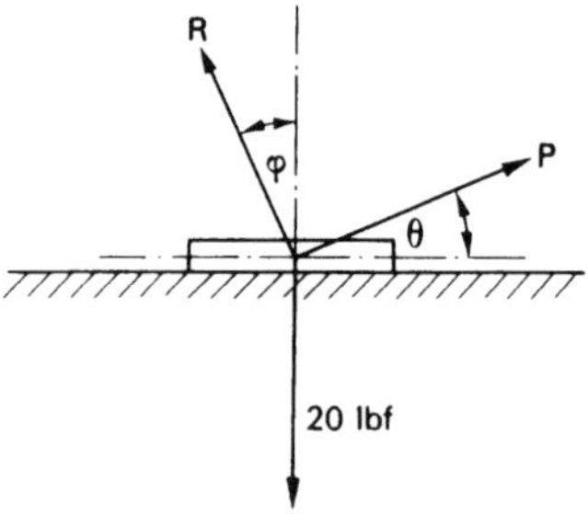

FIG. 1.40

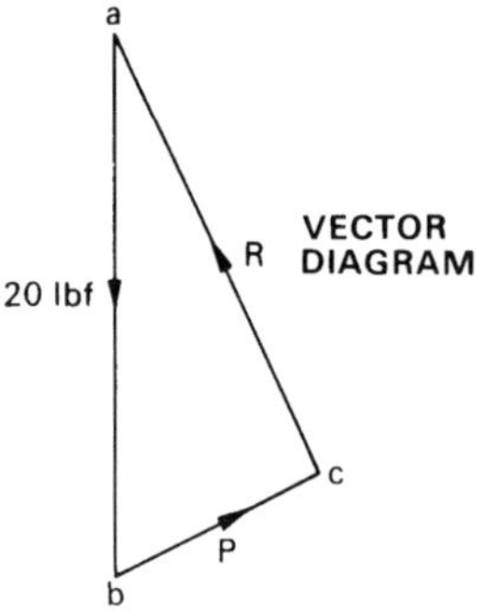

FIG. 1.41

$$P = 20 \sin \phi$$

and $\tan \phi = 0{\cdot}2$,

$$\phi = 11°19',$$

$$\therefore \quad \sin \phi = 0{\cdot}1962,$$

$$P = 20 \times 0{\cdot}1962 \text{ lbf},$$

$$= \mathbf{3{\cdot}92\ lbf.}$$

It should be noticed that the least force to move the block occurs when the angle θ is equal to the angle ϕ (Fig. 1.40).

Problems

1.1. Find the reactions to the beams shown loaded in Fig. 1.42 and supported at A and B.

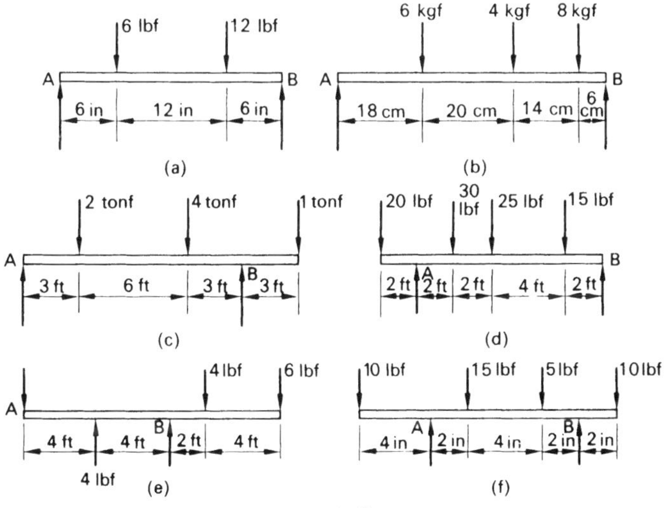

FIG. 1.42

((a) $R_A = 7{\cdot}5$ lbf, $R_B = 10{\cdot}5$ lbf; (b) $R_A = 6{\cdot}3$ kgf, $R_B = 11{\cdot}7$ kgf; (c) $R_A = 2{\cdot}25$ tonf, $R_B = 4{\cdot}75$ tonf; (d) $R_A = 66$ lbf, $R_B = 24$ lbf; (e) $R_A = 7{\cdot}5$ lbf, $R_B = 13{\cdot}5$ lbf; (f) $R_A = 25$ lbf, $R_B = 15$ lbf.)

1.2. A load of 2 tonf is being lifted by a crane using sling chains as shown in Fig. 1.43. Find the load in each leg of the sling chain for values of θ of 30°, 60° 90° and 120°. What conclusion can you draw from your answers?

(1·104 tonf, 1·154 tonf, 1·415 tonf, 2 tonf.)

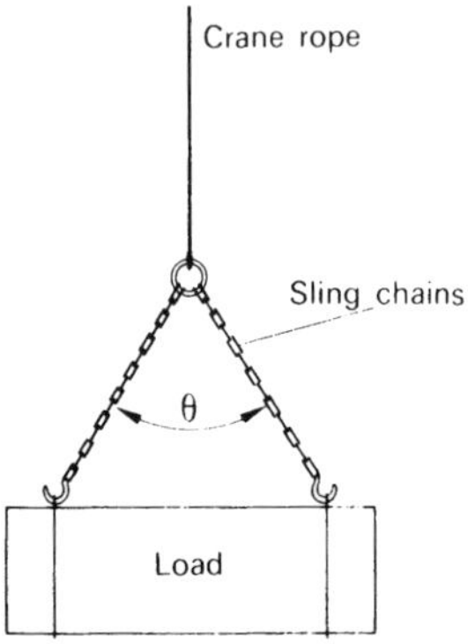

FIG. 1.43

1.3. Two horizontal wires at right angles to each other are attached to the top of a pole. The pull in one wire is 240 lbf and in the other 280 lbf. What is the resultant pull on the pole and what would be the best angular position for a stay wire to be fixed?

(370 lbf; 130° from the 240 lbf in the opposite direction to the 90° angle.)

1.4. Find the horizontal and vertical components of a force of 250 lbf acting in a direction 30° upwards from the horizontal.

(Horizontal component 217 lbf, vertical component 125 lbf.)

1.5. Find the magnitude and direction of the equilibrant of the two forces shown in Fig. 1.44. What alteration must be made to this equilibrant in order that it may become the resultant of the two forces?

(36 lbf 145° from the 15 lbf in the opposite direction to the 60° angle. Change the sense.)

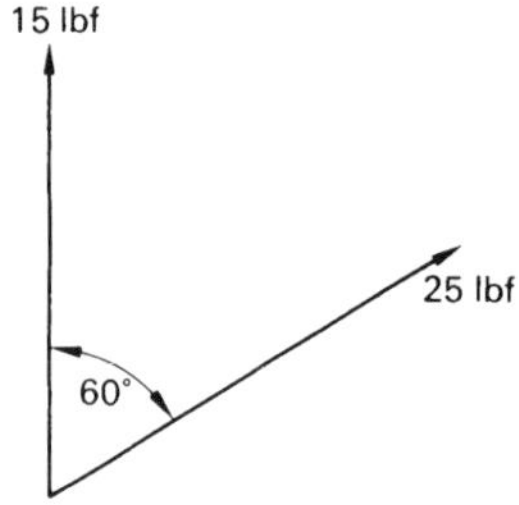

FIG. 1.44

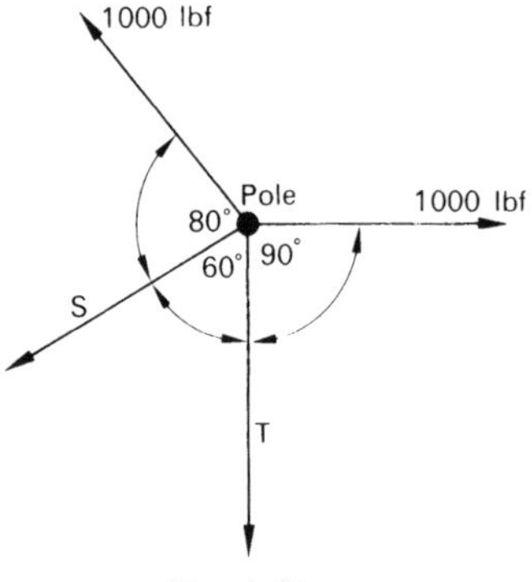

FIG. 1.45

1.6. Figure 1.45 shows the plan of a pole supporting two conductors each having a tension of 1000 lbf. Find the tension in the two horizontal stay wires S and T.

($S=405$ lbf, $T=550$ lbf.)

1.7. Four parallel forces act vertically downwards and are at equal distances of 3 in. apart. If the magnitude of the forces are 2, 4, 5 and 8 lbf respectively calculate the magnitude and position of the resultant force.

(19 lbf acting in the same line as the 5 lbf.)

1.8. Figure 1.46 shows a bell-crank lever pivoted at O. Forces of 10 lbf and 15 lbf are applied on the vertical portion. Find the position of the footstep to enable the lever to be operated by a force of 20 lbf.

($a=20{\cdot}25$ in.)

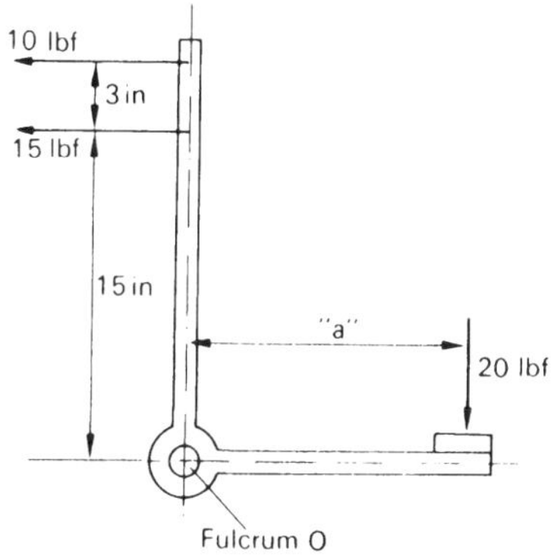

FIG. 1.46

1.9. Find the effort required to operate the lever shown loaded in Fig. 1.47.

(15 lbf.)

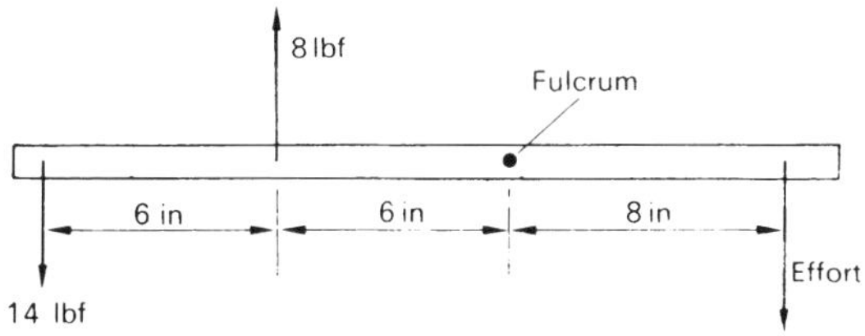

FIG. 1.47

1.10. A "weightless" beam 15 ft long is simply supported at each end and is loaded as follows:

(a) a load of 3 tonf acting vertically downwards at a point 6 ft from the left-hand end;

(b) a load of 8 tonf acting vertically downwards at a point 6 ft from the right-hand end.

Draw the beam showing how it is loaded and find the upward force at each support. (U.E.I.)

(Right-hand support 6 tonf. Left-hand support 5 tonf.)

Answers are not given to the following questions:

1.11. State clearly how a straight line may be used to represent the various features of a force. What force acting at right angles to one of 8 lbf will give a resultant of 12 lbf? State clearly the direction of the resultant force with reference to that of the 8 lbf.

1.12. Explain what is meant by a Polygon of Forces.

Four forces are in equilibrium and act at a common point *O*. The forces are:

(ii) Force *A*: 10 lbf acting horizontally left to right;

(ii) Force *B*: 15 lbf acting at 60° anticlockwise to *A*;

(iii) Force *C*: 15 lbf acting at 60° clockwise to *A*;

(v) Force *X*: to maintain equilibrium.

Draw the force diagram to a scale of 1 inch to 5 lbf and hence find the magnitude of force *X* and its direction with respect to force A. (U.E.I.)

1.13. A load of 2000 lbf is suspended by two chains which are both fastened to the same point on the load. One chain makes an angle of 30° to the horizontal, and the other an angle of 45° to the horizontal. Find the force in each chain. Force scale 1 in. to 400 lbf. (U.L.C.I.)

1.14. A simple wall crane is shown in Fig. 1.48. Find the forces in the jib and tie when lifting a load of 500 lbf.

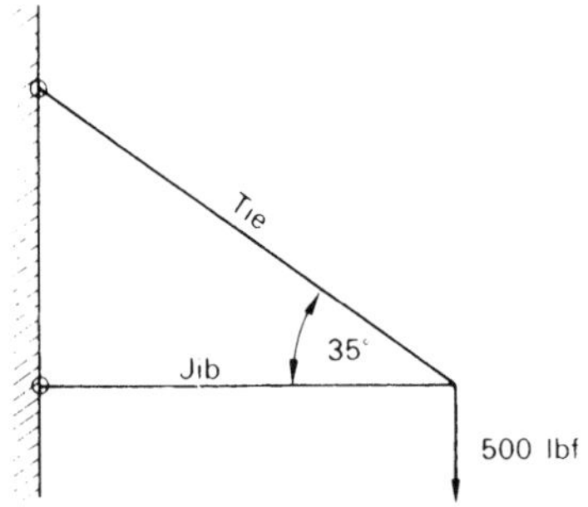

FIG. 1.48

1.15. A beam *AB*, 10 ft long, is simply supported at each end. A force of 30 lbf acts vertically downwards at a point 4 ft from *A*, a force of 20 lbf acts vertically downwards midway along the beam, and a force of 8 lbf acts vertically downwards at a point 2 ft from *B*.

Draw a sketch of the beam showing how it is loaded, and calculate the upward reaction forces *A* and *B*. (U.L.C.I.)

1.16. State the Principle of Moments.

A bell-crank lever consists of a long arm, $8\frac{1}{2}$ in. long, and a short arm 4 in. long, at right angles to each other. Calculate the force to be applied at right angles to the end of the long arm to overcome a resistance of 40 lbf acting at 30° to the vertical of the short arm. Make a drawing of the arrangement.

(U.E.I.)

1.17. Draw to scale the skeleton diagram of the bell-crank lever shown in Fig. 1.49 and measure any distance you require to find the effort required to operate the lever.

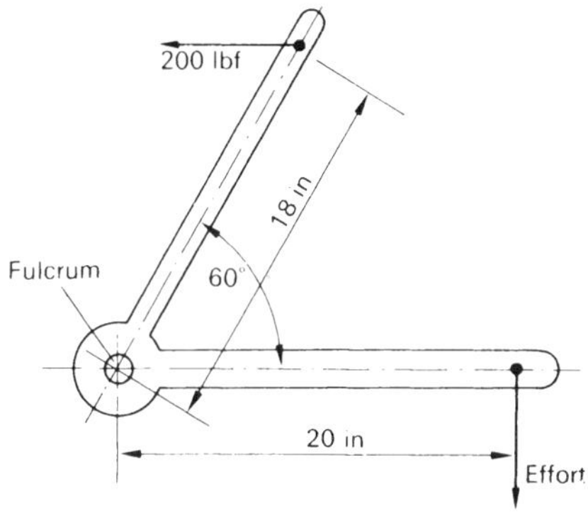

FIG. 1.49

1.18. A load of 50 lbf is supported by two cords, one of which makes an angle of 60° with the vertical. What must be the direction of the second cord so that the tension may be the least possible? Find the tension in the two cords in this case.

1.19. A beam *AB* of uniform cross-section is 20 ft long and weighs 500 lbf. A load of 1000 lbf acts downwards 5 ft from *A* and a load of 750 lbf acts downwards 12 ft from *A*. Find the supporting forces at *A* and *B* assuming that for the purpose of taking moments the weight of the beam acts midway between the supports.

CHAPTER 2

Centre of Gravity

2.1. Definition

Any solid is made up of hundreds of particles and each particle is attracted towards the centre of the earth by the force of gravity. We can consider all these forces to be parallel although in actual fact they will converge at the centre of the earth. The resultant of all the forces acting on the body will pass through a particular point called the Centre of Gravity of the body (Fig. 2.1).

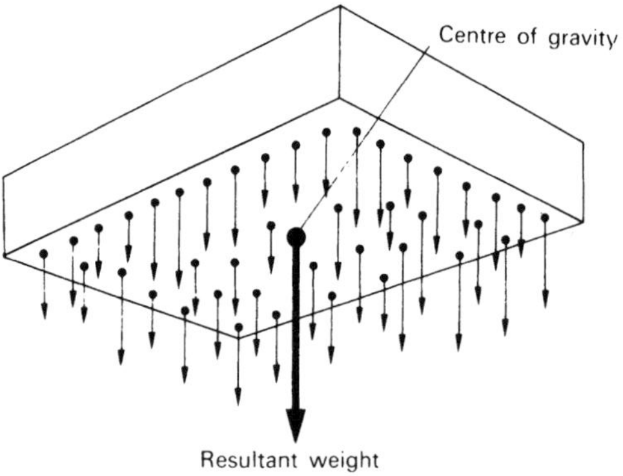

FIG. 2.1

"The centre of gravity of a body is that point through which the resultant weight acts in whatever position the body is placed." See Fig. 2.2.

A solid made from a piece of thin sheet metal or other material of uniform thickness is called a lamina. Now if a lamina is suspended from a corner by a piece of string the centre of gravity of the lamina will move to such a position as to be directly under-

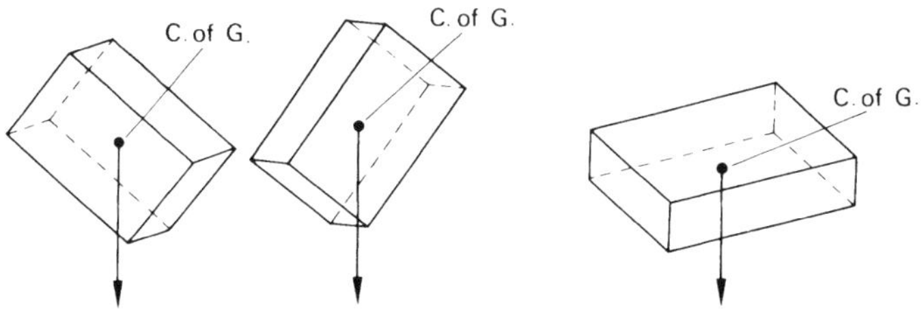

FIG. 2.2. Centre of gravity remains at the same point in the body in whatever position the body is placed.

neath the point of suspension. When this happens the weight of the lamina acting through the centre of gravity and the supporting force are in the same straight line and any tendency to turn will have disappeared. See Fig. 2.3.

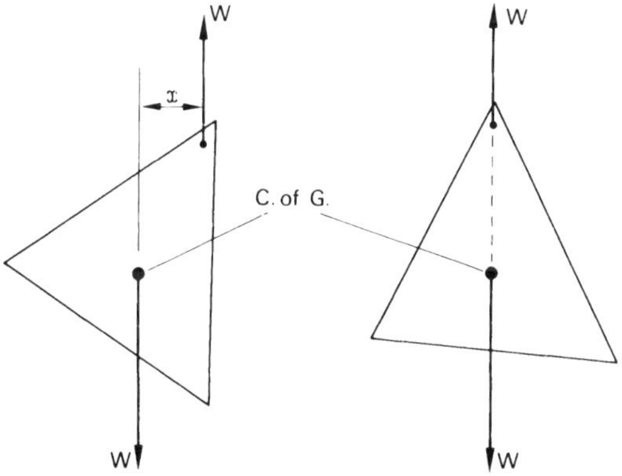

FIG. 2.3. If the centre of gravity is not under the point of suspension a turning moment Wx is introduced causing the body to rotate. This turning moment disappears when the centre of gravity lies under the point of suspension and the body will then remain in equilibrium.

EXPERIMENTAL DETERMINATION OF THE CENTRE OF GRAVITY OF A LAMINA

Equipment required. Irregular shaped piece of cardboard or sheetmetal, piece of string and plumb-bob. See Fig. 2.4.

Pierce three small holes in the lamina at *A*, *B* and *C*. Attach the string at *A* and allow the plumb-bob and lamina to hang from *A*. When they are at rest carefully mark the line of the string on the

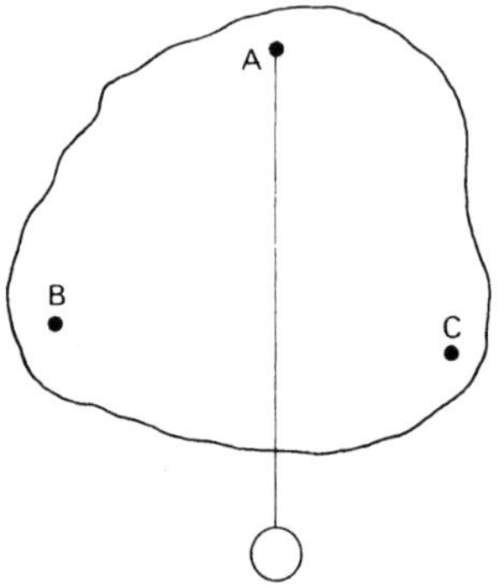

FIG. 2.4

lamina. Repeat this from *B* and *C*. The point where the lines intersect is the centre of gravity of the lamina. To test this, support the lamina on a pin point where the lines intersect and the lamina should balance on the point.

It is possible that the centre of gravity of some solids lies outside the outline (Fig. 2.5).

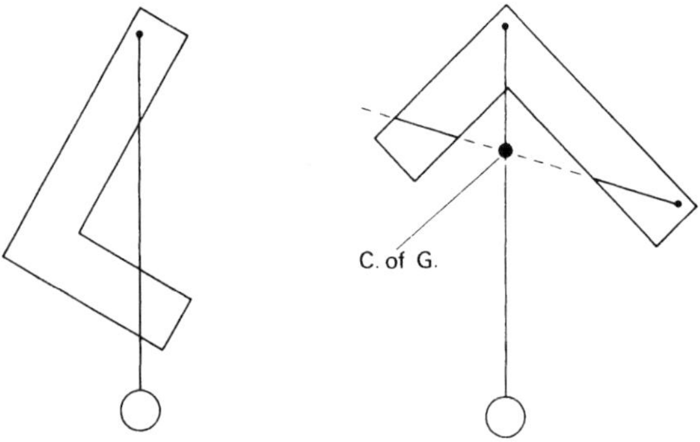

FIG. 2.5. For some bodies the centre of gravity lies outside the outline.

The condition that the centre of gravity lies under the point of suspension applies to all bodies which hang freely.

The positions of the centre of gravity of some simple-shaped laminae are given below:

(a) Triangle: one-third the distance along a median (Fig. 2.6a).

(b) Semicircle: at a point on the radius perpendicular to the bounding diameter and distant $4R/3\pi$ from the diameter (Fig. 2.6b).

(c) Rectangle and square: at the intersection of the diagonals.

(d) Circle: at the centre of the circle.

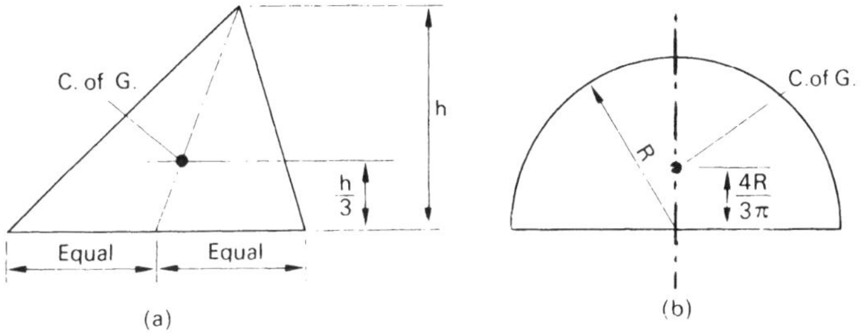

FIG. 2.6

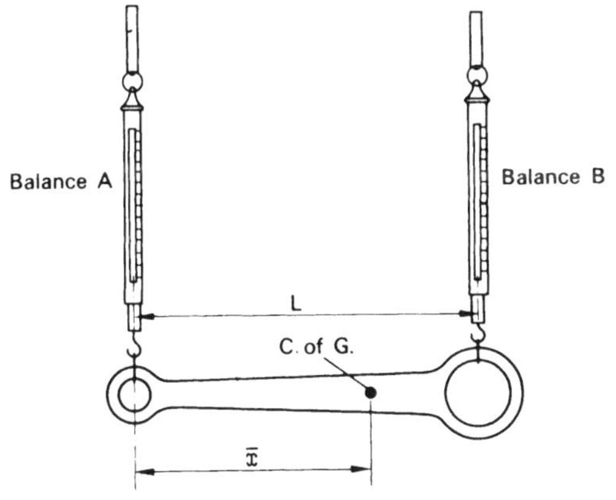

FIG. 2.7

Experimental determination of the centre of gravity of a solid

Equipment required. Solid, two spring balances, metre rule, two stands.

Support the solid from the two spring balances as shown in Fig. 2.7 and adjust the spring balances until the solid is suspended in a horizontal position. Take a reading of balance A and of balance B. Measure the distance apart of the balances, L.

The reading of balance A plus the reading of balance B will be equal to the weight of the solid.

The total weight of the solid can be considered to act through the centre of gravity distance $\bar{x}$ from balance A.

Take moments about balance A.

$$(\text{Force at } B) \times L = \text{Total weight} \times \bar{x},$$

$$\therefore \quad \bar{x} = \frac{(\text{Force at } B) \times L}{\text{Total weight}}.$$

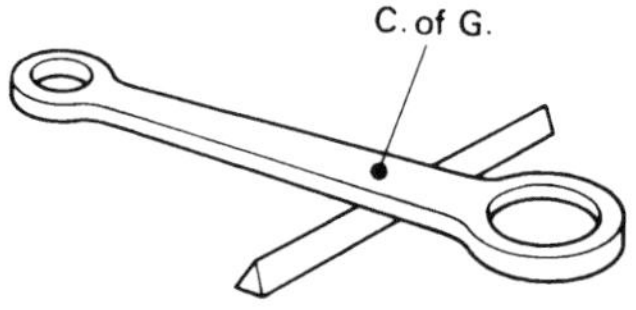

Fig. 2.8

Make a mark on the solid at $\bar{x}$ from end A. Then support the solid on a knife edge at the centre of gravity (Fig. 2.8).

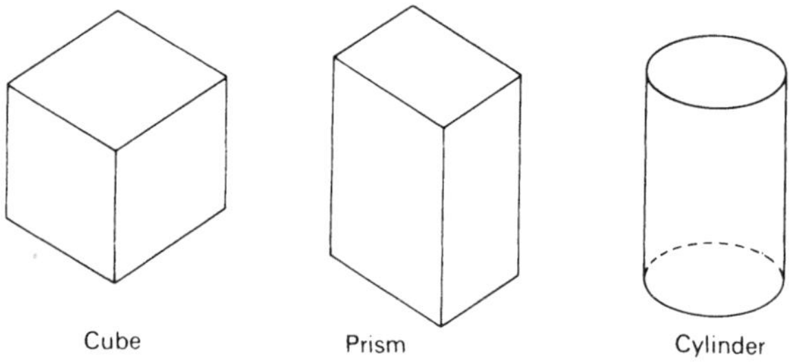

Fig. 2.9. The centre of gravity lies at the mid-point of a line joining the centre of the top and bottom faces.

Balance should be obtained if the experiment has been done carefully.

The position of the centre of gravity of some common solids is given in Figs. 2.9, 2.10, and 2.11.

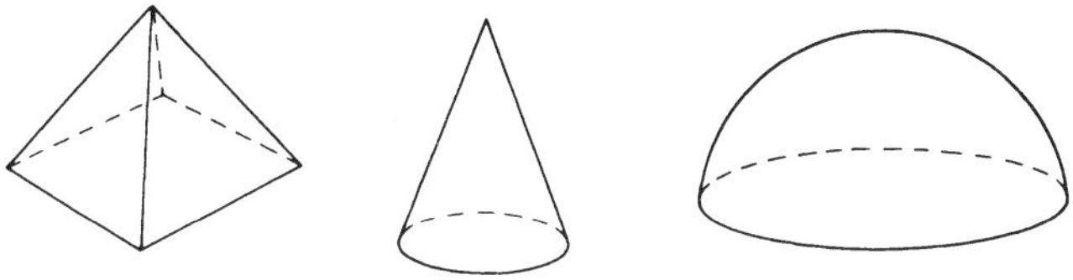

FIG. 2.10. The centre of gravity lies a quarter of the height along a line joining the centre of the base to the apex.

FIG. 2.11. The centre of gravity lies on a radius at right angles to the base and $\frac{3}{8}$ radius from the base.

2.2. Calculation of centre of area

Consider a solid of uniform thickness whose face is made up of regular areas as Fig. 2.12 and let the centre of area be a distance $\bar{y}$ from the axis XX. Now the moment of the whole area about XX

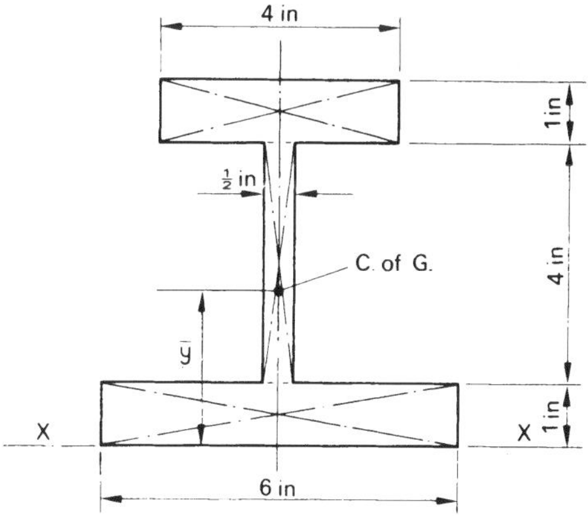

FIG. 2.12

is equal to the sum of the moments of the separate rectangles about XX.

Total area of Fig. 2.12 $= 4\times1+4\times\frac{1}{2}+6\times1=12$ in^2.

Moment of whole area about $XX = 12\bar{y}$.

Moment of bottom rectangle about $XX = 6\times1\times\frac{1}{2}=3$ in^3.

Moment of web about $XX = 4\times\frac{1}{2}\times3=6$ in^3.

Moment of top rectangle about $XX = 4\times1\times5\frac{1}{2}=22$ in^3.

Total moment $= 31$ in^3,

hence $12\bar{y}=31$ in^3,

$$\bar{y}=\frac{31}{12}\text{ in.}$$

$$= \mathbf{2{\cdot}58\ in.}$$

Hence the centre of area is 2·58 in. from the base along the vertical centre line. If the solid is of uniform thickness the centre of gravity will lie midway along the line joining the centre of area of each face.

EXAMPLE. Figure 2.13 shows a piece of 1 in. thick mild steel plate. Find the distance $\bar{x}$ so that when a rope is attached to the plate at B the bottom side remains horizontal.

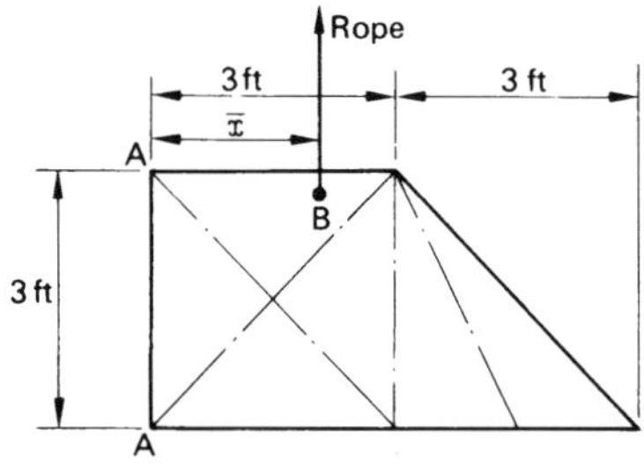

FIG. 2.13

For the bottom side to remain horizontal point B must be directly above the centre of gravity.

Take moments of area about AA.

$$\text{Moment of the whole area about } AA = \left(\frac{3+6}{2}\right)3\bar{x}$$
$$= 13{\cdot}5\bar{x}.$$
$$\text{Moment of square about } AA = 3\times3\times1\tfrac{1}{2} = 13{\cdot}5 \text{ ft}^3.$$
$$\text{Moment of triangle about } AA = 3\times\frac{3}{2}\times4 = 18 \text{ ft}^3.$$
$$\text{Total moment} = 31{\cdot}5 \text{ ft}^3,$$
$$\therefore\ 13{\cdot}5\bar{x} = 31{\cdot}5 \text{ ft}^3,$$
$$\bar{x} = \mathbf{2{\cdot}33\ ft.}$$

EXAMPLE. A casting of uniform thickness has a hole in it 6 in. in diameter positioned as shown in Fig. 2.14. Find the centre of gravity of the solid.

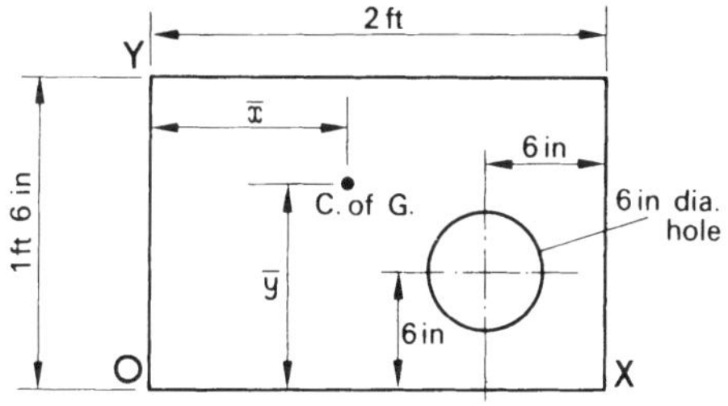

FIG. 2.14

In this problem the surface is not symmetrical about either centre line and therefore it will be necessary to take moments about both axes OX and OY.

(a) Take moments about OX.

$$\text{Moment of the whole area about } OX = (24\times18-\frac{\pi}{4}\times6^2)\bar{y}$$
$$= (432-28{\cdot}3)\bar{y}$$
$$= 403{\cdot}7\bar{y} \text{ in}^3.$$
$$\text{Moment of rectangle about } OX = 24\times18\times9 \text{ in}^3$$
$$= 3888 \text{ in}^3.$$

$$\text{Moment of circle about } OX = \frac{\pi \times 6^2}{4} \times 6 \text{ in}^3$$
$$= 169{\cdot}8 \text{ in}^3.$$

Hence by subtraction the resultant moment

$$= 3888 - 169{\cdot}8 \text{ in}^3$$
$$= 3718{\cdot}2 \text{ in}^3.$$
$$\therefore\ 403{\cdot}7\bar{y} = 3718{\cdot}2 \text{ in}^3$$
$$\bar{y} = \mathbf{9{\cdot}2\ in.}$$

(b) Take moments about OY.

$$\text{Moment of whole area about } OY = 403{\cdot}7\bar{x} \text{ in}^3.$$
$$\text{Moment of rectangle about } OY = 24 \times 18 \times 12 \text{ in}^3$$
$$= 5184 \text{ in}^3.$$
$$\text{Moment of circle about } OY = \frac{\pi \times 6^2}{4} \times 18 \text{ in}^3$$
$$= 509{\cdot}4 \text{ in}^3.$$

Hence by subtraction the resultant moment

$$= 5184 - 509{\cdot}4 \text{ in}^3$$
$$= 4674{\cdot}6 \text{ in}^3.$$
$$403{\cdot}7\bar{x} = 4674{\cdot}6 \text{ in}^3$$
$$\bar{x} = \mathbf{11{\cdot}6\ in.}$$

Hence the position of the centre of gravity is 11·6 in. from OY and 9·2 in. from OX, midway between the faces.

EXAMPLE. A packing case contains equipment weighing 50 lbf and 100 lbf situated as shown in Fig. 2.15. Find the position of the rope sling so that when the case is lifted it remains horizontal.

$$\text{Total force in hoisting rope} = 100 + 50 \text{ lbf}$$
$$= 150 \text{ lbf}.$$

Taking moments about A, $150\bar{x} = 100 \times 6$

$$\bar{x} = \frac{600}{150} \text{ ft}$$
$$= \mathbf{4\ ft.}$$

The sling should be placed 4 ft from A.

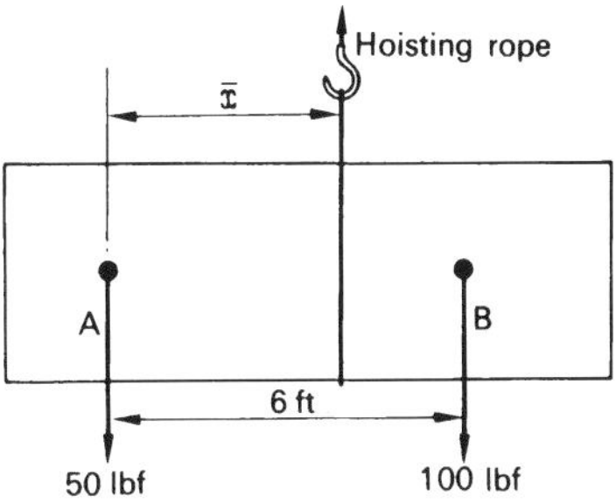

FIG. 2.15

2.3. Stability

There are three conditions of equilibrium known respectively as stable, unstable and neutral. A cone can be used to demonstrate these conditions, as shown in Fig. 2.16. If we consider bodies resting on a horizontal plane, then we can define the conditions of equilibrium as follows:

Stable equilibrium occurs if after a body has been displaced slightly it returns to its original position when the displacing force has been taken away.

Unstable equilibrium occurs if after a body has been displaced slightly it moves farther away from its original position.

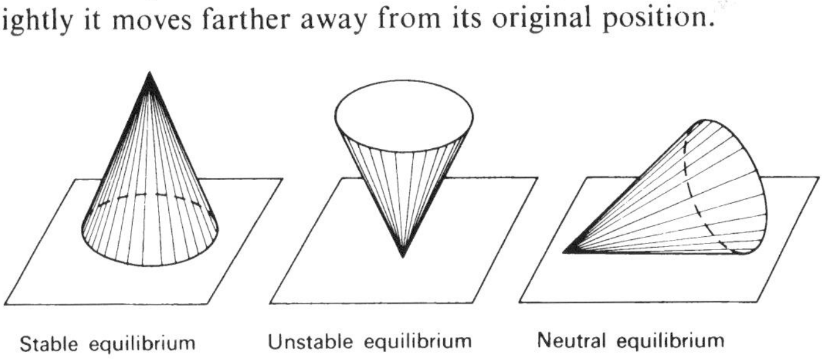

FIG. 2.16

Neutral equilibrium occurs if after a body has been displaced slightly it comes to rest in its new position.

It should be noticed that for stable equilibrium the centre of

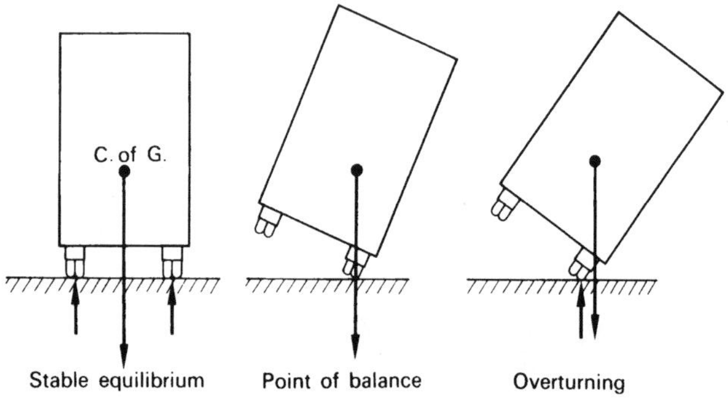

FIG. 2.17

gravity rises relative to the horizontal plane; for unstable equilibrium it falls and for neutral equilibrium it remains the same height.

In the above definitions of equilibrium it was stated that the bodies were displaced slightly. This process of displacement can be described as the overturning of the body. Figure 2.17 shows a

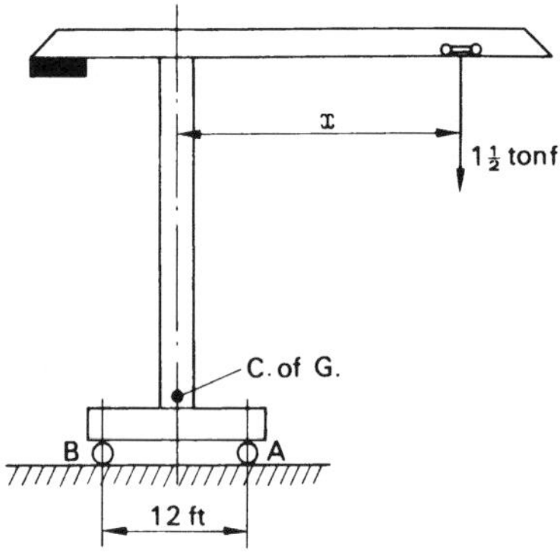

FIG. 2.18

vehicle in the three states of equilibrium. It is important that overturning should not take place and a state of stable equilibrium should prevail. If a vertical line is drawn through the centre of gravity and it passes through the base of the object then it will be stable.

EXAMPLE. Figure 2.18 shows a travelling tower crane weighing 15 tonf. At what radius can a load of $1\frac{1}{2}$ tonf be lifted before the crane is on the point of overturning?

If the crane is on the point of overturning the total reaction will be on wheel A and wheel B will just be leaving the ground.

Take moments about wheel A.

$$\text{Righting moment} = \text{overturning moment}$$

$$15 \times 6 = 1\tfrac{1}{2}(x-6)$$

$$90 = 1\tfrac{1}{2}x - 9$$

$$\therefore x = \frac{99}{1\frac{1}{2}} \text{ ft}$$

$$= \mathbf{66 \text{ ft.}}$$

The crane will be on the point of overturning when the load of $1\frac{1}{2}$ tonf is being lifted at a radius of 66 ft.

Problems

2.1. Calculate the position of the centre of gravity of the thin lamina as shown in Fig. 2.19. Explain how you would find by experiment the position of the centre of gravity of an irregularly shaped lamina.
($2\frac{2}{3}$ ft from top, $2\frac{1}{3}$ ft from left-hand side.) (C. & G.)

2.2. A thin rectangular sheet of metal is 6 ft by 4 ft. A square having sides 2 ft long is cut off from one corner. Calculate the position of the centre of gravity of the remaining piece of metal.
(2 ft 6 in. from 4 ft side, 1 ft 8 in. from 6 ft side.) (N.C.T.E.C.)

2.3. A circular disc of 6 in. radius has a circular hole 3 in. diameter cut from it. The centre of the hole is 3 in. from the centre of the disc. Find the distance of the centre of gravity from the centre of the disc.
(0·2 in.)

2.4. A square prism 6 in. high and 4 in. side of base is tilted on one of the 4-in. edges. At what angle will the prism fall over?
(33°42′ to the vertical.)

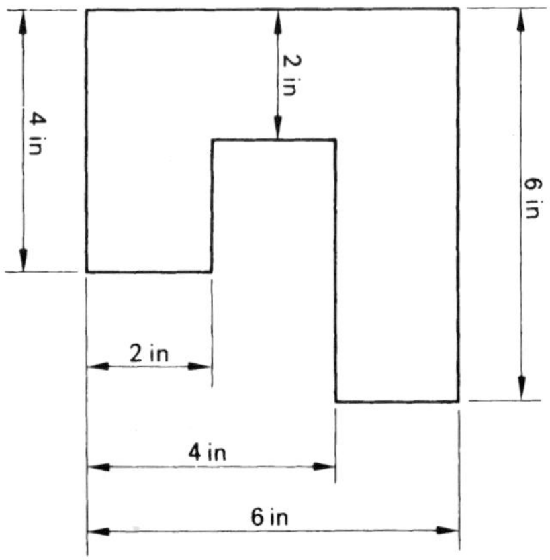

FIG. 2.19

2.5. A square *ABCD* of 10 in. side has weights of 4 lbf, 3 lbf, 6 lbf, and 5 lbf hung from corners *A*, *B*, *C*, and *D* respectively. Find the centre of gravity of the system with reference to sides *AB* and *BC*.

(6·1 in., 5 in.)

2.6. A stepped shaft consists of three coaxial portions. The left-hand portion is 5 in. long and 3 in. diameter, the central portion is 6 in. long and 8 in. diameter and the right-hand portion is 9 in. long and 3 in. diameter. Determine the position of the centre of gravity of the shaft. (E.M.E.U.)

(8·71 in. from left-hand end.)

2.7. A bar of uniform cross-section is 8 ft long. Find the position of the centre of gravity when

(a) the bar is bent into two equal legs of 4 ft at right angles;

(b) the bar is bent into a 6 ft leg and a 2 ft leg at right angles.

((a) 1 ft from each leg; (b) 0·25 ft from 2 ft leg, 2·25 ft from 6 ft leg.)

2.8. A machine consists of two parts which give load concentrations of 14 cwtf and 9 cwtf situated 3 ft 4 in. apart. Find the position in which an eye bolt must be attached to the machine in order that it could be lifted directly above its centre of gravity.

(15·65 in. from the 14 cwtf load.)

Answers are not given to the following questions:

2.9. A rectangular sheet of metal 8 in. long by 6 in. wide is divided into four equal rectangles by lines passing through the centre parallel to the sides. One of these small rectangles is cut away. Determine the position of the centre of gravity of the remainder. How would you check your results experimentally? (U.L.C.I.)

2.10. Distinguish between stable, unstable, and neutral equilibrium, giving an example of each.

A cylinder, 4 in. diameter and 8 in. high, rests on an inclined plane in such a way that it cannot slip. If the inclination of the plane is gradually increased, find the angle at which the cylinder will just topple over. (N.C.T.E.C.)

2.11. A steel bar circular in section, 8 in. overall length, is $\frac{3}{4}$ in. diameter for a length of 4 in. and tapers to a point in the length of the remaining 4 in. Find the position of the centre of gravity.

If you were supplied with a metal bar tapering uniformly from one end to the other and two spring balances, explain carefully how you would determine the position of the centre of gravity of the bar. If the bar weighs 35 lbf, is 28 in. long and has its centre of gravity 12 in. from one end, what would be the readings on the spring balances?

CHAPTER 3

Work, Power and Energy

3.1. Work

When a force acts on a body and causes it to move then work is said to be done and energy is expended. The amount of work is the product of the force and the distance moved in its own line of action.

If a man moves a trolley at constant speed along a level floor and the resistance to motion is equal to 15 lbf then the force exerted by the man will be 15 lbf. If he pulls the trolley 20 ft then the work done is equal to

$$15 \times 20 = 300 \text{ ft-lbf.}$$

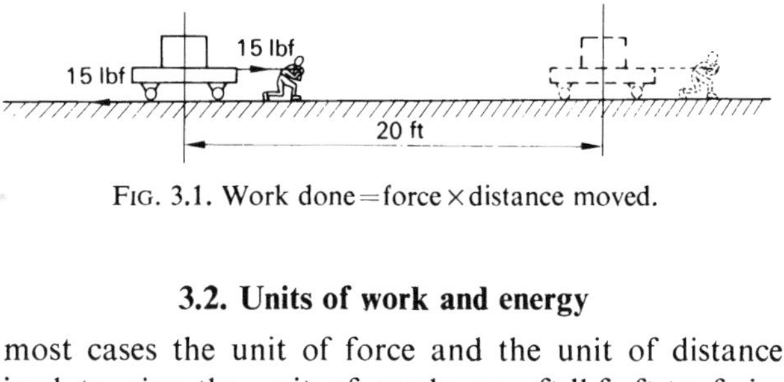

FIG. 3.1. Work done = force × distance moved.

3.2. Units of work and energy

In most cases the unit of force and the unit of distance are combined to give the unit of work, e.g. ft-lbf, ft-tonf, in.-lbf, m-kgf, etc. (See also Section 5.8.)

EXAMPLE. A hammer weighing 4 lbf is raised 3 ft above the bench. How much work has been done on the hammer?

$$\begin{aligned} \text{Work done} &= \text{force} \times \text{distance} \\ &= 4 \times 3 \text{ ft-lbf} \\ &= \mathbf{12\ ft\text{-}lbf.} \end{aligned}$$

EXAMPLE. A wagon is pulling a trailer along a level road at 15 m.p.h. The resistance to motion of the trailer is 60 lbf. How much work does the wagon do on the trailer in 10 min?

$$\text{Distance travelled in 10 min} = \frac{15 \times 5280 \times 10}{60} = 13{,}200 \text{ ft.}$$

$$\begin{aligned}\text{Work done} &= \text{force} \times \text{distance} \\ &= 60 \times 13{,}200 \text{ ft-lbf} \\ &= \mathbf{792{,}000\ ft\text{-}lbf.}\end{aligned}$$

EXAMPLE. How much work is done pumping 50 gallons of water to the top of a building 150 ft high? (1 gallon of water weighs 10 lbf.)

$$\text{Weight of water} = 50 \times 10 = 500 \text{ lbf.}$$

$$\begin{aligned}\text{Work done} &= \text{force} \times \text{distance} \\ &= 500 \times 150 \text{ ft-lbf} \\ &= \mathbf{75{,}000\ ft\text{-}lbf.}\end{aligned}$$

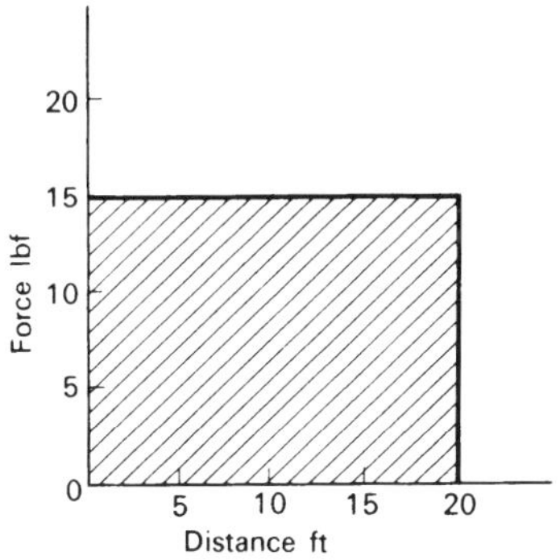

FIG. 3.2. Force–displacement graph. The work done is given by the area under the graph.

3.3. Work diagram

If the first example of the man pulling the trolley by a force of 15 lbf through a distance of 20 ft is plotted on a graph (Fig. 3.2), then the graph becomes a straight horizontal line. The force of 15 lbf was constantly applied throughout the 20 ft.

The area under the force–displacement graph between 0 and 20 ft, taking the scales into account, is equal to 15 lbf × 20 ft, giving a value of 300 lbf-ft, the same as that previously calculated.

Hence, work done = area under the force–displacement graph.

EXAMPLE. A cable 100 ft long weighing 1 lbf per ft of length is being hauled up a shaft 60 ft high. If initially one end of the cable is at the top of the shaft how much work is done in hauling the cable to the top of the shaft?

For the first 40 ft of cable a length of 60 ft will be hanging down the shaft, but then for the last 60 ft of cable the force will gradually decrease as more cable is taken up the shaft. Plot this information on a work diagram (see Fig. 3.3).

Force exerted when the cable is the full length of shaft

$$= 60 \times 1 = 60 \text{ lbf}.$$

Therefore for the first 40 ft the force is uniform at 60 lbf. When

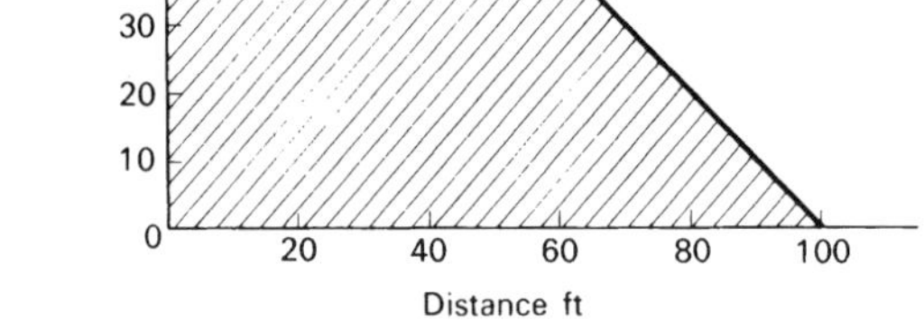

FIG. 3.3

50 ft of cable have been hauled the length hanging down the shaft will be 50 ft.

$$\text{The force required} = 50 \times 1 = 50 \text{ lbf.}$$

When 60 ft have been hauled the length hanging down the shaft is 40 ft.

$$\text{The force required} = 40 \times 1 = 40 \text{ lbf}$$

and the values continue to decrease until all the cable has been hauled up the shaft.

The work done during this operation is the area under the force–displacement graph

$$= 60 \times 40 + \frac{60}{2} \times 60 \text{ ft-lbf}$$
$$= 2400 + 1800 \text{ ft-lbf}$$
$$= \mathbf{4200 \text{ ft-lbf.}}$$

3.4. Work done in rotation

When a force is applied tangentially to a wheel (Fig. 3.4) and the wheel moves through one revolution, then the force has travelled a distance equal to the circumference of the wheel.

Work done = force × circumference

$$= P \times 2\pi r$$
$$= 12 \times 2\pi \times 2 \text{ ft-lbf}$$
$$= \mathbf{48\pi \text{ ft-lbf.}}$$

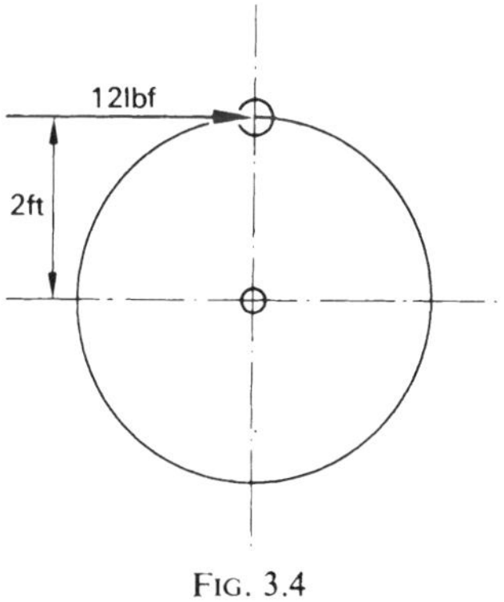

FIG. 3.4

3.5. Radian

A very useful measure of angular distance is the radian. "A radian is the angle subtended at the centre of a circle by an arc equal in length to the radius."

In one revolution of a shaft the angle turned through will be 2π radians, therefore

$$1 \text{ radian} = \frac{360}{2\pi} = 57{\cdot}3^\circ \text{ approx.}$$

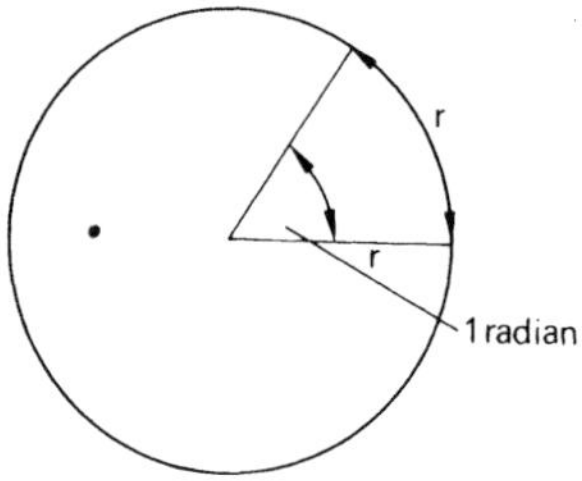

FIG. 3.5

EXAMPLE. Convert 5 radians to degrees.

$$5 \times \frac{360}{2\pi} = \mathbf{286{\cdot}5^\circ.}$$

EXAMPLE. Convert 270° to radians.

$$270 \times \frac{2\pi}{360} = \mathbf{4{\cdot}712 \text{ radians.}}$$

Let us have another look at the work done in rotation.

Figure 3.6 shows a lever r ft long with a force P lbf turning the lever through an angle θ rad.

Now work done = force × distance

$$= P \times \text{length of arc } AB$$
$$= P \times r\theta,$$

but $P \times r$ measures torque (torque being another name for turning moment). Therefore

Work done in rotation = torque × angle turned through in radians

$$= T \times \theta.$$

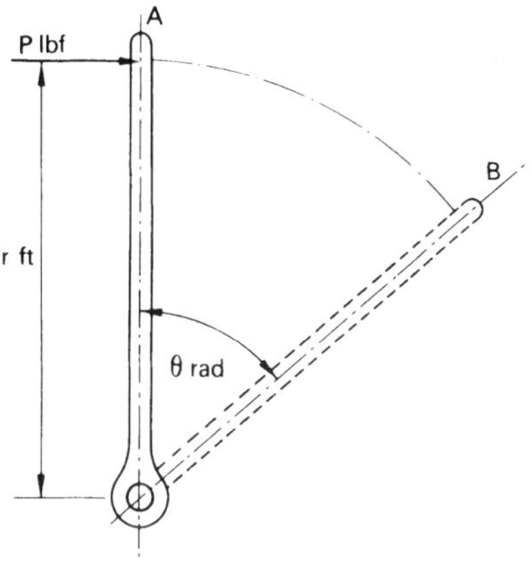

FIG. 3.6

EXAMPLE. How much work is done in 150 revolutions of an electric motor if the torque on its shaft is 1600 lbf-ft?

Angle turned through in radians $= 150 \times 2\pi$.

$$\text{Work done} = 1600 \times 150 \times 2\pi \text{ ft-lbf}$$
$$= \mathbf{1{,}508{,}000 \text{ ft-lbf.}}$$

3.6. Power

The concern of an engineer is not only the amount of work done but also the time in which the work is done or the rate of doing work.

"Power is the rate of doing work."

One horsepower is a rate of working of 33,000 ft-lbf/min or 550 ft-lbf/sec.

One watt is a rate of working of one newton-metre/second
or one joule/second.

This is explained in more detail in Section 10.5.

EXAMPLE. A wagon pulls a trailer weighing 10 tonf at 20 m.p.h. against frictional resistance of 10 lbf/tonf. Find the horsepower required.

Total frictional resistance $= 10 \times 10 = 100$ lbf.

$$\text{Speed of wagon} = \frac{20 \times 5280}{60} = 1760 \text{ ft/min.}$$

Rate of doing work $= 100 \times 1760 = 176{,}000$ ft-lbf/min.

$$\text{Horsepower} = \frac{176{,}000}{33{,}000} = \mathbf{5{\cdot}34\ hp.}$$

EXAMPLE. The hoisting mechanism of a crane is driven by a 10-hp electric motor. If 75% of the power output of the motor is available for lifting the load, find how long it takes to lift a load weighing 8000 lbf a distance of 15 ft.

$$\text{Effective horsepower of crane} = 10 \times \frac{75}{100} = 7{\cdot}5 \text{ hp.}$$

$$\text{Horsepower} = \frac{\text{force (lbf)} \times \text{distance (ft)}}{\text{time (sec)} \times 550}$$

$$\therefore \text{ time} = \frac{8000 \times 15}{7{\cdot}5 \times 550} \text{ sec}$$

$$= \mathbf{29{\cdot}1\ sec.}$$

3.7. Horsepower transmitted in rotation

Work done in rotation
= torque (lbf-ft) × angle turned through (rad).

Hence horsepower transmitted in rotation

$$= \frac{\text{torque (lbf-ft)} \times \text{angle turned through per minute (rad/min)}}{33{,}000}.$$

Now suppose that the shaft or wheel is rotating at a speed of N rev/min, then the number of rad/min would be $2\pi N$.

$$\text{Hence hp} = \frac{2\pi NT}{33{,}000}$$

where T is the torque lbf-ft and N is the speed of shaft in rev/min.

EXAMPLE. A shaft has a torque of 12,600 lbf-in. applied to it and rotates at 500 rev/min. Calculate the horsepower being transmitted through the shaft.

$$\text{Horsepower} = \frac{2\pi NT}{33{,}000}$$

$$\text{and} \qquad T = \frac{12{,}600}{12} \text{ lbf-ft.}$$

$$\text{Horsepower} = \frac{2\pi \times 500 \times 12{,}600}{33{,}000 \times 12}$$

$$= \mathbf{100\ hp.}$$

3.8. Measurement of brake horsepower for an engine or motor

An electric motor, petrol engine, diesel engine or gas engine is a machine whereby energy contained in the form of electricity, petrol, oil or gas can be converted into mechanical energy. During the process of conversion some of the energy is lost in overcoming friction in the machine and also through other heat losses. The power available at the output shaft of an engine is less than the power input to the machine. In practice it is very useful to be able to measure the power at the output shaft and this can be done by means of a dynamometer or brake, hence the term "brake horsepower".

3.9. Simple rope brake

A fairly large pulley is attached to the output shaft of the motor (Fig. 3.7) and around the pulley a double rope is arranged to support weights of W lbf at one end. The other end is attached to a spring balance supported from a hanger.

When the motor pulley is stationary the reading on the spring balance will be the same as the value of W lbf hung on the other end of the rope. If now the motor is started up on the direction of the arrow on the pulley the friction between the rope and pulley will cause the weights of W lbf to be lifted slightly with a consequent reduction in the reading on the spring balance.

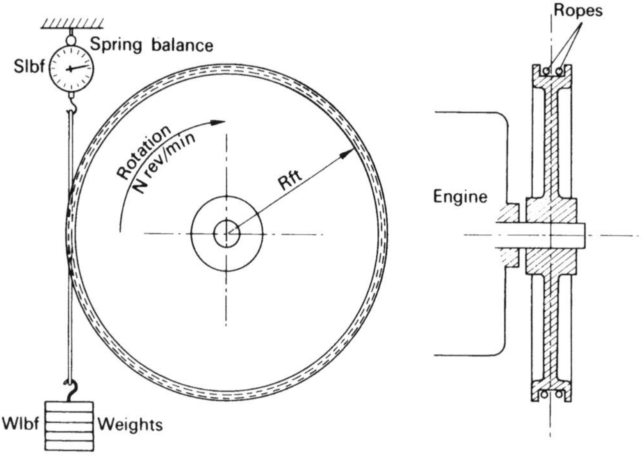

FIG. 3.7. A simple rope brake; $\text{bhp} = \dfrac{2\pi N(W-S)R}{33{,}000}$.

Now let S lbf be the reading on the spring balance and W lbf be the weights on the end of the rope.

Then $(W-S)$ lbf will be the effective force at the circumference of the pulley due to friction between the rope and the pulley tending to stop the motor.

If R ft is the radius of the pulley, then

$$\text{Torque on the shaft} = (W-S)R \text{ lbf-ft},$$

$$\therefore \text{ brake horsepower} = \frac{2\pi N(W-S)R}{33{,}000},$$

where N is the speed of the shaft in revolutions per minute.

An alternative method of measuring brake horsepower is to use two spring balances attached to a hanger as shown in Fig. 3.8.

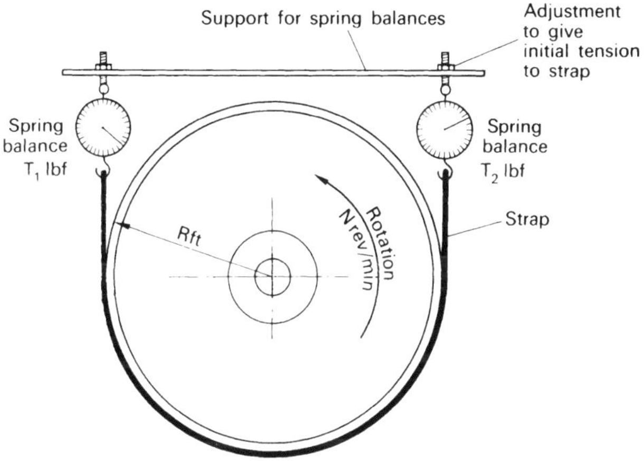

FIG. 3.8. A brake formed of a strap; $\text{bhp} = \dfrac{2\pi N(T_1-T_2)R}{33{,}000}$.

The strap is in contact with the pulley over half the circumference. With the pulley stationary, an initial tension is introduced into the strap giving the same reading on both balances. As the wheel rotates in the direction of the arrow, the reading on balance T_1 is increased and that on T_2 decreased. The brake horsepower is measured as

$$\frac{2\pi N(T_1-T_2)R}{33{,}000}.$$

EXAMPLE. Determine the brake horsepower of an engine from the following information: brake diameter 3 ft, load on brake 400 lbf, spring balance reading 20 lbf, average speed of brake wheel = 220 rev/min.

Resultant force at circumference of wheel = (400 − 20) lbf.

$$\text{Braking torque} = 380 \times \frac{3}{2}\ \text{lbf-ft}$$

$$= 570\ \text{lbf-ft.}$$

$$\text{Brake horsepower} = \frac{2\pi NT}{33{,}000}$$

$$= \frac{2\pi \times 220 \times 570}{33{,}000}\ \text{hp}$$

$$= \mathbf{23{\cdot}9\ hp.}$$

3.10. Energy

When a body is capable of doing work it is said to possess energy. For example, a hammer falling from a height on to an object will cause the object either to be deformed or fractured. Similarly, water stored in a reservoir can pass through pipes to a lower level and be made to drive water turbines which are coupled to electric generators to produce electricity. Petrol and air drawn into a cylinder, compressed and exploded can be made to push a piston down the cylinder to drive a vehicle.

There are other forms of energy, e.g. solar cells attached to satellites collect energy from the sun called solar energy, energy can be found in sound and radio waves, and of course the biggest source of energy discovered this century, nuclear energy.

3.11. Potential energy

The energy stored in objects due to their position above a given datum is called potential energy. Raised weights such as hammers, pile drivers, vehicles at the top of inclines all possess potential energy by virtue of their position.

3.12. Kinetic energy

The energy stored in an object due to its motion is called kinetic energy. Fluids flowing along a channel or pipe, a wheel revolving on a shaft, all possess kinetic energy. A vehicle in motion possesses kinetic energy due to its linear speed and also due to the motion of all the rotating parts such as the gears in the gear-box, back axle and the road wheels.

3.13. Conservation of energy

From the results of experiments and observations it has been concluded that, within the universe, energy can neither be created nor destroyed but may be transformed from one form into another.

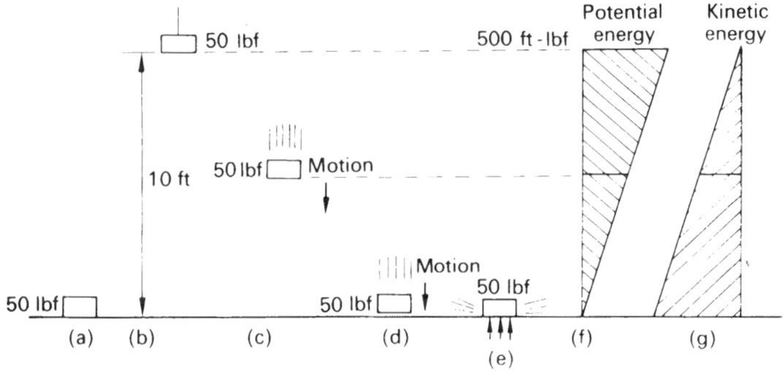

FIG. 3.9

Let us consider a mass weighing 50 lbf resting on the ground (Fig. 3.9(a)). It has no potential or kinetic energy. This mass is now lifted through a height of 10 ft by a crane and in doing this $10 \times 50 = 500$ ft-lbf of work has been done on the mass. This work is stored in the mass as potential energy (Fig. 3.9(b)). If this mass were lowered gently to the ground the work stored or potential energy would be converted into heat energy in the lowering brake and would be lost to the surroundings. Suppose the weight were allowed to fall freely from the height of 10 ft; the potential energy would diminish as the weight fell and at the same time the kinetic energy would increase as the weight gained speed. At any point (Fig. 3.9(c)) the sum of the potential energy and the kinetic energy would be constant. When the mass was just on the point of striking the ground (Fig. 3.9(d)) all the potential energy would have been converted into kinetic energy and the mass would have 500 ft-lbf of kinetic energy. Figures 3.9(f) and 3.9(g) are two graphs showing the relationship between the potential energy and the kinetic energy of the weight at different heights above the datum. At any height the sum of the two energies will be 500 ft-lbf.

When the weight strikes the ground (Fig. 3.9(e)) the energy is dissipated in the form of noise and heat.

3.14. Efficiency of conversion

It is impossible in practice to convert energy completely from one form into another without waste. In mechanical conversions, friction in the bearings and gears causes heat to be generated which is lost to the surroundings.

In the previous example the crane would require more than 500 ft-lbf of energy to lift the weight a height of 10 ft but the energy stored in the weight would be exactly 500 ft-lbf.

In a power station where the energy content of the fuel, oil, coal, etc., is converted to electrical energy, a very careful check is made on the efficiency of energy conversion, i.e.

$$\frac{\text{electrical energy output}}{\text{fuel energy input}} \times 100\%.$$

Even efficient power stations have only an overall efficiency of between 30% and 35%, which means that less than half of the potential energy of the fuel is converted into electrical energy.

3.15. Evaluation of potential energy (P.E.)

Let W = the weight of the body (lbf)
and h = the height above a datum (ft),
then P.E. = Wh ft-lbf.

3.16. Evaluation of kinetic energy (K.E.) due to linear motion

It is found in Section 5.4 that if a body is allowed to fall freely the acceleration is g ft/s^2 and the velocity at any height h is given by

$$v=\sqrt{(2gh)}, \quad \therefore h=\frac{v^2}{2g}.$$

Multiply both sides by W; then $Wh=\dfrac{Wv^2}{2g}$ ft-lbf.

Thus the potential energy, Wh, is all converted into kinetic energy, $Wv^2/2g$, just before hitting the ground.

The K.E., $Wv^2/2g$, measures the energy due to motion in a straight line.

EXAMPLE. A truck weighs 1·5 tonf and moves with a velocity of 45 m.p.h. Calculate its kinetic energy.

If the engine is suddenly stopped, calculate the average force to bring the truck to rest in 200 ft. ($g=32{\cdot}2$ ft/s^2) (1 mile = 5280 ft).

$$45 \text{ m.p.h.}=\frac{45\times 5280}{60\times 60} \text{ ft/s}=66 \text{ ft/s},$$

$$\text{K.E.}=\frac{Wv^2}{2g}=\frac{1{\cdot}5\times 2240\times 66^2}{2\times 32{\cdot}2} \text{ ft-lbf}$$

$$=\mathbf{228{,}000 \text{ ft-lbf.}}$$

Force × distance = change in K.E.

$\therefore$ since the truck comes to rest all the kinetic energy will be lost and the final energy will be zero.

$\therefore$ force × 200 = 228,000 ft-lbf.

$$\text{Force to retard the truck}=\frac{228{,}000}{200} \text{ lbf}$$

$$=\mathbf{1140 \text{ lbf.}}$$

3.17. Evaluation of kinetic energy due to rotation

A rotating wheel is a source of kinetic energy.

The mass of the wheel is considered to be concentrated at a particular radius called the radius of gyration of the wheel.

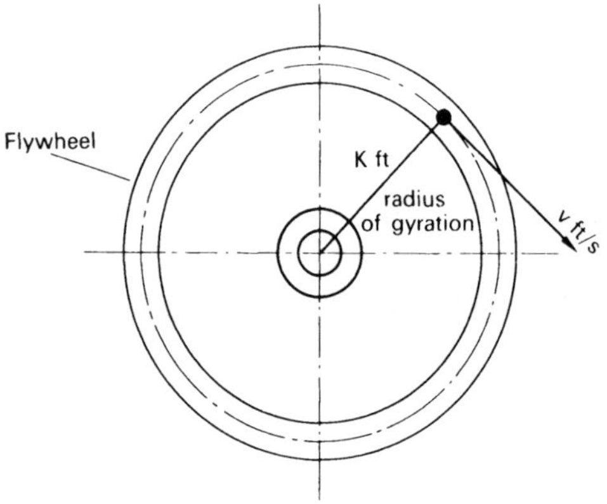

FIG. 3.10

Let k = radius of gyration (ft),
v = velocity of the mass at the radius of gyration (ft/s),
ω = angular velocity of wheel (rad/s),

then
$$\text{K.E.} = \frac{Wv^2}{2g} \text{ ft-lbf.}$$

But
$$v = \omega k \text{ ft/s,}$$

$$\therefore \text{ K.E. due to rotation} = \frac{W}{2g}\omega^2 k^2 \textbf{ ft-lbf.}$$

EXAMPLE. A flywheel has a mass of 5000 lb assumed concentrated at a radius of 4 ft. Calculate the kinetic energy at 150 rev/min.

If the flywheel slows down from 150 to 148 rev/min find the energy abstracted from the wheel. (U.E.I.)

$$\text{K.E.} = \frac{W}{2g}\omega^2 k^2$$

and
$$\omega = 150 \times \frac{2\pi}{60} \text{ rad/s}$$
$$= 5\pi \text{ rad/s}$$
$$\therefore \text{ K.E.} = \frac{5000}{2 \times 32{\cdot}2} \times (5\pi)^2 \times 4^2 \text{ ft-lbf}$$
$$= \frac{5000 \times 25\pi^2 \times 16}{2 \times 32{\cdot}2} \text{ ft-lbf}$$
$$= \mathbf{307{,}000 \text{ ft-lbf.}}$$

Hence K.E. at 150 rev/min is 307,000 ft-lbf.

K.E. at 148 rev/min:
$$\omega = \frac{148 \times 2\pi}{60}$$
$$= 4{\cdot}93\pi \text{ rad/s}$$
$$\therefore \text{ K.E.} = \frac{5000}{2 \times 32{\cdot}2} \times (4{\cdot}93\pi)^2 \times 4^2 \text{ ft-lbf}$$
$$= 299{,}000 \text{ ft-lbf.}$$
$$\text{Energy abstracted} = 307{,}000 - 299{,}000 \text{ ft-lbf}$$
$$= \mathbf{8000 \text{ ft-lbf.}}$$

Problems

3.1. Find the work done in each of the following cases:
(a) A load of 500 lbf raised a vertical distance of 5 ft.
(b) A cycle and rider weighing 150 lbf pedalled up a hill to a height of 70 ft.
(c) 2000 gallons of water pumped to a height of 350 ft.
(2500 ft-lbf, 10,500 ft-lbf, 7,000,000 ft-lbf.)

3.2. What power would be required to lift a load weighing 30 tonf through a vertical height of 40 ft in 15 min?
(5·43 hp.)

3.3. In a brake-horsepower test the circumference of the pulley was 20 ft. The effective load on the brake ($W-S$) was 795·4 lbf. If the average speed was 209·5 rev/min, find the bhp.
(102·1 bhp.)

3.4. A pit cage weighing 3 tonf is suspended by means of a rope weighing 5 lbf/ft. Calculate the power required to lift it at the rate of 2 ft/s, if the efficiency of the gearing is 85% and the cage is at the bottom of a pit 350 ft deep.
(36·2 hp.) (U.L.C.I.)

3.5. A shaft 80 yd deep, 7 ft in diameter, is half full of water. How much work is done when half this water is pumped out assuming that it is delivered at surface level? How long will it take a 5-hp pump to remove all the water? Take the weight of 1 ft^3 of water as 62·5 lbf.

(21,650,000 ft-lbf, 315 min.) (E.M.E.U.)

3.6. The force required to move a body over a specified distance is related to the distance as follows: 0–20 ft, a constant force of 55 lbf; 20–60 ft, a force decreasing uniformly from 55 lbf to 35 lbf; 60–80 ft a force decreasing uniformly from 35 lbf to zero. Give a graphical representation of force and distance and determine the work done by the force.

(3250 ft-lbf.) (U.E.I.)

3.7. An electric tram starts from rest at stop *A* and is driven to stop *B*. The tractive effort of the motors (i.e. the force available) acting on the wheel treads and the corresponding distances from *A* are as follows:

Tractive effort (lbf)	250	750	1250	1340	1360	1330	1250	1120	600	0
Distance from *A* (yd)	0	50	100	120	140	160	180	200	250	300

Plot the curve of tractive effort against distance from *A*. Find (a) the average tractive effort; (b) the total work done in ft-lbf.

(890 lbf, 801,000 ft-lbf.) (N.C.T.E.C.)

No answers are given to the following questions:

3.8. The water for a hydroelectric power station falls through a height of 2400 ft. If one gallon of water weighs 10 lbf how many gallons of water per day must flow through the turbines to generate 4000 horsepower continuously?

3.9. The following table gives the force *F* lbf acting on a moving body at different distances *x* ft from an initial position:

F lbf	10	17	25	38	45	50
x ft	0	7	15	28	35	40

Draw to scale the diagram of work and find the number of ft-lbf of work the diagram represents.

3.10. Explain what you understand by a "foot-pound-force of work".

A man performed 21 in.-tonf of work in raising a lorry by means of a jack. How many ft-lbf of work were done? If the man applied a force of 32 lbf throughout each working stroke of length 15 in., assuming no work lost, how many working strokes did he make? (U.E.I.)

3.11. A test on an engine gave the following results:

Diameter of brake wheel	2 ft $10\frac{1}{2}$ in.
Dead load	87 lbf.
Spring balance reading	10 lbf.
Speed	357 rev/min.

Find the brake horsepower of the engine.

CHAPTER 4

Simple Machines

4.1. Mechanical advantage, velocity ratio and efficiency

A device whereby a load at one point can be moved by a force applied at a different point is called a machine. Usually machines are constructed so that some advantage is gained by the operator. It may be that a small effort is required to overcome a large load or the effort may be applied in a more convenient position than by direct application to the load.

Machines can be very simple such as a rigid lever or single pulley, or very complicated such as automatic machines for mass-production operations. However complicated a machine may be it can be analysed into simple elements such as levers, pulleys, and gears.

There are certain fundamental properties regarding machines which enable the usefulness of the machine for a particular application to be assessed. Consider a machine illustrated in Fig. 4.1. A load W is being lifted by an effort P. The ratio of load to effort is called the mechanical advantage.

$$\text{Mechanical advantage (M.A.)} = \frac{\text{load}}{\text{effort}}.$$

Now some of the effort will be required to overcome internal resistances in the machine such as friction in bearings and gears, and therefore the mechanical advantage will vary depending on the magnitude of the load being lifted.

The effort P will move through a distance d when lifting the load W a height h.

The ratio, $\dfrac{\text{distance moved by effort}}{\text{distance moved by load in the same time}}$, is called the velocity ratio (V.R.) of the machine.

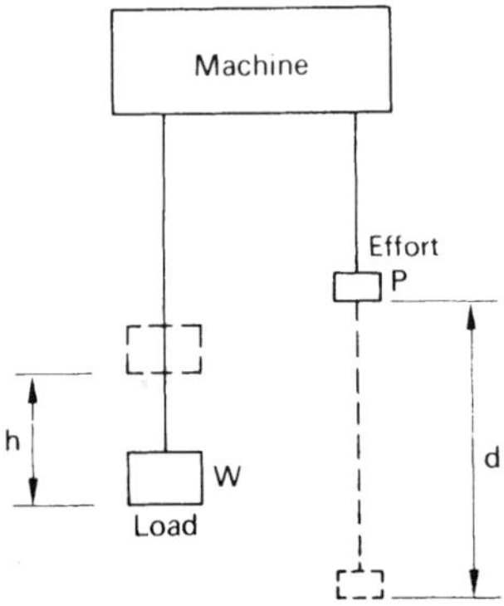

FIG. 4.1. Fundamental properties of a simple machine.

$$\text{Mechanical advantage} = \frac{\text{load}}{\text{effort}},$$

$$\text{Velocity ratio} = \frac{\text{distance moved by effort}}{\text{distance moved by load}}$$

The velocity ratio for a given machine will be constant no matter what load is being lifted. The only way to change the velocity ratio is to change the gears or alter the levers or pulleys in the machine. There are occasions where different velocity ratios can be obtained from a machine by incorporating a gear-box where a lever will select certain gears to be in mesh at a given time. The most common example is the automobile gear-box.

In practice it is found that the work put into a machine is greater than the work available at the output shaft of the machine due to the internal frictional resistances mentioned previously.

$$\text{The efficiency of a machine is given by } \frac{\text{work output}}{\text{work input}}$$

$$\text{and percentage efficiency} = \frac{\text{work output}}{\text{work input}} \times 100\%.$$

$$\text{Work output} = \text{load} \times \text{distance moved by load}$$

$$\text{and work input} = \text{effort} \times \text{distance moved by effort}$$

$$\therefore \text{ efficiency} = \frac{\text{load} \times \text{distance moved by load}}{\text{effort} \times \text{distance moved by effort}},$$

but $\frac{\text{load}}{\text{effort}} = \text{mechanical advantage (M.A.)}$

and $\frac{\text{distance moved by load}}{\text{distance moved by effort}} = \frac{1}{\text{velocity ratio}} \left(\frac{1}{\text{V.R.}}\right)$

$\therefore$ **efficiency** $= \frac{\textbf{M.A.}}{\textbf{V.R.}} \times \mathbf{100}\,\%$.

In a test on a machine having a velocity ratio of 25 the following values of load and effort were obtained:

Load, lbf	0	5	10	15	20	25	30	35
Effort, lbf	0·1	0·5	0·95	1·3	1·65	2·1	2·5	2·95

These values are plotted on a graph (Fig. 4.2).

Since the graph in which the values of effort, P, are plotted vertically and values of load, W, are plotted horizontally, is seen to be a straight line, then P and W must be related by an equation of the form $P = aW + b$.

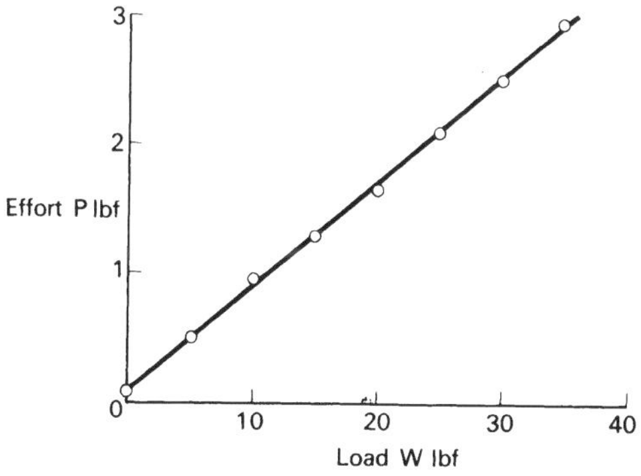

FIG. 4.2. Effort–load graph.

This is the general expression for the Law of a Machine. From the graph (Fig. 4.2) the values of a and b can be found for the particular machine.

a represents the gradient or slope of the graph,

b represents the intercept on the effort axis and is the effort required to overcome frictional resistances at no load.

From Fig. 4.2, $$a=\frac{2}{25}=0{\cdot}08,$$

$$b=0{\cdot}1,$$

$\therefore$ the Law of the Machine can be expressed as $\boldsymbol{P=0{\cdot}08W+0{\cdot}1}$.

Once the law of a machine has been determined it is possible to find the effort required to lift a given load or what load can be lifted by a given effort provided the load is within the capacity of the machine. Since the mechanical advantage varies with the load lifted and the velocity ratio remains constant, the efficiency M.A./V.R. will vary.

The efficiencies of the machine being considered based on the actual loads and efforts are shown in Table 4.1.

TABLE 4.1

Load (lbf)	Effort (lbf)	M.A.	V.R.	Efficiency $=\frac{\text{M.A.}}{\text{V.R.}}\times 100\%$
0	0·1	0	25	0
5	0·5	10	25	40·0
10	0·95	10·5	25	42·0
15	1·3	11·5	25	46·0
20	1·65	12·1	25	48·5
25	2·1	11·9	25	47·6
30	2·5	12	25	48·0
35	2·95	11·9	25	47·6

The graph of efficiency plotted against load Fig. 4.3 shows that the efficiency increases rapidly from zero for small loads and then the efficiency tends to become steady as larger loads are lifted. The maximum efficiency is approximately 48% for this machine within the range of loads lifted.

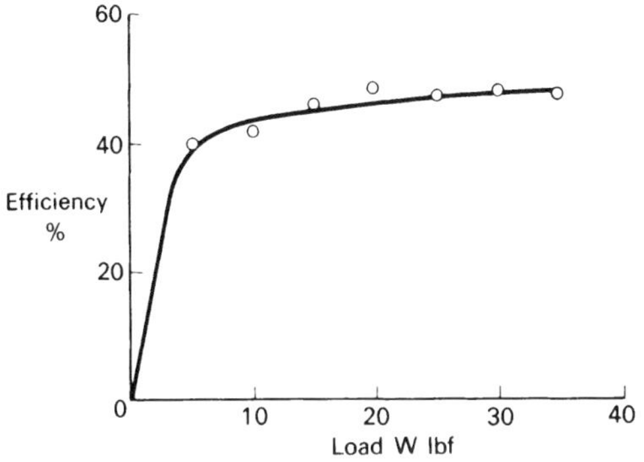

FIG. 4.3. Efficiency–load graph.

4.2. Simple machine elements

Levers. A lever of the first order (Fig. 4.4) has the effort applied on one side of the fulcrum and the load on the other side.

$$\text{V.R.} = \frac{\text{distance moved by effort}}{\text{distance moved by load}} = \frac{b\theta}{a\theta} = \frac{b}{a}.$$

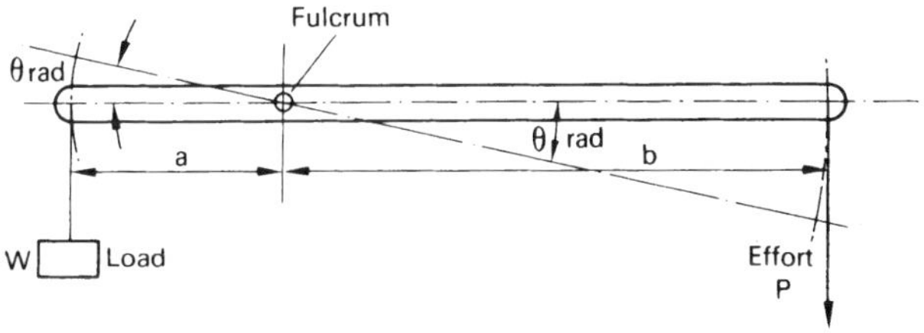

FIG. 4.4

A lever of the second order (Fig. 4.5) has the effort and load on the same side of the fulcrum, the effort being farther away from the fulcrum.

$$\text{V.R.} = \frac{\text{distance moved by effort}}{\text{distance moved by load}} = \frac{b\theta}{a\theta} = \frac{b}{a}.$$

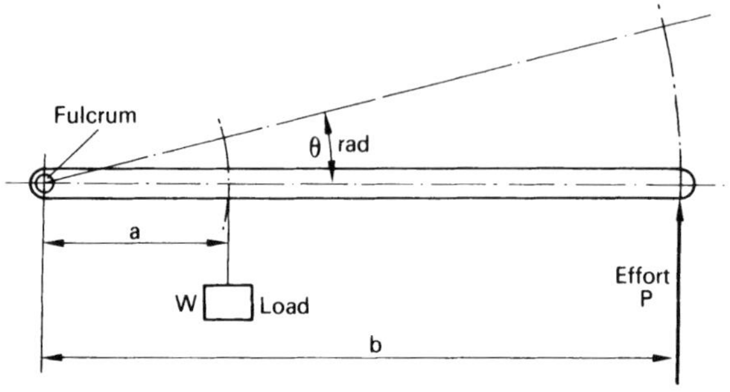

FIG. 4.5

A lever of the third order (Fig. 4.6) has the effort and load on the same side of the fulcrum but the effort is nearer to the fulcrum than the load.

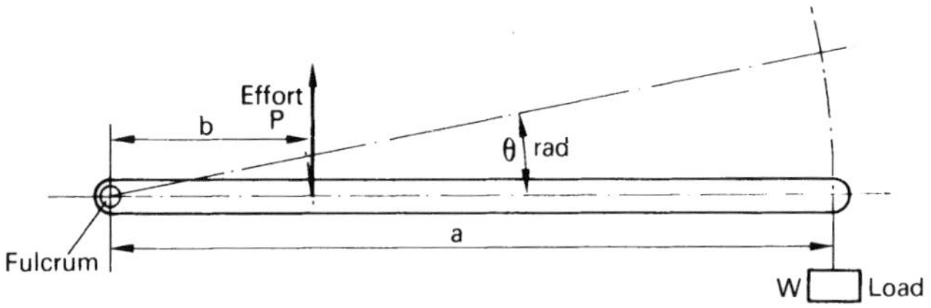

FIG. 4.6

$$\text{V.R.} = \frac{\text{distance moved by effort}}{\text{distance moved by load}} = \frac{b\theta}{a\theta} = \frac{b}{a}.$$

Pulleys. If it is assumed that a machine has no friction in it, then the efficiency would be 100%, in which case $\frac{\text{M.A.}}{\text{V.R.}} = 1$, i.e.

$$\text{M.A.} = \text{V.R.}$$

When calculating the velocity ratio of pulley systems it is easier to make the above assumption and find the mechanical advantage, which will then be equal to the velocity ratio.

Figure 4.7 shows a single pulley supported from a hanger. The pulley serves only to reverse the direction of the effort and the tension in the rope at P will be the same as that at W,

$$\therefore\ P = W,$$

$$\text{V.R.} = \text{M.A.} = \frac{W}{P} = 1.$$

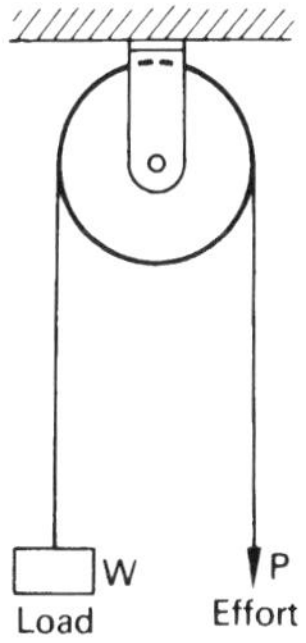

FIG. 4.7

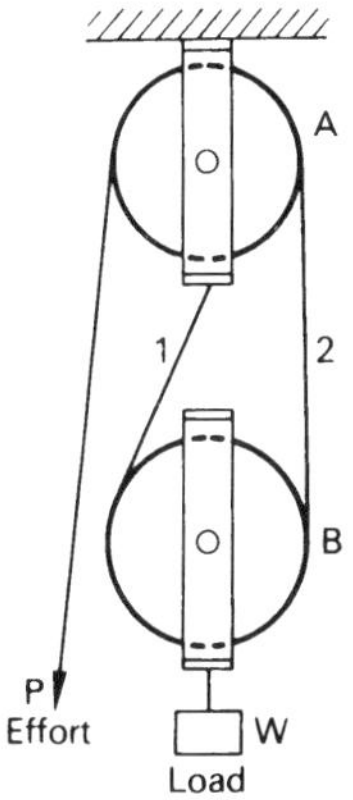

FIG. 4.8

A pulley block system where a single rope passes over pulley A then under pulley B and is fastened to A (Fig. 4.8) causes the load W to be supported by two portions of rope (1) and (2). The tension in the rope will then be $W/2$,

$$\therefore\ P = \frac{W}{2},$$

$$\text{V.R.} = \text{M.A.} = \frac{W}{W/2} = 2.$$

Figure 4.9 shows a three-pulley-block system in which the load is supported by three portions of rope at (1), (2) and (3). The tension in the rope will then be $W/3$,

$$\therefore P = \frac{W}{3},$$

$$\text{V.R.} = \text{M.A.} = \frac{W}{W/3} = 3.$$

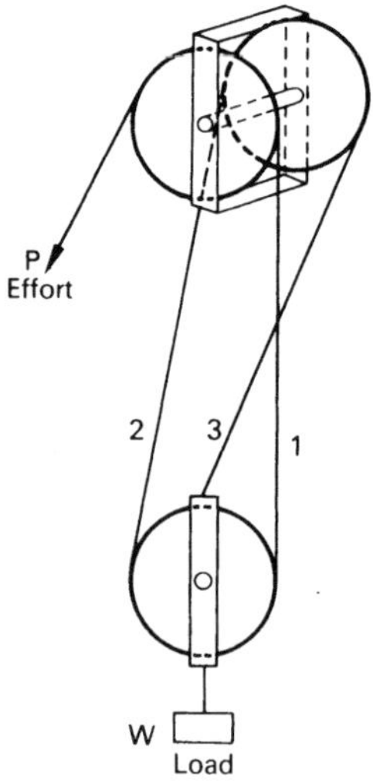

FIG. 4.9

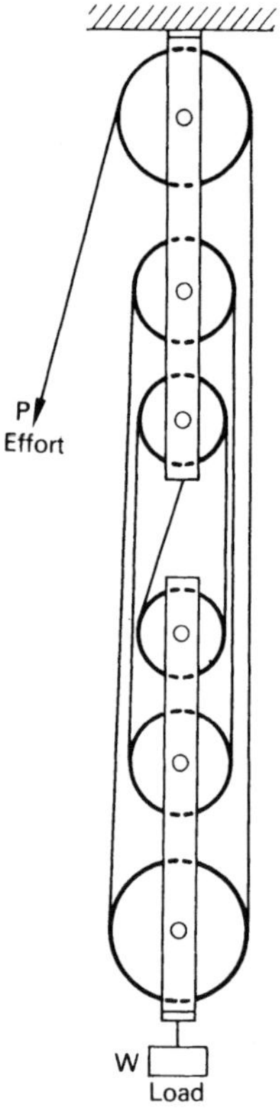

FIG. 4.10

A six-pulley-block system is shown in Fig. 4.10. The load is supported by six portions of rope, therefore the tension in the rope will be $W/6$,

$$\therefore P=\frac{W}{6},$$

$$\text{V.R.}=\text{M.A.}=\frac{W}{W/6}=6.$$

In general for pulley-block systems using a single rope the velocity ratio of the system is given by the number of pulleys in the system.

Gears. Gear wheels are so made that the two wheels can turn together without slipping. A tooth or projection on one wheel enters a recess on the opposite wheel. Figure 4.11 shows two

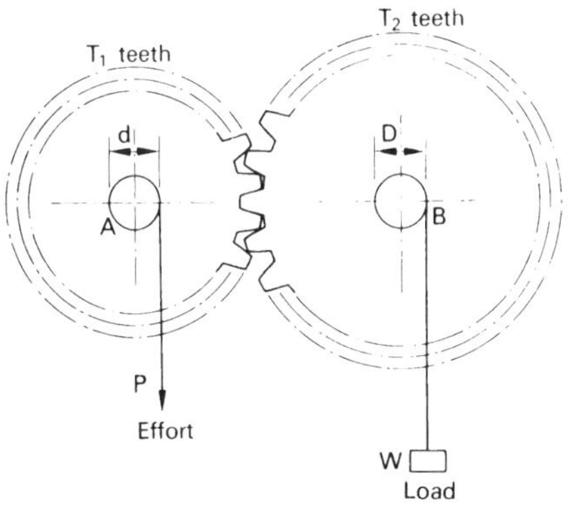

Fig. 4.11

wheels having T_1 teeth and T_2 teeth respectively. If the effort is applied by a rope over shaft A and the load is lifted by a rope over shaft B, then for one revolution of shaft B shaft A will move T_2/T_1 revolutions.

If shafts A and B are the same diameter, then

$$\text{V.R.} = \frac{\text{distance moved by effort}}{\text{distance moved by load}} = \frac{T_2}{T_1}.$$

Winch. A simple arrangement of a hand winch is shown in Fig. 4.12. The effort is applied at the hand lever radius R which rotates the shaft A to which is attached a gear wheel having T_1 teeth. This wheel engages with another gear having T_2 teeth fastened to shaft B. Also on shaft B is a rope barrel, diameter D, on to which the hoisting rope is wound.

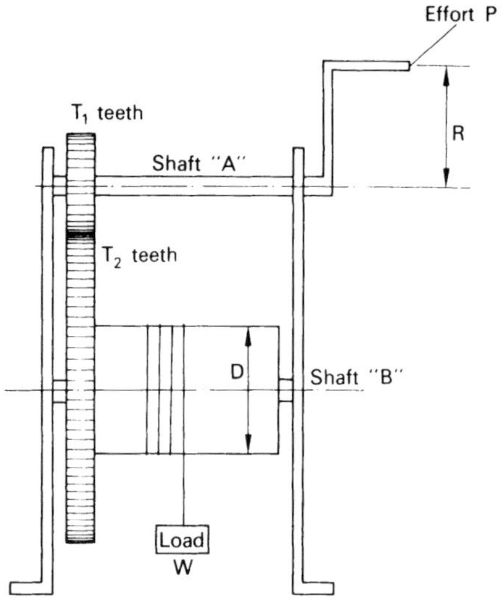

FIG. 4.12

The V.R. of the winch may be found by assuming the shaft B moves through one revolution, the distances moved by the load and effort being calculated. Distance moved by load in one revolution of shaft B equals πD.

Shaft A moves through $\frac{T_2}{T_1}$ revolutions,

therefore distance moved by effort $= 2\pi R \times \frac{T_2}{T_1}$.

$$\text{V.R.} = \frac{\text{distance moved by effort}}{\text{distance moved by load}} = \frac{2\pi R \times (T_2/T_1)}{\pi D}$$

$$= \frac{2R}{D} \times \frac{T_2}{T_1}.$$

Worm and wheel. A worm consists of a thread in the form of a helix cut from a piece of steel or bronze. This thread engages with a special gear wheel having T teeth. If the worm has only one thread running around the core it is called a single-thread worm; if there are two threads it is a double-thread worm and so on. For a single-thread worm (Fig. 4.13) one revolution of shaft A will rotate shaft B a distance corresponding to one tooth; a double-

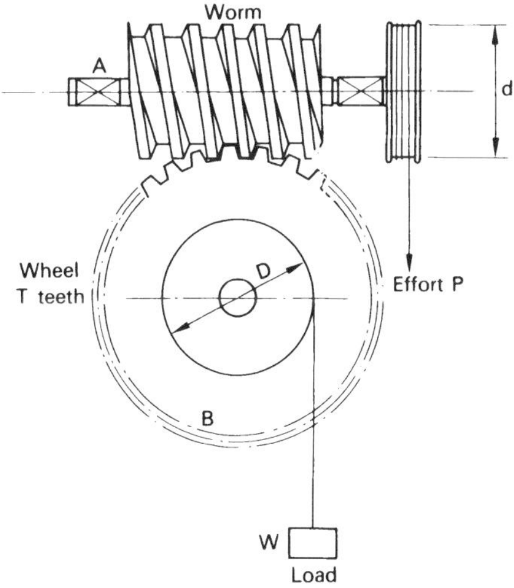

FIG. 4.13

threaded worm for one revolution of shaft *A* will move shaft *B* a distance corresponding to two teeth.

$$\text{V.R. for single-threaded worm} = \frac{\text{distance moved by effort}}{\text{distance moved by load}}$$

$$= \frac{T \times \pi d}{\pi D}$$

$$= \frac{\boldsymbol{d}}{\boldsymbol{D}} \times \boldsymbol{T}.$$

$$\text{V.R. for a double-threaded worm} = \frac{\boldsymbol{d}}{\boldsymbol{D}} \times \frac{\boldsymbol{T}}{\boldsymbol{2}}.$$

Wheel and axle. Figure 4.14 shows an arrangement of a wheel and axle. If D is the diameter of the effort wheel and d is the diameter of the axle, then $\text{V.R.} = \frac{\text{distance moved by effort}}{\text{distance moved by load}}$

$$= \frac{\pi d}{\pi D} = \frac{d}{D}.$$

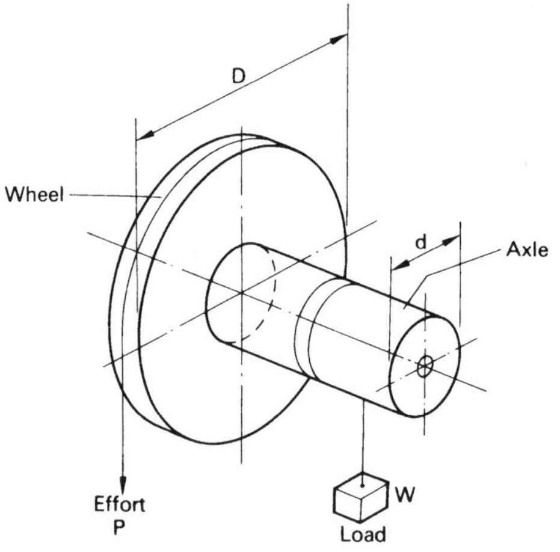

FIG. 4.14

EXAMPLE. A screw of $\frac{3}{4}$ in. pitch is operated by a lever 20 in. long. A force of 30 lbf is applied to the end of the lever to lift a load of 2400 lbf. Find the efficiency of the screw.

$$\text{Efficiency} = \frac{\text{M.A.}}{\text{V.R.}}$$

$$\text{M.A.} = \frac{\text{load}}{\text{effort}} = \frac{2400}{30} = 80.$$

$$\text{V.R.} = \frac{\text{distance moved by effort}}{\text{distance moved by load}}$$

$$= \frac{2\pi \times 20}{\frac{3}{4}} = 167{\cdot}5.$$

$$\text{Efficiency} = \frac{80}{167{\cdot}5} \times 100\%$$

$$= \mathbf{47{\cdot}8}\%.$$

EXAMPLE. An experimental worm and worm wheel consists of a single-threaded worm *A* and on the worm shaft is secured a pulley *B*, 4 in. diameter, around which passes a cord to which the effort is applied. The worm wheel *C* has eighty teeth and on the same shaft is secured a pulley *D*, 5 in. diameter, around which passes a rope with which the load is lifted. Make a sketch of the machine and find its velocity ratio. If the mechanical efficiency is 28% when a load of 60 lbf is being lifted, find the effort.

$$\text{Velocity ratio} = \frac{\text{distance moved by effort}}{\text{distance moved by load}}$$

$$= \frac{80 \times \pi \times 4}{\pi \times 5}$$

$$= \mathbf{64.}$$

$$\text{Efficiency} = \frac{\text{M.A.}}{\text{V.R.}}$$

$$\therefore \ \text{M.A.} = \frac{28}{100} \times 64 = 17{\cdot}9$$

$$\text{and M.A.} = \frac{\text{load}}{\text{effort}}$$

$$\therefore \text{ effort} = 60/17{\cdot}9 = \mathbf{3{\cdot}35\ lbf.}$$

EXAMPLE. In a test on a set of pulley blocks with a velocity ratio of 5 the following values of load and effort were obtained:

Load, lbf	0	14	28	42	56	70
Effort, lbf	3·5	7	11	15	19	24

Draw the graphs of load–effort, load–friction and load–efficiency and determine the law of the machine.

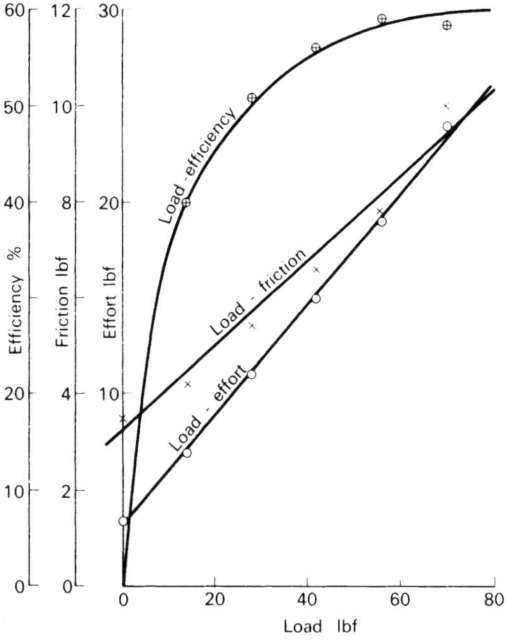

FIG. 4.15

The effect of friction in the machine is the difference between the actual effort and the ideal effort.

$$\text{The ideal effort} = \frac{\text{load}}{\text{V.R.}} = \frac{W}{\text{V.R.}}.$$

$$\text{Effort spent overcoming friction} = \text{actual effort} - \text{ideal effort}$$

$$= P - \frac{W}{\text{V.R.}}.$$

$$\text{Efficiency} = \frac{\text{M.A.}}{\text{V.R.}}.$$

Load (W) (lbf)	Actual effort (P) (lbf)	M.A.	V.R.	Friction (lbf)	Efficiency (%)
0	3·5	0	5	3·5−0=3·5	0
14	7	2	5	7−2·8=4·2	40·0
28	11	2·54	5	11−5·6=5·4	50·8
42	15	2·8	5	15−8·4=6·6	56·0
56	19	2·95	5	19−11·2=7·8	59·0
70	24	2·92	5	24−14=10·0	58·4

The law relating P and W is of the form $P = aW + b$.

From the load–effort graph the value of $a = \dfrac{20{\cdot}5}{70} = 0{\cdot}293$

and the value of $b = 3{\cdot}5$,

$\therefore$ the law of the machine is given by $\boldsymbol{P = 0{\cdot}293W + 3{\cdot}5}$.

Problems

4.1. A load of 65 lbf requires an effort of 18 lbf to lift it using a rope lifting block having six pulleys. What is the velocity ratio? Calculate the efficiency at this load.

(6, 60·2%.)

4.2. In a lifting machine the velocity ration is 30. An effort of 10 lbf is required to raise a load of 35 lbf and an effort of 25 lbf raises a load of 260 lbf. Find the effort required to raise a load of 165 lbf and the efficiency at this load.

(18·67 lbf, 29·4%.) (N.C.T.E.C.)

4.3. A machine raises a load of 360 lbf a distance of 10 in. An effort of 48 lbf is required which moves a distance of 90 in. Calculate the mechanical advantage, velocity ratio, effect of friction and efficiency at this load.

(7·5, 9, 8 lbf, 83·3%.)

4.4. A lever in which the fulcrum is placed at a point one-quarter of the length of the lever from the load has an efficiency of 89 per cent. If the available effort is 28 lbf find the load the lever could raise.

(74·8 lbf.)

4.5. What is meant by the "efficiency of a machine"?

A screw, having 8 threads per inch, is tightened up by a spanner 8 in. long. What is the velocity ratio? If a force of 25 lbf applied at the end of the spanner causes the screw to exert a force of 1200 lbf, what is the mechanical advantage? Calculate the efficiency.

(402, 48, 11·9%.) (N.C.T.E.C.)

4.6. A load of $\frac{1}{4}$ tonf is lifted by a worm wheel hoist with an efficiency of 28%. The hoist consists of a single-threaded worm driven at 120 rev/min through a pulley 10 in. diameter. The worm wheel has forty teeth and the load rope drum is 9 in. diameter. Calculate the velocity ratio and the horsepower of the motor required to drive the hoist.

(44·4, 0·428 hp.)

No answers are given to the following questions:

4.7. The following data were obtained from a small winch having a velocity ratio of 11. An effort of 5·7 lbf lifted a load of 40 lbf and an effort of 19·9 lbf lifted a load of 150 lbf. The effort–load graph is a straight line. Find the law of the machine, the effort required for a load of 200 lbf, and the efficiencies of the winch for loads of 40 lbf, 150 lbf, and 200 lbf. Plot an efficiency curve on a load base. (U.L.C.I.)

4.8. A hoisting winch has a handle 18 in. long. There are twenty-five teeth on the gear pinion and ninety-five teeth on the gear wheel. The diameter of the hoisting drum is 6 in. and the lifting rope 0·5 in. Calculate the velocity ratio.

If an effort of 25 lbf is applied to the handle, what load can be lifted if the efficiency is 78%?

4.9. The law of a machine taken from the load (W) and effort (P) graph where W and P are in lbf is $P = 0{\cdot}343W + 5{\cdot}63$.

Find (a) the effort which can raise a load of 200 lbf, (b) the effort required to operate the machine at no load, (c) the load which an effort of 22 lbf will lift, (d) the efficiency for a load of 120 lbf if the velocity ratio is 4.

4.10. The table of a pillar drilling machine weighs 120 lbf and requires a force of 8 lbf at the end of the elevating handle to raise it. One turn of the handle, which is 9 in. long, raises the table 0·25 in. Find the mechanical advantage, velocity ratio and efficiency of the mechanism.

CHAPTER 5

Motion and Inertia

5.1. Motion

When a body is moving it is said to have speed. To determine the speed it is necessary to know the distance travelled by the body and the time taken.

"Speed is the rate of change of distance."

If a body always moves over equal distances in equal lengths of time however small, then it is said to be moving at a constant speed.

In considering the average speed of a body it does not follow that the body is moving at a constant speed. It might be that the first half of the distance travelled was at a higher speed than the second half in which case the speed is not constant. The average speed is given by

$$\text{Average speed} = \frac{\text{distance moved}}{\text{time taken}}$$

Speed is a scalar quantity and has magnitude only. The unit of speed is feet per second (ft/s), kilometres per hour (km/h), miles per hour (m.p.h.). Any unit of distance divided by time will specify speed.

Velocity is a vector quantity and involves direction, and therefore to express velocity we need to know the speed in a given direction. We can change velocity by changing the speed or the direction, or by changing both of these.

A motor car is travelling along a straight road at 20 m.p.h. and 30 seconds later at 30 m.p.h. Therefore its speed has changed.

An aeroplane is travelling at 450 m.p.h. due W. and 30 seconds later at 450 m.p.h. due S. Its speed has not changed but its velocity has.

EXAMPLE. A cyclist moves at a constant speed of 12 m.p.h. How many feet will he travel in 10 seconds and how long will it take him to travel 27 miles?

Distance travelled per hour = 12 miles.

$$\text{Distance travelled per second} = \frac{12 \times 5280}{60 \times 60}\ \text{ft} = \frac{88}{5}\ \text{ft} = 17{\cdot}6\ \text{ft}.$$

$$\text{Distance travelled in 10 seconds} = 17{\cdot}6 \times 10\ \text{ft} = \mathbf{176\ ft.}$$

$$\text{Time taken to travel 27 miles} = \frac{27}{12}\ \text{h} = \mathbf{2{\cdot}25\ h,}$$

or **2 h 15 min.**

Whenever the velocity of a body changes, an acceleration is produced. It may be an increase in velocity called acceleration or a reduction in velocity called a retardation or deceleration.

"Acceleration is the rate of change of velocity."

Uniform acceleration is acceleration which does not vary however small the interval of time.

$$\text{Acceleration} = \frac{\text{change in velocity}}{\text{time}}.$$

A basic unit of acceleration is the foot per second per second, usually abbreviated to ft/s^2.

We frequently require to change the units involved in the calculations of velocity and acceleration.

To change miles per hour to feet per second multiply by the factor $\frac{88}{60}$. This is obtained by changing one mile into feet and one hour into seconds as follows:

$$\frac{1 \times 5280}{60 \times 60} = \frac{88}{60}.$$

Conversely, to change ft/s into m.p.h. multiply by $\frac{60}{88}$.

EXAMPLE. Convert 90 yd/min into ft/s.

$$90\ \text{yd/min} = \frac{90 \times 3}{1 \times 60}\ \text{ft/s} = \mathbf{4{\cdot}5\ ft/s.}$$

Convert 45 m.p.h. into ft/s.

$$45\ \text{m.p.h.} = 45 \times \frac{88}{60}\ \text{ft/s} = \mathbf{66\ ft/s.}$$

Convert 22 ft/s into m.p.h.

$$22\ \text{ft/s} = \frac{22 \times 60}{88}\ \text{m.p.h.} = \mathbf{15\ m.p.h.}$$

Convert an acceleration of 30 m.p.h./s into ft/s^2.

$$30\ \text{m.p.h./s} = 30 \times \frac{88}{60}\ \text{ft/s}^2 = \mathbf{44\ ft/s^2.}$$

5.2. Speed–time graphs

Plotting speed on the vertical axis and time on the horizontal axis a graph of uniform speed will look like Fig. 5.1.

Suppose the speed is constant at 4 ft/s and the time interval is 5 sec, then the distance travelled will be $4 \times 5 = 20$ ft. Note that the area of the graph represents this value.

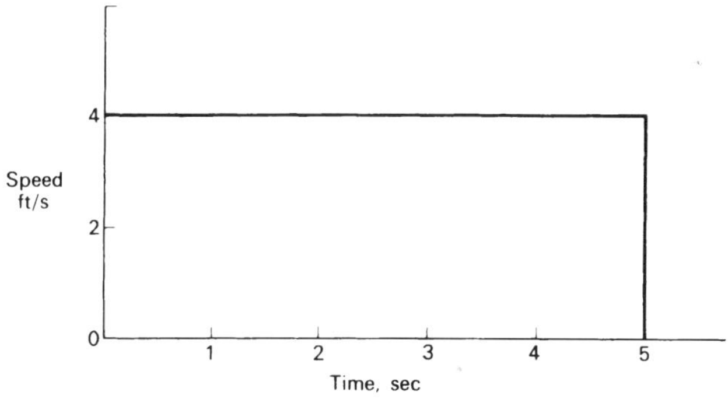

FIG. 5.1

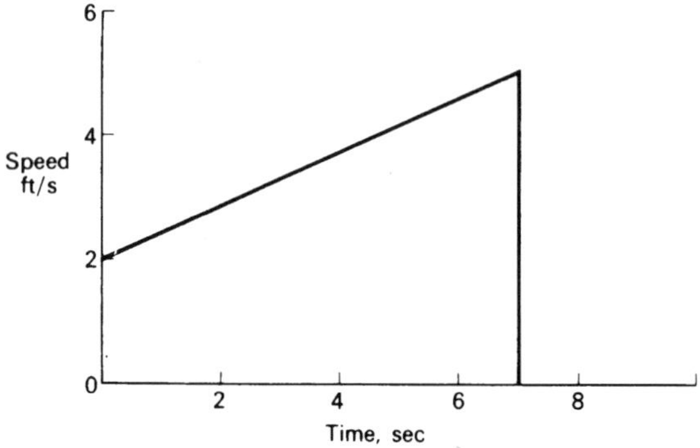

FIG. 5.2

A graph showing uniform increase in speed in equal time intervals is shown in Fig. 5.2. In a time interval of 7 sec the speed has increased from 2 ft/s to 5 ft/s.

$$\text{Distance travelled} = \text{average speed} \times \text{time}$$

$$= \frac{2+5}{2} \times 7 \text{ ft}$$

$$= 24\tfrac{1}{2} \text{ ft.}$$

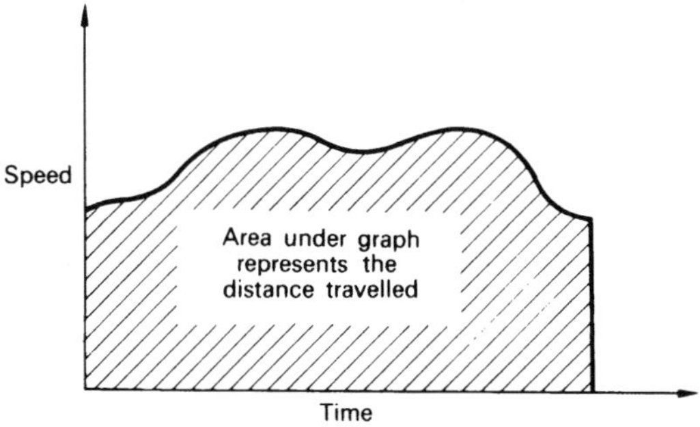

FIG. 5.3

The area under the speed–time graph (Fig. 5.3) again measures the distance travelled.

Any graph showing speed plotted against time will give the distance travelled as the area under the graph, provided the time unit is the same.

EXAMPLE. The following table gives the speed of a car at time intervals. Plot these values on a graph and find the distance travelled by the car.

Speed, ft/s	0	10	20	30	30	30	30	15	0
Time, seconds	0	1	2	3	4	5	6	7	8

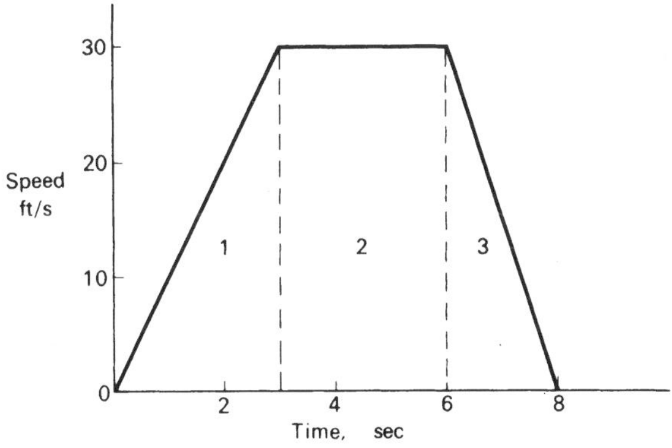

FIG. 5.4

Distance travelled = area under graph.
Divide the area into two triangles and a rectangle.

$$\text{Area of triangle No. } 1 = \frac{30 \times 3}{2} = 45 \text{ ft.}$$

$$\text{Area of rectangle No. } 2 = 30 \times 3 = 90 \text{ ft.}$$

$$\text{Area of triangle No. } 3 = \frac{30 \times 2}{2} = 30 \text{ ft.}$$

Total distance travelled = **165 ft.**

5.3. Equations of uniformly accelerated motion in a straight line

Using the following symbols let

s be the distance travelled (ft),
u the initial velocity (ft/s),
v the final velocity (ft/s),
f the acceleration (ft/s^2),
t the time (sec).

Now distance travelled = average velocity × time,

therefore $$s = \frac{(u+v)}{2} \times t. \qquad (1)$$

Also acceleration $= \dfrac{\text{change in velocity}}{\text{time}}$,

therefore $$f = \frac{v-u}{t}$$

or $$v = u + ft. \qquad (2)$$

Substituting the value of v in equation (2) into equation (1)

$$s = \frac{u+u+ft}{2} \times t$$

$$= \left(\frac{2u+ft}{2}\right)t,$$

$$\therefore s = ut + \tfrac{1}{2}ft^2. \qquad (3)$$

Squaring both sides of (2)

$$v^2 = u^2 + 2u\,ft + f^2t^2$$

$$= u^2 + 2f\,(ut + \tfrac{1}{2}ft^2),$$

and from equation (3) $s = ut + \frac{1}{2}ft^2$,

then $$v^2 = u^2 + 2fs. \qquad (4)$$

These equations enable most of the problems concerning uniformly accelerated motion to be solved. They are summarised as follows:

$$s=\left(\frac{u+v}{2}\right)t$$

$$v=u+ft$$

$$s=ut+\tfrac{1}{2}ft^2$$

$$v^2=u^2+2fs.$$

EXAMPLE. A car is travelling at 20 m.p.h. and 10 sec later its speed is 40 m.p.h. Find (a) the acceleration assuming it to be uniform and (b) the distance travelled during the 10 sec.

(a) Assemble the information down the side of your paper, converting the speeds to feet per second.

$$u=20\times\frac{88}{60}=\frac{88}{3}\text{ ft/s},$$

$$v=40\times\frac{88}{60}=\frac{2\times 88}{3}\text{ ft/s},$$

$$t=10\text{ sec},$$

$$f=?$$

Choose an equation which includes the four symbols u, v, t, and f, i.e.

$$v=u+ft.$$

Change the subject of the equation to f.

$$f=\frac{v-u}{t}$$

and substitute the values.

$$f=\frac{\frac{2\times 88}{3}-\frac{88}{3}}{10}\text{ ft/s}$$

$$=\frac{88}{3\times 10}\text{ ft/s}^2$$

$$=\mathbf{2{\cdot}93\ ft/s^2}.$$

(b) Reassemble the information.

$u = \frac{88}{3}$ ft/s,

$v = 2 \times \frac{88}{3}$ ft/s,

$t = 10$ sec,

$s = ?$

Choose an equation which includes u, v, t, s.

$$s = \frac{(u+v)}{2} \times t$$

$$= \frac{\left(\frac{88}{3} + \frac{2 \times 88}{3}\right)}{2} \times 10 \text{ ft}$$

$$= \frac{88}{2} \times 10 = \mathbf{440 \text{ ft.}}$$

EXAMPLE. A body has an initial velocity of 15 m.p.h. and is uniformly accelerated at the rate of 0·5 yd/s² for 0·5 mile. Find the final velocity in ft/s.

$u = 15 \times \frac{88}{60} = 22$ ft/s,

$f = \frac{1}{2} \times 3 = 1\frac{1}{2}$ ft/s²,

$s = 5280 \times \frac{1}{2} = 2640$ ft,

$v = ?$

$$v^2 - u^2 = 2fs,$$

$$\therefore v^2 = 2fs + u^2$$

$$= 2 \times 1\tfrac{1}{2} \times 2640 + 22^2$$

$$= 7920 + 484$$

$$= 8404,$$

$$\therefore v = \sqrt{8404}$$

$$= \mathbf{91 \cdot 7 \text{ ft/s.}}$$

EXAMPLE. A train reduces its speed from 60 m.p.h. to 20 m.p.h. in a distance of $\frac{1}{8}$ mile. Assuming uniform retardation, find how much farther the train will run before coming to rest and the time taken.

To find the retardation of the train.

$$u = 60 \times \frac{88}{60} = 88 \text{ ft/s},$$

$$v = 20 \times \frac{88}{60} = \frac{88}{3} \text{ ft/s},$$

$$s = \tfrac{1}{8} \times 5280 = 660 \text{ ft},$$

$$f = ?$$

$$v^2 - u^2 = 2fs,$$

$$\therefore f = \frac{v^2 - u^2}{2s}$$

$$= \frac{\left(\frac{88}{3}\right)^2 - 88^2}{2 \times 660} \text{ ft/s}$$

$$= \mathbf{-5 \cdot 22 \ ft/s^2}.$$

The negative sign indicates negative acceleration or retardation.

To find the distance travelled to rest.

$$u = 20 \times \frac{88}{3} = \frac{88}{3} \text{ ft/s},$$

$$v = 0,$$

$$f = -5 \cdot 22 \text{ ft/s}^2,$$

$$s = ?$$

$$v^2 - u^2 = 2fs$$

$$s = \frac{v^2 - u^2}{2f}$$

$$= \frac{0 - \left(\frac{88}{3}\right)^2}{-2 \times 5 \cdot 22} \text{ ft}$$

$$= \mathbf{82 \cdot 3 \ ft}.$$

To find the time taken.

$u = \frac{88}{3}$ ft/s,

$v = 0$,

$f = -5{\cdot}22$ ft/s^2,

$t = ?$

$$v = u + ft$$

$$t = \frac{v-u}{f}$$

$$= \frac{0 - \frac{88}{3}}{-5{\cdot}22}$$

$$= \mathbf{5{\cdot}61\ sec.}$$

5.4. Motion under the action of gravity

If a body is allowed to fall freely it moves towards the centre of the earth. It is found that the time taken to fall a given distance is the same for all sizes and shapes of objects and for bodies made from different materials providing they are free to fall and that there is no resistance to motion.

The acceleration due to gravity, usually given the symbol "g", varies slightly over the earth's surface. The accepted average value is 32·2 ft/s^2 or 981 cm/s^2.

If a body is thrown vertically upwards its retardation will be g ft/s^2 until it reaches the highest point when it will reverse its direction and start falling with an acceleration of g ft/s^2. The acceleration g is always acting downwards and we can put a negative sign in front of the g to signify downwards direction. This is equivalent to changing the signs in equations (2), (3) and (4) and regarding g as being numerically positive. This is shown in equations (5), (6) and (7) below.

Similarly if the initial velocity of the object is upwards we can call this a positive velocity and a downwards velocity negative.

Distances measured upwards from the point of projection are positive and downwards distances negative.

The distance is always measured from the point of projection and represents the distance from the starting-point which is not necessarily the same as the distance travelled by the object.

The equations used previously for uniformly accelerated linear motion can be modified to incorporate g as follows:

$$s = ut - \tfrac{1}{2}gt^2, \tag{5}$$

$$v^2 = u^2 - 2gs, \tag{6}$$

$$v = u - gt. \tag{7}$$

EXAMPLE. A body is allowed to fall freely from the top of a building 130 ft high. (a) How long will it take to reach the ground? (b) What will be its velocity on reaching the ground? (c) What will be the distance fallen after 1·25 sec?

Proceed with the solution as explained in the previous section.

(a) $u = 0$,
$s = 130$ ft,
$g = 32{\cdot}2$ ft/s^2,
$t = ?$

$$s = ut - \tfrac{1}{2}gt^2$$

$$\therefore -130 = 0 - \tfrac{1}{2} \times 32{\cdot}2t^2$$

$$\therefore \quad t^2 = \frac{130 \times 2}{32{\cdot}2}$$

$$\therefore \quad t = \sqrt{8{\cdot}07} \text{ sec}$$

$$= \pm \mathbf{2{\cdot}84 \text{ sec.}}$$

The positive answer signifies the time passing from when the body was allowed to fall.

(b) $u = 0$,
$s = 130$ ft,
$g = 32{\cdot}2$ ft/s^2,
$v = ?$

$$v^2-u^2=-2gs$$
$$\therefore\ v^2-0=2(-32{\cdot}2)(-130)$$
$$\therefore\ v^2=8372$$
$$\therefore\ v=\pm\mathbf{91{\cdot}5\ ft/s.}$$

The negative sign would be the correct one signifying downward velocity.

(c) $u=0$,

$g=32{\cdot}2$ ft/s^2,

$t=1{\cdot}25$ sec,

$s=?$

$$s=ut-\tfrac{1}{2}gt^2$$
$$=0-\tfrac{1}{2}\times 32{\cdot}2\times 1{\cdot}25^2$$
$$=-\mathbf{25{\cdot}2\ ft.}$$

The negative sign means downward distance.

EXAMPLE. A body is projected upwards from the top of a tower 96 ft high with a velocity of 64·4 ft/s. (a) How high does the body travel above the top of the tower? (b) What velocity has the object on reaching ground level?

(a) At the maximum height the velocity will be zero.

$u=+64{\cdot}4$ ft/s,

$v=0$,

$s=?$

$g=32{\cdot}2$ ft/s^2.

$$v^2-u^2=-2gs,$$
$$s=\frac{0-64{\cdot}4^2}{-2\times 32{\cdot}2}=\mathbf{64{\cdot}4\ ft.}$$

(b) $u=+64{\cdot}4$ ft/s,

$v=?$

$s=-96$ ft,

$g=32{\cdot}2$ ft/s^2.

$$v^2 - u^2 = -2gs,$$

$$v^2 - 64{\cdot}4^2 = 2(-32{\cdot}2)(-96),$$

$$v^2 = 6182{\cdot}4 + 4147{\cdot}4,$$

$$v = \sqrt{10{,}329{\cdot}8} \text{ ft/s}$$

$$= \pm \mathbf{101{\cdot}4 \text{ ft/s}}.$$

The negative sign will indicate downward velocity.

5.5 Angular motion

A set of equations similar to those given for linear motion can be derived for angular motion. It has been mentioned in Section 3.5 that the unit of angular displacement is the radian and this is used in calculations involving rotating bodies.

Let θ = angular displacement (radians),
ω_1 = initial angular velocity (rad/s),
ω_2 = final angular velocity (rad/s),
α = angular acceleration (rad/s^2),
t = time (sec).

Then angular displacement = average angular velocity × time

$$\therefore \theta = \left(\frac{\omega_1 + \omega_2}{2}\right) t. \tag{8}$$

Also angular acceleration $= \dfrac{\text{change in angular velocity}}{\text{time}}$

$$\therefore \alpha = \frac{\omega_2 - \omega_1}{t}$$

or

$$\omega_2 = \omega_1 + \alpha t. \tag{9}$$

Substituting the value of ω_2 in equation (9) into equation (8)

$$\theta = \left(\frac{\omega_1 + \omega_1 + \alpha t}{2}\right) t,$$

$$\therefore \theta = \omega_1 t + \tfrac{1}{2}\alpha t^2. \tag{10}$$

Squaring both sides of equation (9)

$$\omega_2{}^2 = \omega_1{}^2 + 2\omega_1 \alpha t + \alpha^2 t^2$$
$$= \omega_1{}^2 + 2\alpha(\omega_1 t + \tfrac{1}{2}\alpha t^2).$$

Substituting from equation (10) for $\omega_1 t + \frac{1}{2}\alpha t^2$

$$\omega_2{}^2 = \omega_1{}^2 + 2\alpha\theta. \qquad (11)$$

EXAMPLE. An electric motor has its speed increased from 120 rev/min to 240 rev/min in 30 sec. Calculate (a) the average angular acceleration, (b) the number of revolutions turned through by the motor shaft.

To convert rev/min to rad/s use a multiplying constant.

There are 2π radians in one revolution and 60 sec in 1 min.

$$\therefore \text{rev/min} \times \frac{2\pi}{60} = \text{rad/s}.$$

(a) $\omega_1 = 120 \times \dfrac{2\pi}{60} = 4\pi$ rad/s,

$\omega_2 = 240 \times \dfrac{2\pi}{60} = 8\pi$ rad/s,

$t = 30$ sec,

$\alpha = ?$

$$\omega_2 = \omega_1 + \alpha t,$$

$$\therefore \alpha = \frac{\omega_2 - \omega_1}{t}$$
$$= \frac{8\pi - 4\pi}{30}$$
$$= \frac{4\pi}{30}$$
$$= \mathbf{0{\cdot}419\ rad/s^2}.$$

(b) $\omega_1 = 4\pi$ rad/s,

$\omega_2 = 8\pi$ rad/s,

$t = 30$ sec,

$\theta = ?$

$$\theta = \frac{(\omega_1 + \omega_2)}{2} t$$

$$= \left(\frac{4\pi + 8\pi}{2}\right) \times 30 \text{ rad}$$

$$= \mathbf{180\pi \text{ rad.}}$$

There are 2π radians in the angle turned through in one revolution.

$$\text{Number of revolutions} = \frac{180\pi}{2\pi} = \mathbf{90.}$$

EXAMPLE. A flywheel is uniformly accelerated from rest and reaches a speed of 30 rev/min in 2 min. Find (a) the acceleration in rad/s², (b) the number of revolutions made by the wheel when the speed has reached 120 rev/min. (N.C.T.E.C.)

(a) $\omega_1 = 0$,

$\omega_2 = 30 \times \frac{2\pi}{60} = \pi$ rad/s,

$t = 2 \times 60 = 120$ sec,

$\alpha = ?$

$$\omega_2 = \omega_1 + \alpha t,$$

$$\alpha = \frac{\omega_2 - \omega_1}{t}$$

$$= \frac{\pi - 0}{120} \text{ rad/s}^2$$

$$= \mathbf{0{\cdot}0262\ rad/s^2}$$

(b) $\omega_1 = 0,$

$\alpha = 0{\cdot}0262 \text{ rad/s}^2,$

$\omega_2 = \dfrac{120 \times 2\pi}{60} = 4\pi \text{ rad/s},$

$\theta = ?$

$$\omega_2^2 - \omega_1^2 = 2\alpha\theta,$$

$$\theta = \frac{\omega_2^2 - \omega_1^2}{2\alpha}$$

$$= \frac{(4\pi)^2 - 0}{2 \times \dfrac{\pi}{120}} \text{ rad}$$

$$= \frac{16\pi^2 \times 120}{2 \times \pi} \text{ rad},$$

$$\therefore \theta = 960\pi \text{ rad.}$$

$$\text{Number of revolutions} = \frac{960\pi}{2\pi} = \mathbf{480.}$$

5.6. Momentum, inertia, mass, and weight

In Chapter 1 we considered the equilibrium of a set of forces, the forces being either stationary or moving at a constant speed. Let us now investigate the action of a body when a gradually increasing force is applied to it.

Take a 1-lb mass and place it on a smooth table (Fig. 5.5). Attach a piece of string to the mass and let the other end of the string hang over the side of the table with a weight hanger fastened to it. Apply weights to the weight hanger until the 1-lb mass just moves at a constant speed along the table. The system is now in equilibrium, the weights on the weight hanger being sufficient to overcome the frictional resistance of the 1-lb mass on the table.

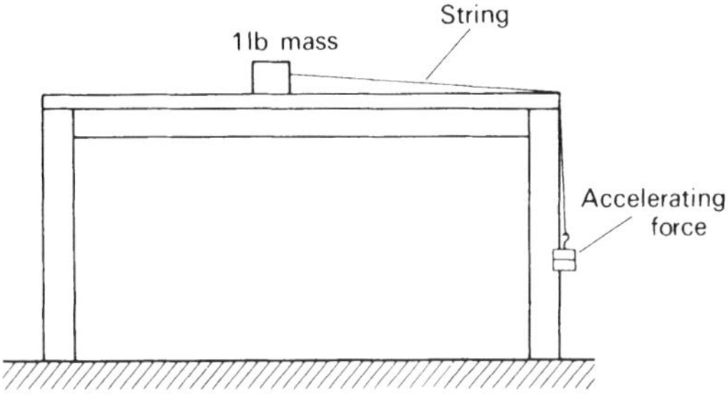

FIG. 5.5

Now add an extra weight on to the hanger and notice the response of the 1-lb mass; it begins to move faster and faster, that is, it accelerates. The extra force on the string has produced an acceleration on the 1-lb mass. Another weight added to the hanger will increase the acceleration of the 1-lb mass.

Sir Isaac Newton stated a second law relating the applied force and the resulting acceleration. "The rate of change of momentum is proportional to the impressed force."

To be able to apply this law requires certain terms to be defined and understood.

MOMENTUM

Momentum is defined as the quantity of motion possessed by a body and is measured by the product of the mass of the body and its velocity.

INERTIA

Inertia is a measure of the resistance a body has to change of motion.

A tanker full of petrol requires a larger force to accelerate it than does an empty tanker. Similarly, to stop a full tanker requires a larger braking force than to stop the empty one, i.e. the full tanker has more inertia than the empty one.

Mass

The mass of a body is the numerical measure of the inertia of the body. The only way to alter the mass of the body is to cut a portion away or to add more matter to the body.

Weight

The weight of a body is a measure of the earth's attraction on the body or the gravitational force exerted on the body.

It was stated in Chapter 1 that the gravitational attraction decreases as the bodies move away from each other; therefore it follows that the weight of a body will alter depending on where the body is weighed. There will be a slight difference in the weight of an object at the top of a high mountain to the weight in a deep pit.

Newton's second law states

$$\text{Force} \propto \text{rate of change of momentum.}$$

Let v_1 = initial velocity (ft/s),

v_2 = final velocity (ft/s),

m = mass (lb),

t = time to change from v_1 to v_2 (sec),

then first momentum $= mv_1$,

second momentum $= mv_2$.

Change in momentum $= m(v_2 - v_1)$.

Rate of change of momentum $= m\dfrac{(v_2 - v_1)}{t}$,

but acceleration is defined as $\dfrac{v_2 - v_1}{t} = f$.

Therefore the rate of change of momentum = mass × acceleration.

From Newton's second law; force $\propto$ rate of change of momentum.

Force $\propto$ mass $\times$ acceleration. (12)

Force = a constant $\times$ mass $\times$ acceleration. (13)

5.7. Systems of units

To be able to use expression (13) requires the value of the constant to be found, or alternatively the constant can equal unity if a force is chosen which will give a unit mass a unit acceleration.

This forms the basis of the different systems of units when dealing with problems involving forces and accelerations.

A force of 1 lbf is the gravitational attraction on a mass of 1 lb. If we take a mass of 1 lb and let it fall freely it will have an acceleration of g ft/s^2 and therefore we can write

1 lbf (force) = k (constant) $\times$ 1 lb (mass) $\times$ g ft/s^2 (acceleration)

$$\therefore\ k = \frac{1\ (\text{lbf})}{1\ (\text{lb}) \times g\ (\text{ft/s}^2)} = \frac{1\ \text{lbf-s}^2}{g\ \text{lb-ft}}.$$

The following systems take the value of $k = 1$ and derive new units for either force or mass.

FOOT-POUND-SECOND (F.P.S.) SYSTEMS

1. *Absolute system of units. New unit of force.*

   Unit of force is the POUNDAL (pdl).

   Unit of mass is the pound (lb).

   Unit of acceleration is the foot per second per second (ft/s^2).

Definition. The POUNDAL is that force which, when acting on a mass of one pound, produces an acceleration of one foot per second per second.

Force = mass $\times$ acceleration

1 (pdl) = 1 (lb) $\times$ 1 (ft/s^2).

Hence g pdl = 1 lbf.

2. *Gravitational or Engineer's system of units. New unit of mass.*

Unit of force is the pound-force (lbf).

Unit of mass is the SLUG.

Unit of acceleration is the foot per second per second (ft/s^2).

Definition. The SLUG is that mass which, when acted on by a force of one pound-force, has an acceleration of one foot per second per second.

$$\text{Force} = \text{mass} \times \text{acceleration}$$
$$1\ (\text{lbf}) = 1\ (\text{slug}) \times 1\ (\text{ft/s}^2).$$
$$\text{Hence}\quad 1\ \text{slug} = g\ \text{lb}$$

or a body weighing W lbf has a mass of W lb or $\frac{W}{g}$ slugs.

METRE-KILOGRAMME-SECOND (M.K.S.) SYSTEM

3. *Absolute system of units. New unit of force.*

Unit of force is the NEWTON (N). See also Section 10.8.

Unit of mass is the kilogramme (kg).

Unit of acceleration is the metre per second per second (m/s^2).

Definition. The NEWTON is that force which, when acting on a mass of one kilogramme, produces an acceleration of one metre per second per second.

$$\text{Force} = \text{mass} \times \text{acceleration}$$
$$1\ (\text{N}) = 1\ (\text{kg}) \times 1\ (\text{m/s}^2).$$

Other systems have been used in the past such as the centimetre-gramme-second (c.g.s.) system, but the m.k.s. system has replaced these.

The International System of units (SI) is to be adopted for use in this country and a gradual change-over will take place in the next few years.

5.8. Comparison of units

The following table gives the basic units of mass, force, energy and power in different systems.

System	Mass	Force	Energy	Power
SI absolute	kilogramme (kg)	newton (N)	joule (J)	watt (W)
M.K.S. absolute	kilogramme (kg)	newton (N)	joule (J)	watt (W)
M.K.S. gravitational	unit of mass $=9{\cdot}81$ kg	kilogramme-force 1 kgf $=9{\cdot}81$ N	kgf-m 1 kgf-m $=9{\cdot}81$ J	kgf-m/s 1 kgf-m/s $=9{\cdot}81$ W
C.G.S. absolute	gramme 1 g $=10^{-3}$ kg	dyne 1 dyne $=10^{-5}$ N	erg 1 erg $=10^{-7}$ J	erg/s 1 erg/s $=10^{-7}$ W
F.P.S. absolute	pound 1 lb $=0{\cdot}4536$ kg (1 kg $=2{\cdot}2$ lb)	poundal 1 pdl $=0{\cdot}1383$ N	ft-pdl 1 ft-pdl $=0{\cdot}04214$ J	ft-pdl/s 1 ft-pdl/s $=0{\cdot}04214$ W
F.P.S. gravitational	slug 1 slug $=32{\cdot}2$ lb $=14{\cdot}59$ kg	pound-force 1 lbf $=32{\cdot}2$ pdl $=4{\cdot}448$ N (1 N $=0{\cdot}225$ lbf)	ft-lbf 1 ft-lbf $=32{\cdot}2$ ft-pdl $=1{\cdot}356$ J	ft-lbf/s 1 ft-lbf/s $=32{\cdot}2$ ft-pdl/s $=1{\cdot}356$ W

EXAMPLE. A motor car weighing 1 tonf has an acceleration of $0{\cdot}3$ ft/s^2. What is the magnitude of the force producing this acceleration (a) in poundals, (b) in pound-force units?

(a) Using the f.p.s. absolute system

$$\text{Force} = \text{mass} \times \text{acceleration},$$

$$\text{Force} = P \text{ pdl},$$

$$\text{Mass} = 2240 \text{ lb},$$

$$\text{Acceleration} = 0{\cdot}3 \text{ ft/s}^2,$$

$$\therefore\ P = 2240 \times 0{\cdot}3 \text{ pdl}$$

$$= \mathbf{672\ pdl.}$$

(b) Using the f.p.s. gravitational system

$$\text{Force} = P \text{ lbf},$$

$$\text{Mass} = \frac{w}{g} = \frac{2240}{32 \cdot 2} \text{ slugs},$$

$$\text{Acceleration} = 0 \cdot 3 \text{ ft/s}^2,$$

$$\therefore P = \frac{2240}{32 \cdot 2} \times 0 \cdot 3 \text{ lbf}$$

$$= \mathbf{20 \cdot 9 \text{ lbf.}}$$

EXAMPLE. A train weighing 330 tonf reaches a speed of 30 m.p.h. from rest in $2\frac{1}{2}$ min. Find the force producing this acceleration (a) in poundals, (b) in pound-force units.

(a) $u = 0$,

$$v = 30 \times \frac{88}{60} = 44 \text{ ft/s},$$

$$t = 2 \cdot 5 \times 60 = 150 \text{ sec},$$

$$v = u + ft,$$

$$\therefore f = \frac{v - u}{t}$$

$$= \frac{44 - 0}{150} = \frac{44}{150} \text{ ft/s}^2.$$

$$\text{Force} = P \text{ pdl},$$

$$\text{Mass} = 330 \times 2240 \text{ lb},$$

$$\text{Acceleration} = \frac{44}{150} \text{ ft/s}^2,$$

$$\text{Force} = \text{mass} \times \text{acceleration}$$

$$= 330 \times 2240 \times \frac{44}{150} \text{ pdl}$$

$$= \mathbf{217{,}000 \text{ pdl.}}$$

(b) $$\text{Force} = P \text{ lbf},$$

$$\text{Mass} = \frac{330 \times 2240}{32 \cdot 2} \text{ slugs},$$

$$\text{Acceleration} = \frac{44}{150} \text{ ft/s}^2,$$

$$\text{Force} = \text{mass} \times \text{acceleration},$$

$$P = \frac{330 \times 2240}{32 \cdot 2} \times \frac{44}{150} \text{ lbf}$$

$$= \mathbf{6740 \text{ lbf.}}$$

EXAMPLE. A mass of 100 kg is supported by a vertical rope. Calculate the rope tension in newtons when the load is being lifted with an acceleration of 2 m/s^2.

Let the tension in the rope be T newtons. The rope tension T is both holding the weight up and also accelerating the weight therefore T must be greater than 100 kgf for the weight to accelerate upwards.

$$\text{Accelerating force} = T - (100 \times 9 \cdot 81) \text{ N},$$

$$\text{Mass} = 100 \text{ kg},$$

$$\text{Acceleration} = 2 \text{ m/s}^2,$$

$$\text{Force} = \text{mass} \times \text{acceleration}.$$

$$\therefore T - (100 \times 9 \cdot 81) = 100 \times 2 \text{ N},$$

$$\therefore T = 981 + 200 \text{ N}$$

$$= \mathbf{1181 \text{ N.}}$$

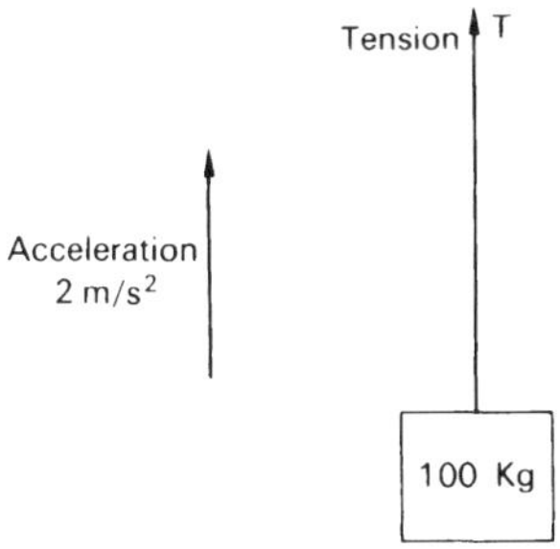

FIG. 5.6

EXAMPLE. Calculate the acceleration produced and the work done in 80 sec from rest when a force of

(a) 30 pdl acts on a mass of 0·75 ton,
(b) 4 kgf acts on a mass of 24 kg,
(c) 10 lbf acts on a mass of 200 lb,
(d) 50 N acts on a mass of 30 kg.

(a)

$$\text{Force} = 30 \text{ pdl},$$
$$\text{Mass} = 0{\cdot}75 \times 2240 \text{ lb},$$
$$\text{Force} = \text{mass} \times \text{acceleration},$$
$$\therefore\ 30 = 0{\cdot}75 \times 2240 \times f,$$
$$\therefore\ f = \frac{30}{0{\cdot}75 \times 2240} \text{ ft/s}^2 = \mathbf{0{\cdot}018\ ft/s^2}.$$

$u = 0$,
$t = 80$ sec,
$f = 0{\cdot}018$ ft/s^2,
$s = ?$

$$s = ut + \tfrac{1}{2} ft^2,$$
$$\therefore\ s = \tfrac{1}{2} \times 0{\cdot}018 \times 80^2 \text{ ft}.$$
$$= 57{\cdot}1 \text{ ft}.$$

$$\text{Work done} = \text{force} \times \text{distance}$$
$$= 30 \times 57{\cdot}1 \text{ ft-pdl}$$
$$= 1713 \text{ ft-pdl}$$
$$= \frac{1713}{32{\cdot}2} \text{ ft-lbf}$$
$$= \mathbf{53{\cdot}2\ ft\text{-}lbf.}$$

(b)

$$\text{Force} = 4 \text{ kgf}$$
$$= 4 \times 9{\cdot}81 \text{ N},$$
$$\text{Mass} = 24 \text{ kg},$$
$$\text{Force} = \text{mass} \times \text{acceleration}$$
$$4 \times 9{\cdot}81 = 24 \times f$$
$$\therefore\ f = \frac{4 \times 9{\cdot}81}{24} \text{ m/s}^2 = \mathbf{1{\cdot}635\ m/s^2}.$$

$u = 0,$
$t = 80$ sec,
$f = 1{\cdot}635$ m/s^2,
$s = ?$

$$s = ut + \tfrac{1}{2} ft^2,$$
$$\therefore\ s = \tfrac{1}{2} \times 1{\cdot}635 \times 80^2 \text{ m}$$
$$= 5240 \text{ m.}$$

Work done = force × distance
$= 4 \times 9{\cdot}81 \times 5240$ Nm
$=$ **205,500 Nm.**

Now 1 Nm = 1 joule,
∴ work done = **205,500 J.**

(c) Force = 10 lbf,

$$\text{Mass} = \frac{200}{32{\cdot}2} \text{ slugs,}$$

Force = mass × acceleration,

$$10 = \frac{200}{32{\cdot}2} \times f,$$
$$\therefore\ f = \frac{10 \times 32{\cdot}2}{200} \text{ ft/s}^2$$
$$= \mathbf{1{\cdot}61\ ft/s^2}.$$

$u = 0,$
$t = 80$ sec,
$f = 1{\cdot}61$ ft/s^2,
$s = ?$

$$s = ut + \tfrac{1}{2} ft^2$$
$$= \tfrac{1}{2} \times 1{\cdot}61 \times 80^2 \text{ ft}$$
$$= 5150 \text{ ft.}$$

Work done = force × distance
$= 10 \times 5150$ ft-lbf
$=$ **51,500 ft-lbf.**

(d)

$$\text{Force} = 50 \text{ N},$$
$$\text{Mass} = 30 \text{ kg},$$

$$\text{Force} = \text{mass} \times \text{acceleration},$$
$$50 = 30 \times f,$$
$$\therefore f = \frac{\mathbf{50}}{\mathbf{30}} \ \mathbf{m/s^2}.$$

$u = 0,$

$t = 80$ sec,

$f = \frac{5}{3}$ m/s^2,

$s = ?$

$$s = ut + \tfrac{1}{2} ft^2$$
$$= \tfrac{1}{2} \times \frac{5}{3} \times 80^2$$
$$= 5350 \text{ m}.$$

$$\text{Work done} = \text{force} \times \text{distance}$$
$$= 50 \times 5350 \text{ Nm}$$
$$= 267{,}500 \text{ Nm}$$
$$= \mathbf{267{,}500\ J.}$$

Problems

5.1. Convert into ft/s: (a) 15 m.p.h., (b) 50 m.p.h., (c) 180 m.p.h. Convert into m.p.h.: (d) 20 ft/s, (e) 66 ft/s, (f) 176 ft/s.

(22 ft/s, 73·3 ft/s, 264 ft/s. 13·6 m.p.h., 45 m.p.h., 120 m.p.h.)

5.2. Convert into rad/s: (a) 15 rev/min, (b) 250 rev/min, (c) 4000 rev/min. Convert into rev/min: (d) 75 rad/s, (e) 350 rad/s, (f) 6500 rad/s.

(1·57 rad/s, 26·2 rad/s, 418 rad/s. 715 rev/m, 3660 rev/m, 62,000 rev/m.)

5.3. A body comes to rest in a distance of 100 ft with a uniform retardation of 8 ft/s^2. What was its initial velocity?

(40 ft/s.)

5.4. A vehicle is accelerated at 4 ft/s^2 from a speed of 15 m.p.h. Calculate (a) the time required to reach a speed of 45 m.p.h., (b) the distance travelled in this time.

If the vehicle is now retarded uniformly until it comes to rest after travelling a further 363 yd, calculate the retardation in ft/s^2.

(11 sec, 484 ft, 2 ft/s^2.) (U.E.I.)

5.5. A body starts from rest and accelerates uniformly for 24 sec by which time a velocity of 32 ft/s has been reached. This velocity is maintained until the body has travelled a distance of 1200 ft from rest. Find the distance travelled in the first 24 sec and the time taken to cover the distance of 1200 ft. If the body is then decelerated at 2 ft/s^2 find the total time taken and the total distance covered from start to rest. (N.C.T.E.C.)

(384 ft, 49·5 sec, 1456 ft.)

5.6. The armature of an electric motor is rotating at 5000 rev/min. The motor comes to rest 10 sec after the current is switched off. Calculate (a) the average angular retardation of the armature, (b) the number of revolutions made by the armature when the current is switched off.

(52·3 rad/s^2, 416·67 rev.) (U.E.I.)

5.7. Water is projected vertically upwards at 100 ft/s. Find (a) the greatest height to which the water rises, (b) the time for the water to reach ground level again.

(155·2 ft, 6·21 sec.)

5.8. Define the Newton in terms of mass and acceleration. A motor car weighs 2000 lbf and on a level road its speed increases from 20 to 50 m.p.h. in 10 sec. Neglecting friction and other resistances to motion, find (a) the constant force necessary, (b) the hp developed at 20 m.p.h. ($g = 32{\cdot}2$ ft/s^2).

(234 lbf, 14·5 hp.) (N.C.T.E.C.)

5.9. What is meant by (a) the weight of a body, (b) the momentum of a body?

The speed of a car at a given instant is 75 m.p.h. It is given a uniform deceleration of 4 ft/s^2 until its speed is reduced to 30 m.p.h. Determine the time taken to reduce its speed and the distance travelled by the car during deceleration.

(16·55 s, sec, 907·5 ft.)

5.10. Define the unit of mass used in the f.p.s. system of units and state the name of the unit.

A heavy load on a crane would accelerate rapidly if the brake were removed. Explain (a) why this occurs and (b) what happens to the energy acquired by the load when the brakes are applied.

5.11. Define the unit of force in the m.k.s. system of units. For what are the letters m.k.s. an abbreviation?

A cyclist mounts his bicycle at the top of a hill and free-wheels down it. With reference to potential and kinetic energy, explain why the cycle descends the hill.

What energy conversion takes place when the cyclist applies his brakes? (N.C.T.E.C.)

5.12. Explain what is meant by (a) momentum (b) inertia.

Describe an application in engineering where inertia is of importance.

5.13. Given that a force of 1 lbf accelerates a mass of 1 lb at 32·2 ft/s^2 define the unit of force in the m.k.s. system.

Show that 1 newton = 0·225 pound-force and 1 foot-pound-force = 1·359 joules.

(1 ft = 30·48 cm, 1 kg = 2·2 lb.)

5.14. Define (i) velocity, (ii) acceleration.

A body changes its speed from 10 m.p.h. to 40 m.p.h. in 1 min. Determine its acceleration and the distance moved in this time.

If the body weighs 20 lbf, calculate the force required to produce the acceleration in (a) pound-force, (b) newtons (1 lb=0·45 kg, g=9·81 m/s^2, 1 ft=0·3 m.)

(0·73 ft/s^2, 2200 ft, 0·45 lbf, 1·98 N.) (N.C.T.E.C.)

No answers are given to the following questions.

5.15. A train is reduced in speed from 28 m.p.h. to 25 m.p.h. in 520 yd after which it is brought to rest in 2 min. Determine the average retardation in ft/s^2 during each stage. (U.L.C.I.)

5.16. A rifle bullet is fired vertically upwards with a velocity of 2500 ft/s. Neglecting air resistance determine the height to which it will rise and the time taken to reach ground level.

5.17. An electric train starting from rest at one station comes to rest at the next station 7200 ft away in 3·5 min, having first a uniform acceleration then a uniform speed for 2·5 min, then a uniform retardation. The time of acceleration is twice the time of retardation. Find (a) the uniform speed in m.p.h. at which the train runs for 2·5 min, (b) the uniform acceleration in ft/s^2, (c) the uniform retardation in ft/s^2, (d) the space passed over in ft during the acceleration period, (e) the space passed over in ft during the retardation period. (U.E.I.)

5.18. A weight of 1000 lbf is accelerated from rest by a constant force of 100 lbf. Calculate the velocity of the body, the distance travelled and the work done by the force after 5 sec.

5.19. A passenger lift has an upward acceleration of 1·5 ft/s^2. What force will a man weighing 140 lbf exert on the floor of the lift? Find the force if the lift is descending with an acceleration of 2 ft/s^2.

5.20. A train weighing 200 tonf has a frictional resistance of 12 lbf/tonf. What average force will be required to give it a speed of 30 m.p.h. in 1½ min from rest on the level? What horsepower will the engine be developing when the speed is 30 m.p.h.?

CHAPTER 6

Elasticity

6.1. Hooke's law

When a force is applied to a body and the body is resistant to movement either by frictional effects or more permanent fastenings such as bolts or rivets, then the body changes shape. The extent to which this takes place depends on the magnitude of the force and also on the material from which the body is made.

The response of the body to the applied force measures its elasticity. A piece of rubber can be stretched easily and as soon as the stretching force is taken away the rubber returns to its original dimensions. On the other hand, a piece of chalk as used in the classroom shows no apparent change in shape before it breaks. All materials deform when loaded although the deformation caused may be so small that it is not noticeable.

Providing the force applied to the body is not excessive compared to the strength of the material, the material will return to its original length when the force is taken away. The extension of the material is then said to be within the "elastic limit". If the force is increased then the material suffers a permanent extension and the elastic limit has been exceeded.

EXPERIMENT. *To determine the extension of a spring as different loads are applied.*

Equipment required. Spring, stand, metre rule, weights, weight hanger. Hang the spring from the stand and to the other end of the spring attach the weight hanger. Take the rule and measure the height from the bench top to the underside of the hanger.

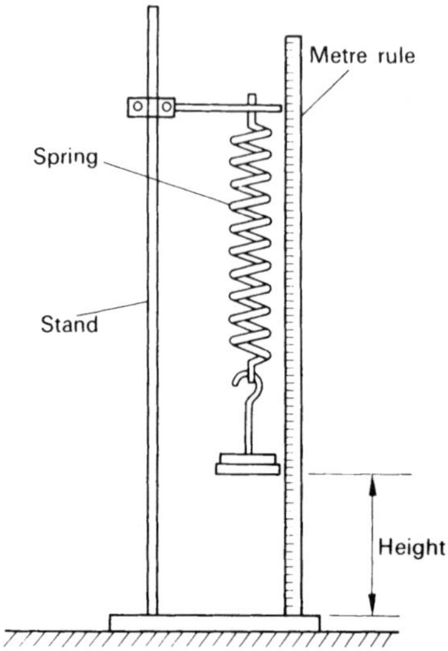

FIG. 6.1

Apply a weight of 1 lbf on to the hanger and remeasure the height. Repeat this until sufficient weight has been applied to the hanger and tabulate the results. Take off the weights and establish whether the spring returns to its original length.

In such an experiment the results shown in Table 6.1 were obtained.

TABLE 6.1

Weight (lbf)	Height above bench (in.)	Extension of spring (in.)
0	15·75	0
1	14·0	1·75
2	12·25	3·5
3	10·5	5·25
4	8·75	7·0
5	7·0	8·75
6	5·25	10·5
7	3·5	12·25
8	1·75	14·0

When all the weights were taken off the spring the distance to the underside of the hanger measured 15·75 in. This showed that the material was not stretched permanently and therefore was working within its elastic limit.

A graph of load against extension is shown in Fig. 6.2.

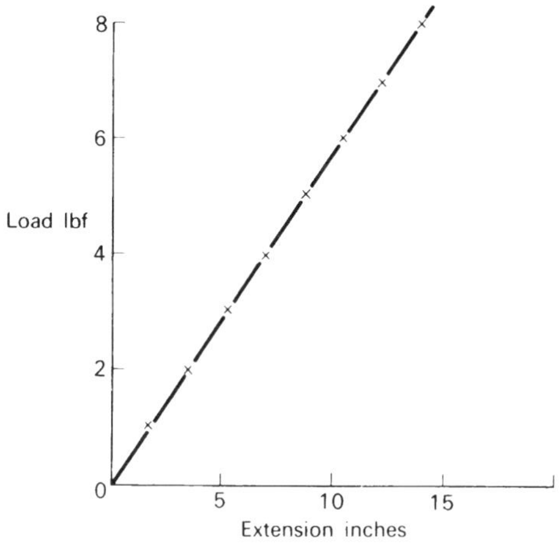

FIG. 6.2

EXPERIMENT. *To determine the deflection of a beam as different loads are applied.*

Equipment required. Steel beam, two knife edge supports, dial gauge and stand, weights and weight hanger.

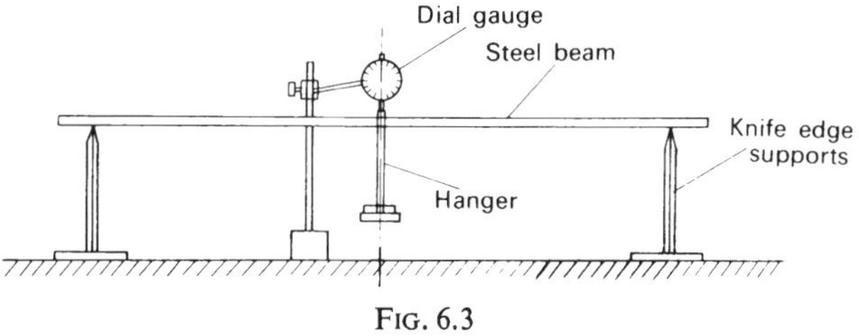

FIG. 6.3

Support the beam on the two knife edges and with the weight hanger in the centre of the beam set the dial gauge over the weight hanger and zero the gauge. Apply loads in 1-lbf increments to the hanger and each time take a reading from the dial gauge. Take off all the weights and establish whether the beam returns to its original position. The results in Table 6.2 were obtained in such an experiment.

TABLE 6.2

Load (lbf)	Deflection (1/1000 in.)
0	0
1	55
2	110
3	165
4	220
5	275
6	335
7	390
8	442

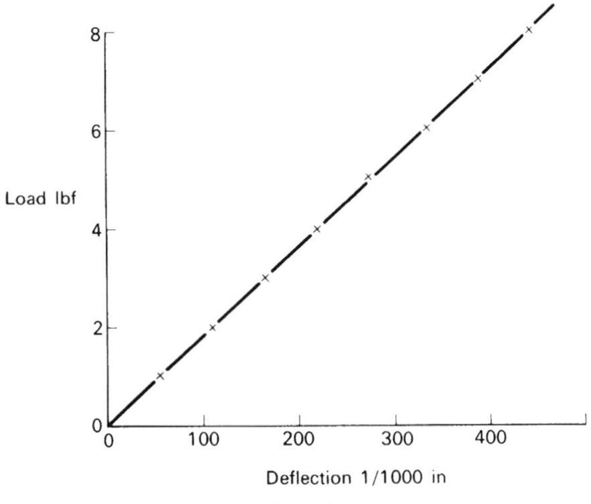

FIG. 6.4

The reading on the dial gauge when all the load was taken off was zero.

A graph of load against deflection is shown in Fig. 6.4.

EXPERIMENT. *To find the relationship between the angle of twist and the torque applied to a circular shaft.*

Equipment required. Steel rod of circular cross-section, two pointers to attach to the rod, pulley, frame, two scales for angular measurement, weight hanger and weights.

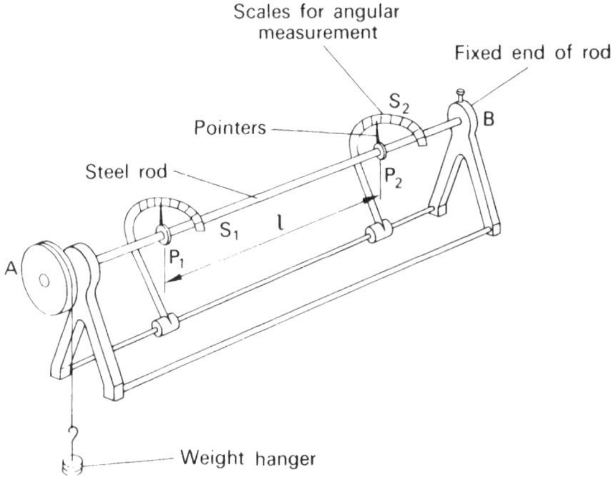

FIG. 6.5

Fix the rod at B and to the other end attach a pulley A. Fasten the two pointers to the shaft a given distance apart to read zero on the scales. Put weights on the hanger in 0·5-lbf increments, taking readings on the scales S_1 and S_2. Now the pointer P_1 will turn more than the pointer P_2 and to measure the angle of twist over length l the reading P_2 must be subtracted from the reading P_1. The torque is measured by multiplying W, the weight attached, by the radius of the pulley. Take off the weights and establish whether the rod returns to its original position.

In such an experiment the results in Table 6.3 were obtained.

TABLE 6.3

Torque	Angular displacement degrees		
$W \times r$ (lbf-in.)	P_2	P_1	$P_1 - P_2$
1	$\frac{1}{8}$	1	$\frac{7}{8}$
2	$\frac{1}{4}$	2	$1\frac{3}{4}$
3	$\frac{3}{8}$	3	$2\frac{5}{8}$
4	$\frac{1}{2}$	4	$3\frac{1}{2}$
5	$\frac{5}{8}$	5	$4\frac{3}{8}$
6	$\frac{3}{4}$	6	$5\frac{1}{4}$
7	$\frac{7}{8}$	7	$6\frac{1}{8}$
8	1	8	7

A graph of torque against angle of twist is shown in Fig. 6.6.

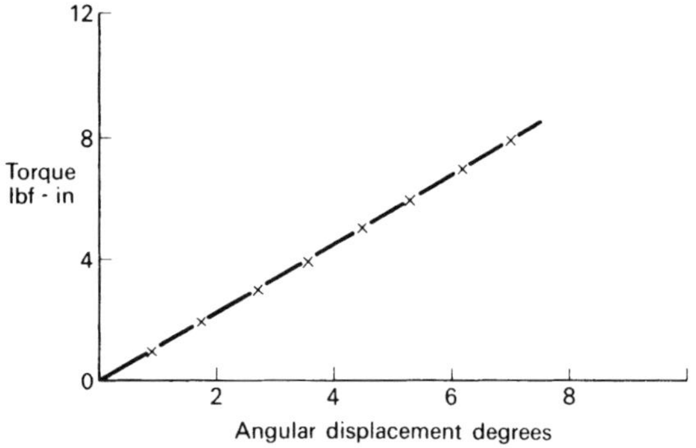

FIG. 6.6

From the results of these three experiments it should be possible to deduce that the relationship in each case between the applied force and the deformation is a linear one. We can say the extension of the spring is proportional to the applied load, the deflection of the beam is proportional to the applied load and the angle of twist of the rod is proportional to the torque applied.

Furthermore, when the load was taken away the material returned to its original dimensions. This indicated the material was loaded within the elastic limit.

To illustrate this point let us take a piece of steel wire and apply loads gradually to it and measure accurately the extension produced.

EXPERIMENT. *To examine the behaviour of a length of steel wire under gradually increasing load.*

Equipment required. Piece of steel wire approximately 4 ft long, vernier measuring attachment fitted to wall, weight hanger and weights, metre rule.

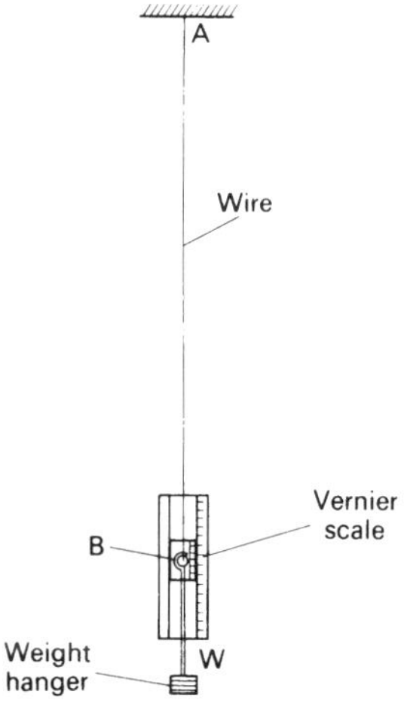

FIG. 6.7

Attach a length of steel wire to the wall at A (Fig. 6.7) and the other end to a sliding vernier scale at B. Measure the length of AB and the diameter of the wire. Apply weights at W each time measuring the extension of the wire.

The readings in Table 6.4 were obtained from a piece of wire 0·055 in. diameter and 66·3 in. long.

TABLE 6.4

Load, lbf	0	10	20	30	40	50	60	70	80	90
Extension, in.	0	0·01	0·021	0·034	0·049	0·061	0·073	0·098	0·19	0·40
Load, lbf	95	100	105	110	115	120	125	130	135	140
Extension, in.	0·55	0·75	1·02	1·26	1·60	1·92	2·28	2·80	3·25	3·82
Load, lbf	145	150	151	Wire broke						
Extension, in.	4·65	5·56	6·90							

A graph of these readings is shown in Fig. 6.8. Notice from the graph that the wire behaves in a different manner for values of loading from 0 to 60 lbf to the loading from 60 to 150 lbf. The first

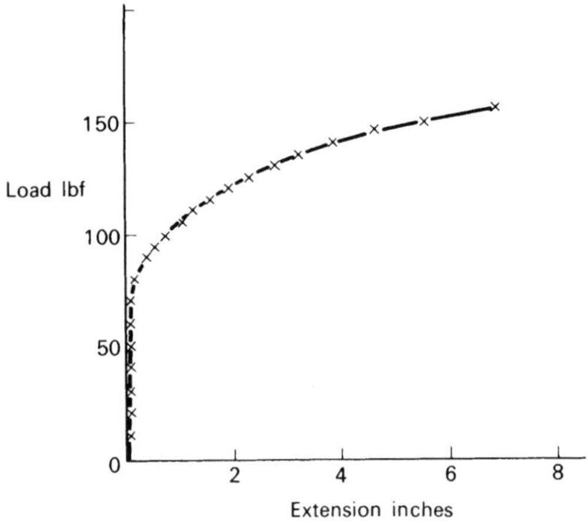

FIG. 6.8

section of the graph shows the linear relationship between load and deflection similar to that obtained from the previous experiments, but then the graph begins to curve and continues to do so until the wire breaks.

The first section of the graph gives the "elastic extension" of the wire, and the second section the "plastic extension".

If the experiment is repeated and after applying a load of 30 lbf the load is removed it will be found that the wire returns to its original length. If a 100-lbf load is applied to the wire and then removed it will be found that the wire does not return to its original length, in which case we say the material has a "permanent set" or a "permanent extension". The material has extended beyond the elastic limit and a permanent extension has taken place.

Most of the work we do in calculations regarding the strength of materials is within the elastic limit, although modern techniques use calculations within the plastic stages of deformation.

Hooke's Law

Robert Hooke, a scientist in the seventeenth century, performed experiments on wires and springs similar to the ones previously mentioned and he established the law relating load and extension. "Hooke's law states that within the elastic limit the deformation produced is proportional to the force producing it."

6.2. Stress

A load applied to a solid causes internal actions between the molecules of the material and when this happens the material is said to be in a state of stress. The more material available for transmitting a load or force the less the reaction between the molecules and therefore the less the stress.

Intensity of stress

The intensity of stress is the load or force per unit area.

$$\text{Intensity of stress} = \frac{\text{load}}{\text{cross-sectional area}}$$

Unit of stress is lbf/in^2, $tonf/in^2$, kgf/mm^2, N/m^2, etc., depending on the unit of force and the unit of cross-sectional area.

DIRECT STRESSES

If a material is pulled we say it is subject to a *Tensile stress*, and if the material is pushed or squeezed then the material is subject to a *Compressive stress*.

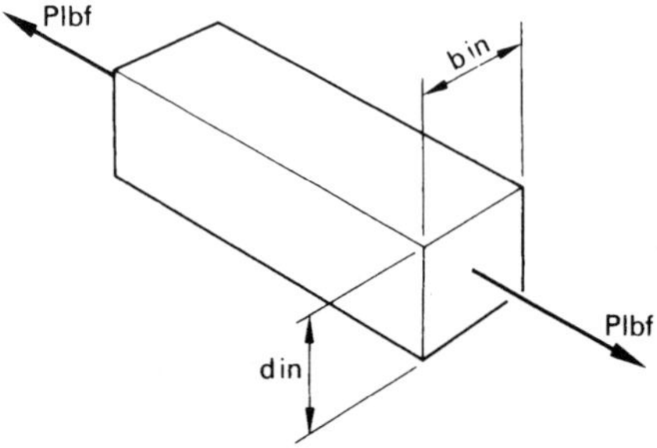

$$\text{Tensile stress} = \frac{P}{bd} \text{ lbf/in}^2$$

FIG. 6.9

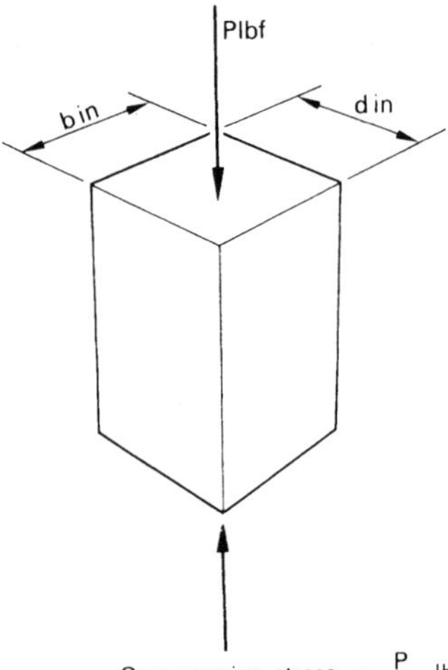

$$\text{Compressive stress} = \frac{P}{bd} \text{ lbf/in}^2$$

FIG. 6.10

SHEAR STRESS

When equal and opposite forces are applied to a material causing the material to slip across a plane in the direction in which the load is applied then a *Shear stress* is induced.

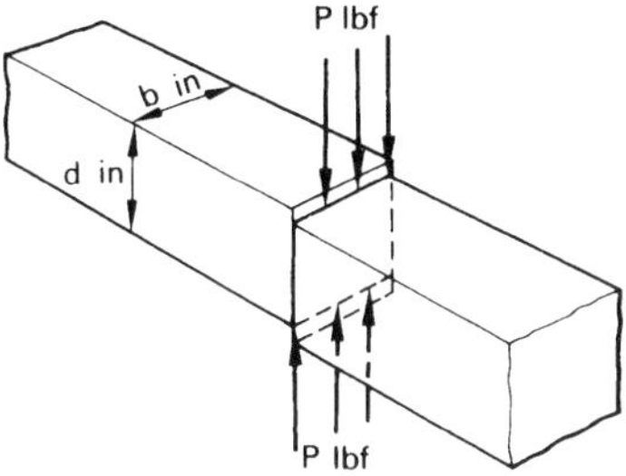

$$\text{Shear stress} = \frac{P}{bd}\ \text{lbf/in}^2$$

FIG. 6.11

6.3. Strain

From the previous experiment it was found that a load applied to a material produces a deformation. Within the elastic limit the relationship between the extension and the original length for a given load is constant and this measures the strain in the material.

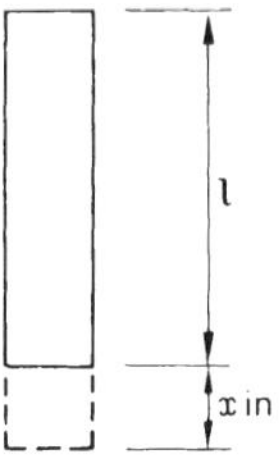

$$\text{Tensile strain} = \frac{x}{l}$$

FIG. 6.12

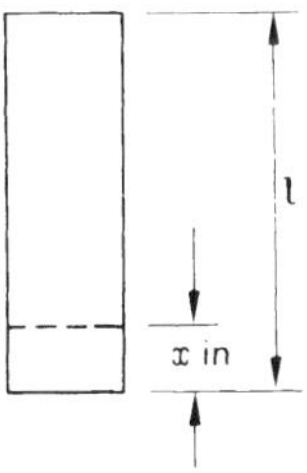

$$\text{Compressive strain} = \frac{x}{l}$$

FIG. 6.13

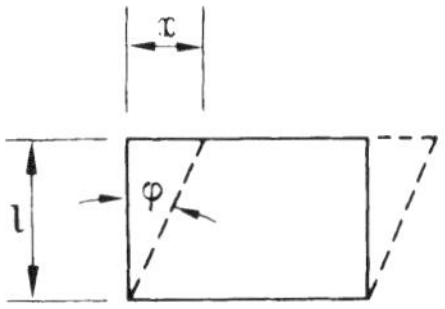

$$\text{Shear strain} = \frac{x}{l} \simeq \varphi\ \text{rad}$$

FIG. 6.14

$$\text{Tensile strain } e = \frac{\text{extension}}{\text{original length}} = \frac{x}{l}.$$

$$\text{Compressive strain } e = \frac{\text{compression}}{\text{original length}} = \frac{x}{l}.$$

$$\text{Shear strain } \phi = \text{angular deformation in radians} = \frac{x}{l}.$$

Notice that the strain is measured by the ratio $\frac{\text{extension}}{\text{original length}}$ and has *no unit.*

6.4. Young's Modulus of Elasticity

Different materials having the same length and cross-sectional area when loaded with the same weight have different extensions. If we consider Hooke's law and write

$$\text{load} \propto \text{extension},$$

then, dividing load by cross-sectional area and extension by original length, we get

$$\frac{\text{load}}{\text{area}} \propto \frac{\text{extension}}{\text{original length}}$$

or $\text{stress} \propto \text{strain};$

then $\text{stress} = E \times \text{strain},$

where E represents *Young's Modulus of Elasticity.*

$E = \frac{\text{stress}}{\text{strain}}$ and is measured in the same unit as stress.

The value of E is constant for a given material but different materials have different values of E.

Typical values of Young's Modulus of Elasticity are given in Table 6.5.

TABLE 6.5

Material	E (tonf/in.^{2})
Mild steel	13,400
Aluminium	4,500
Copper, hard drawn wire	8,250
Copper, hot rolled	6,700
Brass	6,250
Timber	700
Glass	8,600

6.5. Maximum stress

The maximum or ultimate stress in a material is determined from the load–extension graph and is the maximum load carried by the material divided by the original cross-sectional area.

$$\text{Maximum stress} = \frac{\text{maximum load}}{\text{original cross-sectional area}}.$$

Some materials when tested to destruction reach a maximum load and then, due to changes in dimensions of the test piece even if the load is slightly decreased, extension (or yielding) continues and fracture occurs. The breaking load is then less than the maximum load.

6.6. Factor of safety

The designer of a machine must ensure that no part of the machine is stressed anywhere near the maximum stress because if a slight overload occurs the part will break. To help the designer a factor of safety is introduced. This is the ratio of the maximum permissible stress to the working stress.

$$\text{Factor of safety} = \frac{\text{maximum stress}}{\text{working stress}}$$

The value of the factor of safety depends on the material being used and the circumstances under which the material will work. A static structure such as a lamp standard can have a lower factor of safety than, say, the lifting ropes on a crane which are subject to wear and constant variations in loading.

Experimental determination of Young's Modulus of Elasticity and maximum stress

The experiment already performed on the mild steel wire will enable a value for E for mild steel to be found. Remember that we are considering extensions within the elastic limit and therefore it is the straight-line portion of the graph of load–extension which will give us the necessary information.

Let E = Young's Modulus of Elasticity (lbf/in^2),

W = load (lbf),

A = cross-sectional area of wire (in^2),

x = extension (in.),

l = original length of wire (in.),

$$\text{then } E = \frac{\text{stress}}{\text{strain}} = \frac{W/A}{x/l} = \frac{Wl}{Ax}.$$

This can be written $\frac{W}{x} \times \frac{l}{A}$, where W/x represents the slope or gradient of the load–extension graph. Therefore if the slope of the graph is multiplied by the constant length/area the value for Young's Modulus can be found.

$$\text{Slope of graph (Fig. 6.8)} = \frac{53}{0{\cdot}05}.$$

$$\text{Cross-sectional area of wire} = \frac{\pi \times 0{\cdot}055^2}{4} = 0{\cdot}00237 \text{ in}^2.$$

Original length of wire $= 66{\cdot}3$ in.,

$$\therefore\ E = \frac{53}{0{\cdot}05} \times \frac{66{\cdot}3}{0{\cdot}00237}\ \text{lbf/in}^2$$

$$= \mathbf{29{\cdot}65 \times 10^6\ lbf/in^2}.$$

Maximum load supported by wire $= 151$ lbf,

$$\therefore\ \text{maximum stress} = \frac{151}{0{\cdot}00237}\ \text{lbf/in}^2$$

$$= \mathbf{63{,}600\ lbf/in^2}.$$

Example. Find the maximum allowable load that can be carried by a column of outside diameter 5 in. and inside diameter 4 in. if the maximum stress is 28 tonf/in^2 and a factor of safety of 4 is used.

$$\text{Factor of safety} = \frac{\text{maximum stress}}{\text{working stress}}.$$

$$\text{Working stress} = \frac{28}{4} = 7\ \text{tonf/in}^2.$$

$$\text{Cross-sectional area of column} = \frac{\pi \times 5^2}{4} - \frac{\pi \times 4^2}{4}$$

$$= \frac{9\pi}{4} = 7{\cdot}06\ \text{in}^2.$$

$$\text{Stress} = \frac{\text{load}}{\text{area}},\ \therefore\ \text{load} = \text{stress} \times \text{area}$$

$$= 7 \times 7{\cdot}06\ \text{tonf}$$

$$= \mathbf{49{\cdot}42\ tonf.}$$

Example. Find the modulus of elasticity of steel if a bar 10 ft long, $1\frac{1}{2}$ in. diameter stretches 0·04 in. under a load of 8 tonf.

$$E = \frac{\text{stress}}{\text{strain}}.$$

$$\text{Stress} = \frac{\text{load}}{\text{cross-sectional area}} = \frac{8}{\pi \times 1{\cdot}5^2/4} = 4{\cdot}52\ \text{tonf/in}^2.$$

$$\text{Strain} = \frac{\text{extension}}{\text{original length}} = \frac{0{\cdot}04}{10 \times 12} = 0{\cdot}000333.$$

$$E = \frac{4{\cdot}52}{0{\cdot}000333}\ \text{tonf/in}^2 \quad = \mathbf{13{,}580\ tonf/in^2}.$$

EXAMPLE. Find the least force required to cut a steel plate 2 in. wide by $\frac{1}{4}$ in. thick by shears if the maximum shear stress is 26 tonf/in^2.

$$\text{Area of plate resisting shear} = 2 \times \tfrac{1}{4} = 0{\cdot}5\ \text{in}^2,$$

$$\begin{aligned}\text{Shearing force} &= \text{area} \times \text{stress}\\ &= 0{\cdot}5 \times 26\ \text{tonf}\\ &= \mathbf{13\ tonf.}\end{aligned}$$

EXAMPLE. A square hole having 2-in. sides is punched through a steel plate $\frac{1}{4}$ in. thick. Find the force required and the mean intensity of stress in the punch. Maximum shear stress = 26 tonf/in^2.

$$\begin{aligned}\text{Area of metal resisting shear} &= \text{perimeter of hole} \times \text{thickness of metal}\\ &= 2 \times 4 \times \tfrac{1}{4} = 2\ \text{in}^2\end{aligned}$$

$$\therefore\ \text{force required} = 2 \times 26\ \text{tonf} = 52\ \text{tonf}.$$

$$\text{Cross-sectional area of punch} = 2 \times 2 = 4\ \text{in}^2,$$

$$\therefore\ \text{mean compressive stress in punch} = \frac{52}{4} = \mathbf{13\ tonf/in^2}.$$

6.7. Material behaviour

The experiment on the mild steel wire showed the elastic region and plastic region for this particular material. The method of applying the load and measuring the extension was relatively simple and more elaborate machines and measuring instruments are used to give a more precise relationship between load and extension.

A specimen is made by turning the material in a lathe to the shape shown in Fig. 6.15. The diameter is made a given proportion

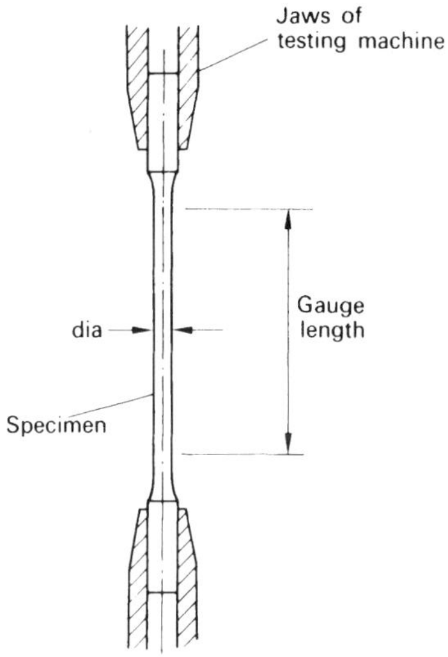

FIG. 6.15

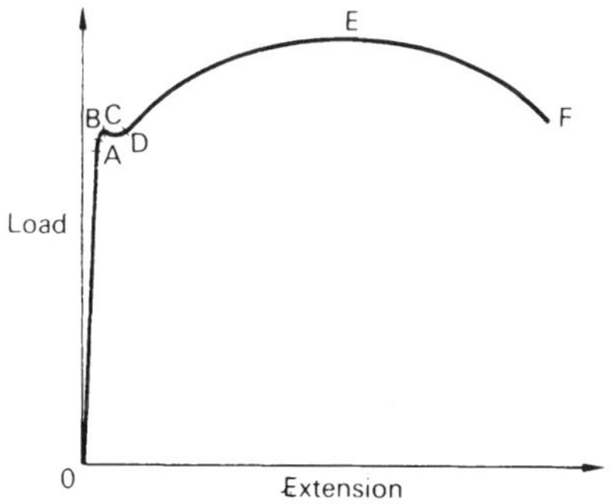

FIG. 6.16

of the gauge length and slightly larger diameters are formed at the ends to allow the jaws of the testing machine to grip the specimen.

If readings are taken of load and extension from a test on this specimen we should find the graph for mild steel to be similar to Fig. 6.16.

A larger scale for extension is shown in Fig. 6.17 to illustrate more clearly what happens after the elastic limit has been passed.

Starting from no load the material extends uniformly as the load increases up to point *A* which is called the *limit of proportionality*. As the name suggests, the limit of proportionality is

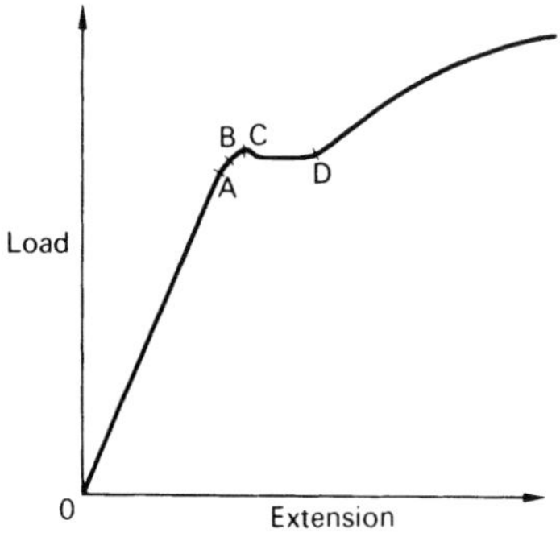

FIG. 6.17

that point at which the stress ceases to be proportional to strain. Slightly beyond the limit of proportionality is the *elastic limit B* which has been mentioned previously as the point at which the material will return to its original dimensions if the load is taken away.

At point *C* a phenomenon occurs known as yielding where the material stretches considerably without increasing the load. In fact some further yielding may occur even if the load is now slightly reduced. Point *C* is called the *yield point* and is very important in materials testing.

When the load is increased the material continues to extend, but not now in proportion to the load, until point *E* (Fig. 6.16) is reached. Up to this point no appreciable change in diameter has taken place in the specimen, but at point *E* the material begins to form a waist or neck (Fig. 6.18) and from *E* to *F*, the point of fracture, the load decreases.

At fracture the specimen is similar to Fig. 6.19.

If after fracture the separation of the original gauge points is

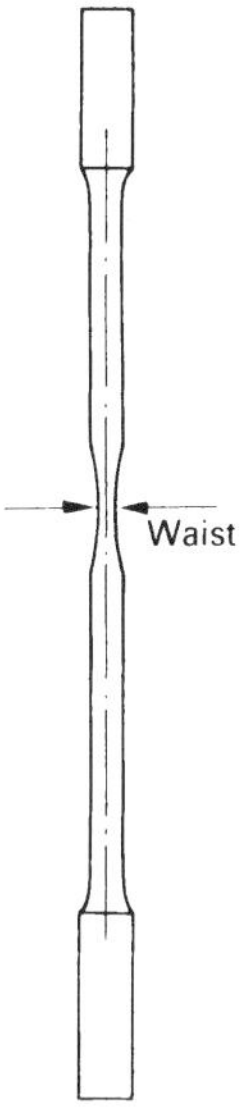

FIG. 6.18

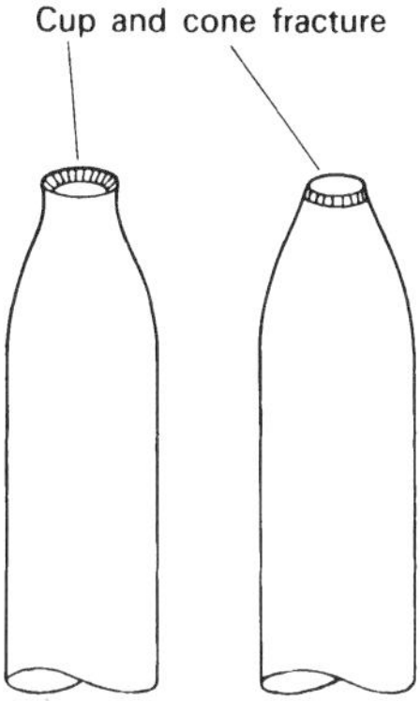

FIG. 6.19

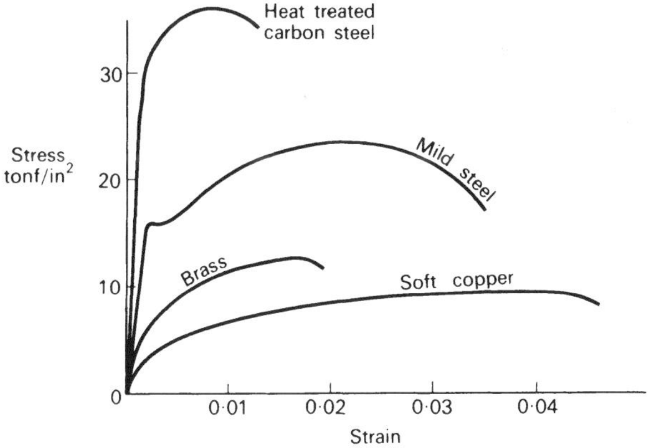

FIG. 6.20

measured it will be possible to form an idea of the ductility of the material. Some materials, e.g. mild steel, will stretch much more than other materials such as cast iron and concrete. Ductility measures the ability of a material to be pulled out or stretched or deformed without breaking. A brittle material fractures with very little deformation.

Figure 6.20 shows graphs of stress–strain for different materials.

From these graphs an idea of the properties of the material can be ascertained. If the graph shows a long extension before fracture, then the material is ductile; if a short extension, then the material is brittle. If the graph has very little straight portion, then the material is not elastic. Some materials have a poorly defined yield point.

Problems

6.1. State the meaning of the following terms: stress, strain, Young's Modulus of Elasticity. A bar 20 ft long and 1 in. in diameter is stretched 0·08 in. when loaded. If the Modulus of Elasticity is 30×10^6 lbf/in^2 calculate (a) the stress in the bar, (b) the load applied.

(10,000 lbf/in^2, 7850 lbf.) (E.M.E.U.)

6.2. A steel tie-rod is of circular section and is required to transmit a pull of 20 tonf. It is to be made of steel of maximum tensile strength 30 tonf/in^2 and the factor of safety is to be 6. Find the diameter required. If the length of the tie-rod is 15 ft, find the extension which would be produced by the 20 tonf load.

Young's Modulus of Elasticity $=$ 13,000 tonf/in^2.

(2·26 in., 0·069 in.)

6.3. What is meant by the term "shearing stress"? Two flat plates are held together by a cylindrical rivet 0·5 in. in diameter. This rivet shears when a force of 1 tonf acts along the axis of the plates. What is the value of the maximum shearing stress the material of the rivet can bear?

If the force tending to pull the plates apart is normally 500 lbf, what is the factor of safety?

(5·1 tonf/in^2, 4·5.)

6.4. A steel bar is 1 in. diameter for 3 ft and $1\frac{1}{2}$ in. in diameter for 9 ft making an overall length of 12 ft. If the maximum tensile stress in the bar is 10,000 lbf/in^2, calculate the force to produce this stress and the total extension of the 12-ft bar when this force is acting.

Take $E = 30 \times 10^6$ lbf/in^2.

(7854 lbf, 0·028 in.)

6.5. A steel tube 10 in. outside diameter, 0·5 in. thick wall and 10 ft long is used as a column. If the maximum allowable compressive stress in the tube is 5·15 tonf/in^2 calculate the axial load which can be placed on the column. Calculate the decrease in length of the column when this load is applied. $E = 13{,}500$ tonf/in^2.

(82·1 tonf, 0·0089 in.)

6.6. A test on a metal specimen, 0·534 in. in diameter and of 8 in. gauge length, gave the following results:

Load lbf	0	200	400	600	800	1000	1200	1400	1600	1800
Extension $\times 10^4$ in.	0	7·3	13·3	19·5	25·8	32·2	39	47	57	70

Note! At 200-lbf load the extension is 0·00073 in. Calculate the stress at the limit of proportionality, and the modulus of elasticity.

(2·39 tonf/in^2, 4950 tonf/in^2.)

No answers are given to the following problems:

6.7. Explain what is meant by the terms "tensile stress", "tensile strain", "Young's Modulus".

During a tensile test on a wire of length 2 m and diameter 0·12 cm, the following figures were obtained:

Load in kgf	2	4	6	8	10	12
Extension in cm	0·010	0·019	0·028	0·037	0·048	0·063

Comment on the shape of the load extension graph, and calculate Young's Modulus for the material of the wire.

How could the measurements of extension and diameter be made?

(C. & G.)

6.8. A rod 12 in. long is of $1\frac{1}{2}$ in. diameter over a length of 10 in. and 1 in. diameter over the remaining 2 in. of its length. The smaller portion is subject to a tensile stress of 9 tonf/in^2. What is the stress intensity in the larger portion? How much will an 8-in. length of the larger portion stretch under the load if the Modulus of Elasticity is 13,500 tonf/in^2? (U.E.I.)

6.9. What is the meaning of the term "factor of safety"? Calculate the diameter of a rod to take a load of 22 tonf (a) if the extension is not to exceed 0·005%, (b) if the factor of safety is 8 and the maximum stress of the rod is 30 tonf/in^2.

Take $E = 30 \times 10^6$ lbf/in^2.

CHAPTER 7

Heat

7.1. Temperature

Temperature measures the degree of hotness of a body. If one body is hotter than another body we say the first body is at a higher temperature than the second. The method of measuring temperature is by a thermometer or pyrometer.

A thermometer makes use of the thermal expansion of a liquid or gas or solid to measure temperature. A common type of thermometer consists of a glass tube with a bulb at one end which is filled with mercury. The other end of the tube is sealed after all the air in the tube has been extracted so that a vacuum is produced. The mercury in the bulb expands when heated, pushing a column of mercury up the tube. A pocket at the top of the tube acts as a safety measure against overheating of the thermometer. When the lower part of the thermometer is placed in melting ice the height of the column of mercury is marked on the tube. This point registers the *freezing point* of water. The tube is then immersed in steam from distilled water boiling at atmospheric pressure, that is from a container open to the atmosphere, and the height of the column of mercury is again marked on the tube. The mark registers the *boiling point* of water. The distance between the freezing point and the boiling point marks is called the *fundamental interval* of the temperature scale.

The *Fahrenheit* scale of measurement takes the fundamental interval and divides it into 180 equal divisions. The freezing point is taken as 32 degrees and boiling point as 212 degrees.

The *Celsius* or *Centigrade* scale of measurement divides the fundamental interval into 100 equal divisions. The freezing point is taken as 0 degrees and boiling point as 100 degrees.

A comparison of the two scales is shown in Fig. 7.1.

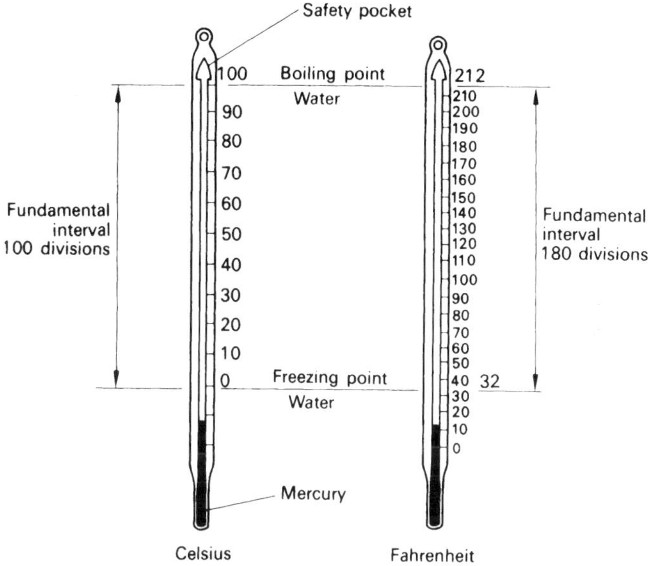

FIG. 7.1. Comparison of thermometer scales.

7.2. Conversion of scales

$$180 \text{ deg F} = 100 \text{ deg C},$$

$$\therefore\ 1 \text{ deg F} = \frac{5}{9} \text{ deg C}.$$

If F represents a reading on the Fahrenheit thermometer and C a reading on the Centigrade thermometer, then

$$F = \frac{9}{5}C + 32 \text{ degrees}$$

$$C = (F - 32)\frac{5}{9} \text{ degrees}.$$

It is usual to write deg F or deg C when indicating a temperature difference and °F and °C when indicating an actual temperature.

Specially constructed thermometers using mercury as the measuring fluid can read up to about 1000°F or 540°C, but for measurements of higher temperatures other instruments known as pyrometers are used. Some of these pyrometers incorporate thermocouples.

7.3. Thermocouples

A thermocouple consists of two wires of different metals joined at each end. One junction is at the point where the temperature is to be measured and the other junction is kept at a lower fixed temperature. Owing to this difference in temperature at the

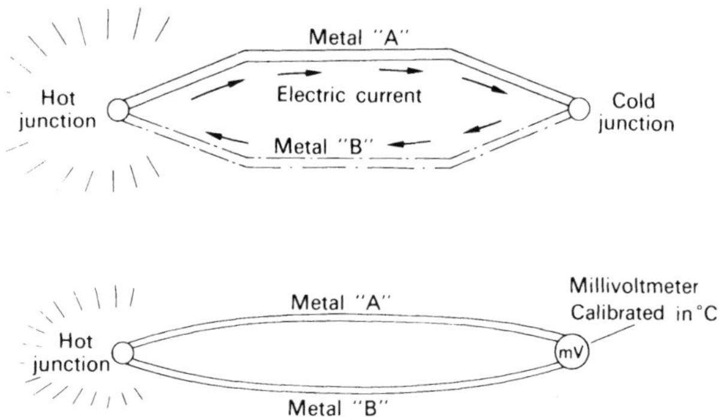

FIG. 7.2. Principle of a thermocouple.

junctions a tiny electromotive force is set up which drives an electric current around the circuit. The bigger the difference between the temperatures of the junctions the larger the electromotive force. This electromotive force can be measured by means of a galvanometer or a potentiometer.

Typical combinations of metals used for thermocouples are:

(a) copper and constantan (60% copper and 40% nickel),

(b) platinum and platinum-rhodium alloy.

7.4. Radiation pyrometer

If the measurement of the temperature has to be made without contact with the hot material then a radiation pyrometer can be used. This type of pyrometer measures the intensity of heat radiated from the hot body by concentrating the rays on to the hot junction of a thermocouple using a lens and then measuring the electromotive force produced. See Fig. 7.3.

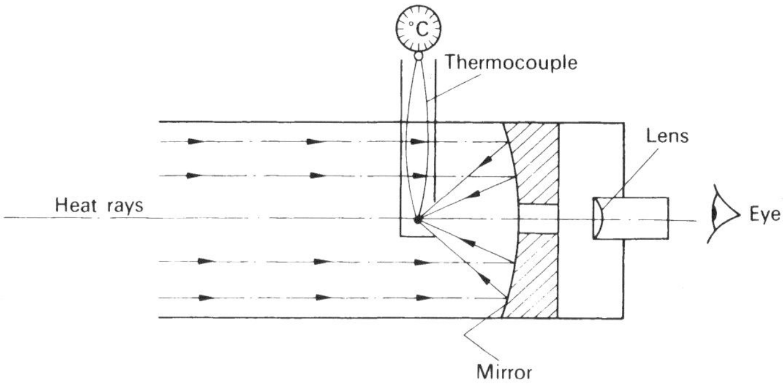

FIG. 7.3. A radiation pyrometer.

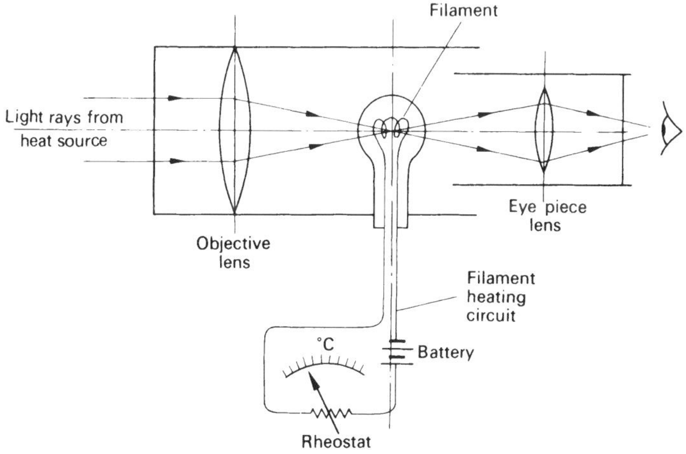

FIG. 7.4. An optical pyrometer.

7.5. Optical pyrometer

The disappearing filament pyrometer is a typical pyrometer of this type. A small telescope containing a metal filament is pointed at the source of heat, e.g. a furnace. The rheostat is then adjusted until the metal filament glows the same colour as the furnace. When this happens the filament will tend to disappear. The current needed for this can be made to give a measure of the furnace temperature.

7.6. Expansion of solids by heat

Nearly all materials expand when heated although the amount of expansion for a given temperature rise varies with the type of material. The increase in dimensions takes place in all directions. We call the increase in length linear expansion, the increase in area superficial expansion, and the increase in volume cubical expansion. The ratio $\frac{\text{increase in length}}{\text{original length}}$ for 1 degree rise in temperature is called the coefficient of linear expansion,

the ratio $\frac{\text{increase in area}}{\text{original area}}$ for 1 degree rise in temperature is called the coefficient of superficial expansion,

and the ratio $\frac{\text{increase in volume}}{\text{original volume}}$ for 1 degree rise in temperature is called the coefficient of cubical expansion.

It is generally accepted that if the coefficient of linear expansion is the same in all directions for a given material, then the coefficient of superficial expansion is approximately twice the coefficient of

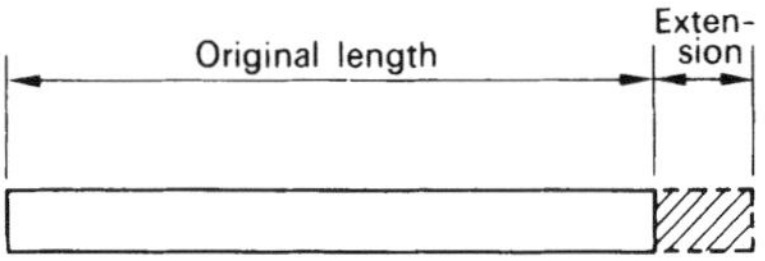

FIG. 7.5. Linear expansion.

linear expansion and the coefficient of cubical expansion three times the coefficient of linear expansion.

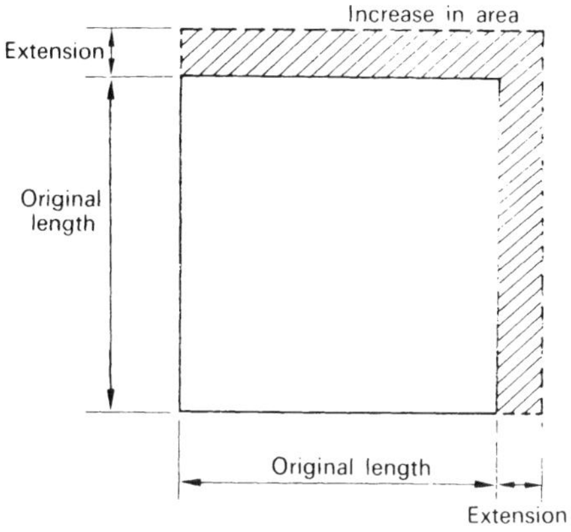

FIG. 7.6. Superficial expansion.

Values of coefficients of linear expansion for various materials are given in Table 7.1. It should be noticed that the coefficient of

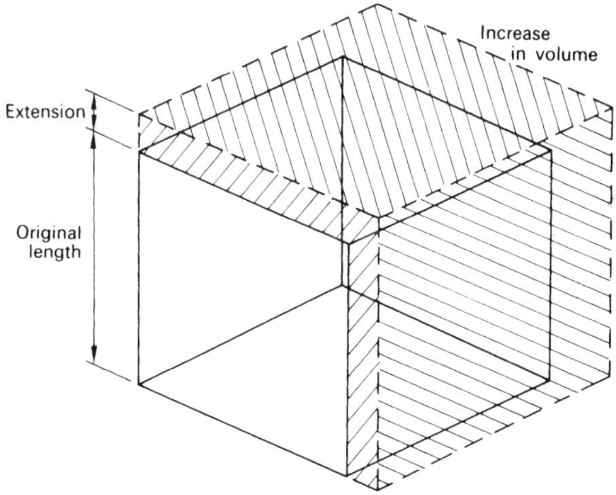

FIG. 7.7. Cubical expansion.

expansion of a material should be qualified by the degree of temperature increase, i.e. deg F or deg C.

The coefficient of expansion per deg F $= \frac{5}{9}$ coefficient of expansion per deg C.

TABLE 7.1

COEFFICIENTS OF LINEAR EXPANSION (α)

Material	per deg F	per deg C
Aluminium	0·0000128	0·000023
Brass	0·0000104	0·0000188
Copper	0·0000093	0·0000167
Cast iron	0·0000061	0·000011
Lead	0·0000161	0·000029
Monel	0·0000076	0·0000137
Nickel	0·0000071	0·0000128
Platinum	0·00000495	0·0000089
Cast steel	0·0000061	0·000011
Tin	0·0000117	0·000021
Glass	0·0000047	0·0000085
Graphite	0·0000044	0·0000079

EXAMPLE. (a) At 0°C a brass rod is 100 in. long. If the coefficient of linear expansion of brass is 0·0000188 per deg C, what will be its length at 80°C? (b) A square sheet of brass 10 in. length of side at 0°C is heated to 80°C. What will then be its area?

(a) Coefficient of expansion

$$= \frac{\text{increase in length}}{\text{original length}} \text{ per 1 deg C temperature rise.}$$

Increase in length

$$= 0{\cdot}0000188 \times 100 \times 80 \text{ in.}$$
$$= 0{\cdot}1504 \text{ in.}$$

Final length

$$= 100 + 0{\cdot}1504 \text{ in.}$$
$$= \mathbf{100{\cdot}1504\ in.}$$

(b) Coefficient of superficial expansion

$=2\times 0{\cdot}0000188$

$=0{\cdot}0000376.$

Increase in area

$=0{\cdot}0000376\times 10\times 10\times 80 \text{ in}^2$

$=0{\cdot}3008 \text{ in}^2.$

Final area

$=100+0{\cdot}3008 \text{ in}^2$

$=\mathbf{100{\cdot}3008\ in^2}.$

EXAMPLE. What increase in length will take place in a steam pipe 120 ft long, the temperature of the steam being 450°F, if the atmospheric temperature is 50°F and the coefficient of linear expansion of the material is 0·0000061 per deg F?

Coefficient of expansion

$$=\frac{\text{increase in length}}{\text{original length}} \text{ per deg F temperature change.}$$

Increase in length

$=0{\cdot}0000061\times 120\times(450-50) \text{ ft}$

$=0{\cdot}0000061\times 120\times 400 \text{ ft}$

$=0{\cdot}2928 \text{ ft}$

$=\mathbf{3{\cdot}5136\ in.}$

EXPERIMENTAL VERIFICATION OF COEFFICIENT OF LINEAR EXPANSION

Equipment required. Copper rod, steam jacket, boiler, thermometer, dial gauge, and bunsen burner.

The length of the rod is measured from the free end to the centre of the fastening at the fixed end. The dial gauge measuring to 1/1000 in. is zeroed at room temperature and this temperature is read off the thermometer. Steam produced in the boiler is then passed through the steam jacket in which the rod is enclosed. After a few minutes the rod will have reached the same temperature as the steam and this temperature is read off the thermometer. The expansion is measured on the dial gauge.

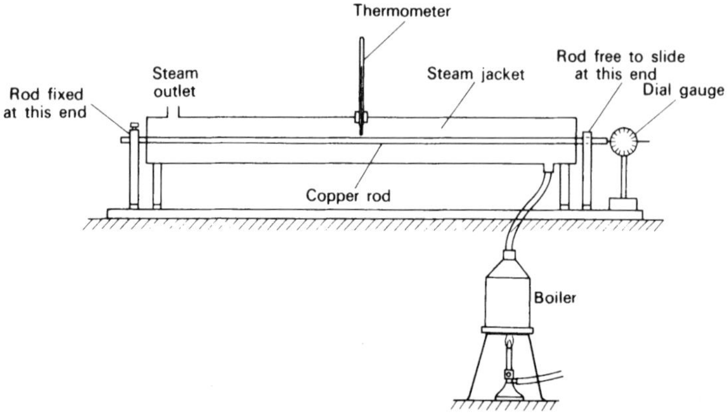

FIG. 7.8

Original length of copper rod

$$= 30{\cdot}25 \text{ in.}$$

Room temperature

$$= 65°\text{F}.$$

Expansion of rod (from dial gauge reading) = 0·041 in.

Final temperature = 212°F.

Coefficient of linear expansion

$$= \frac{\text{increase in length}}{\text{original length}} \text{ per deg temperature rise}$$

$$= \frac{0{\cdot}041}{30{\cdot}25\,(212-65)}$$

$$= \frac{0{\cdot}041}{30{\cdot}25 \times 147}$$

$$= \mathbf{0{\cdot}0000093 \text{ per deg F.}}$$

7.7. Thermostats

A thermostat is an instrument allowing the temperature to be maintained within given limits by the use of a device which cuts off the supply of heat when the required temperature is exceeded and automatically restores the supply when the temperature falls below that required. A bi-metal strip is often used to control current supply. This consists of two different metals having different coefficients of expansion, metal *B* having a larger coefficient of expansion than metal *A* (Fig. 7.9). If the temperature falls, then rod *B* contracts more than rod *A*, causing the two to bend and close the contacts at *C*, thus making the circuit to restore the heat supply and increase the temperature.

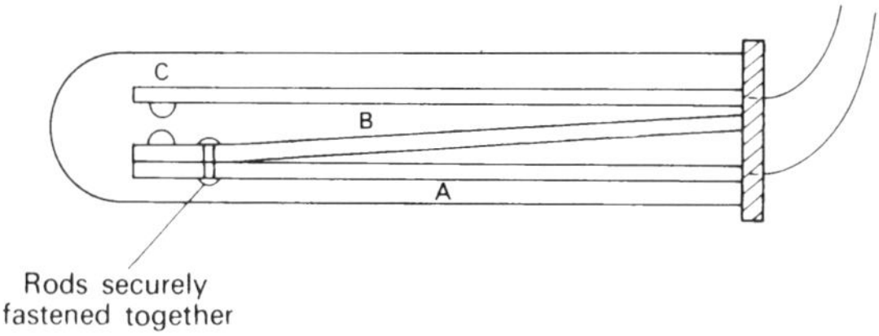

FIG. 7.9

7.8. Transfer of heat

The transfer of heat can take place in three ways—namely, radiation, conduction, and convection.

RADIATION

Radiant heat is transferred from one body to another without contact between the bodies. The sun is the greatest source of radiant heat that we know and this heat travels through space in the form of electromagnetic waves, reaching the earth and warming it during the daylight hours. If you stand in front of a fire then radiant heat is transmitted to you, but if someone comes between you and the fire then the heat supply is blocked and you cease to feel the benefit of the heat from the fire. The radiant heat waves travel in straight lines unless deflected by some reflecting surface.

A thermopile (Fig. 7.10) is an instrument by which radiation can be detected. It consists of a heat-sensitive surface *A* formed by a number of thermocouples connected in series and coupled to a sensitive galvanometer *B* to measure the tiny electromotive forces made by the heat energy. Using the instrument it is possible to compare the heat radiating properties of different surfaces. It is well known that the radiating power of a surface depends not only on the temperature but also on the nature and colour of the surface. A black surface will absorb and give out heat more quickly than a light coloured surface. A dull matt surface will have different radiating qualities from a highly polished surface.

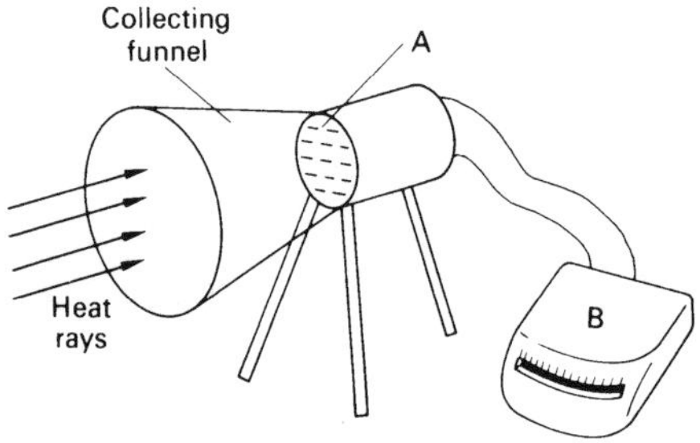

FIG. 7.10. A thermopile.

If you examine an electric fire, the reflector behind the element is made from a highly polished piece of metal so that very little heat is absorbed but a lot of heat is reflected and concentrated where it is required.

In countries where the sunshine is abundant, steam may be raised by reflecting the sun's rays on to a boiler by means of a large mirror.

CONDUCTION

When heat is transferred by conduction, the bodies must be in contact. Conduction is also the means of transferring heat from one part of a solid to another part of the same solid. If a hot body is placed in contact with a cooler body then the heat will flow from the hot body to the cooler body until both reach the same temperature. Some materials such as silver, copper, and aluminium are good conductors of heat whilst materials such as glass, wood, cork, and asbestos are poor conductors of heat. Hence good saucepans are made from copper or aluminium whilst asbestos can be used for fire-fighting suits; wool is a good material for blankets and certain plastic materials are used for pan handles.

Liquids and gases are generally poor conductors of heat.

CONVECTION

When heat is transferred by convection it is necessary that the medium through which the heat is transferred shall be in motion, and therefore convection is only possible in liquids and gases. If one part of a liquid or gas in a container is heated it expands and becomes less dense and therefore rises above the surrounding fluid. After reaching the top it is cooled and sinks again and this cycle is repeated continuously causing convection currents to be formed in the fluid.

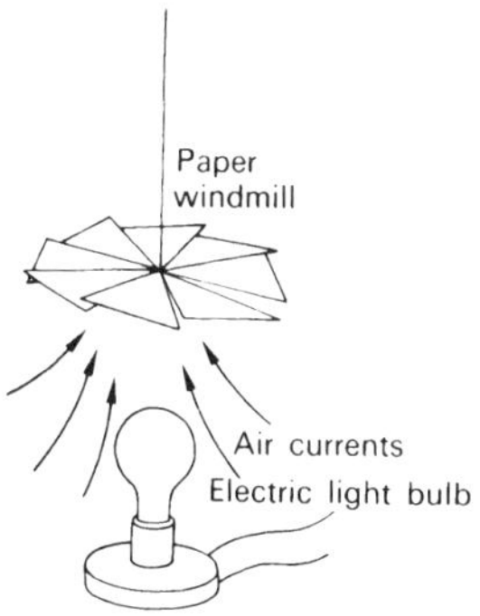

FIG. 7.11. Convection in air.

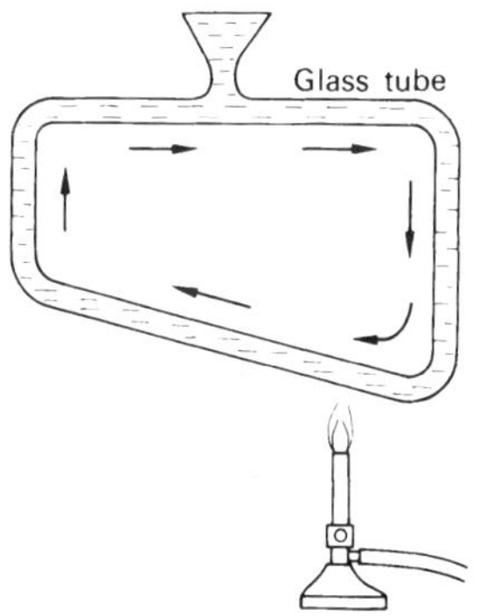

FIG. 7.12. Convection in water.

Figure 7.11 illustrates convection in air which is shown by making a paper windmill and suspending it over an electric light bulb. The air in contact with the bulb is heated and rises, this air being replaced by cool air which in turn is heated and rises, causing a continuous upward flow of air and thereby turning the paper windmill.

Figure 7.12 is a simple piece of apparatus consisting of a glass tube filled with water. The bottom leg of the apparatus is heated causing the water to flow up the inclined leg. When the system is working a drop of blue ink introduced into the water will show clearly the circulation. Domestic hot-water systems use the principle of convection to circulate water around the boiler and hot-water storage cylinder. Figure 7.13 shows a typical arrange-

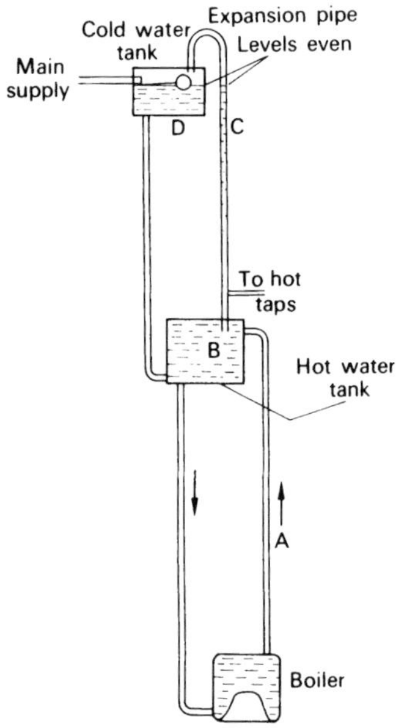

FIG. 7.13. Domestic hot-water system.

ment where the water is heated in a boiler situated on the lowest floor of the house and rises through pipe *A* to the hot-water storage cylinder *B*. From the top of the cylinder an expansion pipe *C* is taken to the cold water storage tank *D* situated at the highest point in the system. This pipe *C* is a safety device in case the water boils. It should be noticed that the level of the water in pipe *C* is the same as that in the cold water storage tank *D*. Providing the tank *D* is above the level of the highest tap, then sufficient pressure will be available to force the water out of the taps. Cold water is led into the bottom of the hot-water cylinder and from there into the bottom of the boiler. The arrows on Fig. 7.13 show the direction of the convection currents.

7.9. Measurement of heat energy

To supply the means whereby a body can become hotter requires energy to be given to the body to speed up the movement of its molecules. The energy given to a body to increase its temperature is heat energy. Heat energy is measured in a variety of units.

One Centigrade heat unit (Chu) is the amount of heat energy required to raise the temperature of a mass of one pound of water 1 deg C.

One British thermal unit (Btu) is the amount of heat energy required to raise the temperature of a mass of one pound of water 1 deg F.

One therm is equal to 100,000 Btu.

One gram calorie (cal) is the amount of heat energy required to raise the temperature of a mass of one gramme of water 1 deg C.

One kilocalorie (kcal) is the amount of heat energy required to raise the temperature of a mass of one kilogramme of water 1 deg C.

Since 1 lb $= 454$ g,
then 1 Chu $= 454$ cal
and 1 Btu $= 454 \times \frac{5}{9} = 252$ cal.

In 1843 the scientist Joule performed some experiments to determine the relationship between heat energy and mechanical energy. A sketch of the apparatus is shown in Fig. 7.14. The weight W is allowed to fall through a height h and in doing so rotates the paddles P immersed in a given quantity of water. Joule found that 778 ft-lbf of work was required to raise the temperature of 1 lb of water 1 deg F or 1400 ft-lbf of work was required to raise the temperature of 1 lb of water 1 deg C. These values are known as Joule's equivalents.

1 Btu $= 778$ ft-lbf,
1 Chu $= 1400$ ft-lbf.

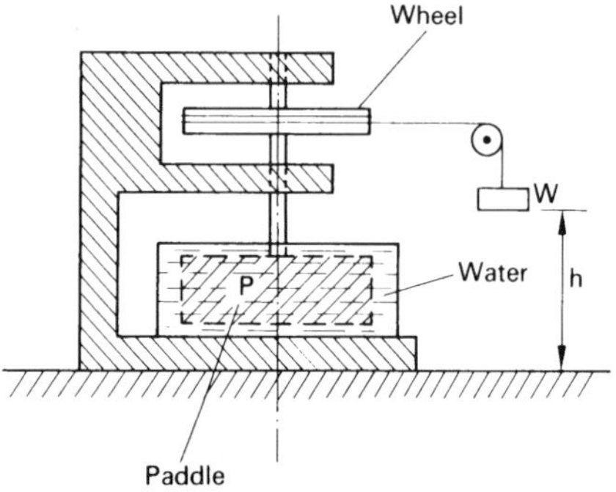

FIG. 7.14

7.10. Quantity of heat

From our definition of heat units it is easy to see that if we have a container holding 1 lb of water and another container holding 4 lb of water, then it will require four times as much heat to raise the temperature of the 4 lb of water through the same temperature range as the 1 lb of water. If a 3-pint electric kettle is filled it will take longer to boil the water than if it is only half filled. The quantity of heat required to raise the temperature of a given quantity of water through a given range is given by:

Heat required = mass of water × temperature rise.

If, on the other hand, the water is cooling,

Heat given out = mass of water × temperature fall.

EXAMPLE. Determine the amount of heat in calories required to raise the temperature of 1215 g of water 30 deg C.

Heat required = 1215 × 30 = **36,450 cal.**

EXAMPLE. A radiator containing a mass of 30 lb of water gives out 450 Btu of heat. If initially the water was at a temperature of 180°F, what is the final temperature of the water?

Heat given out = mass of water × temperature fall

$$450 = 30(t_1 - t_2), \quad \text{where } t_1 = \text{initial temp.}$$
$$= 30(180 - t_2) \qquad t_2 = \text{final temp.}$$
$$\therefore \quad \frac{450}{30} = 180 - t_2,$$
$$\therefore \; t_2 = 180 - 15,$$
$$t_2 = \mathbf{165°F.}$$

EXAMPLE. How much heat is lost by one gallon of water in cooling from 85°C to 15°C? Give an answer in Chu, Btu and cal. (1 gallon of water has a mass of 10 lb, 1 lb = 454 g.)

(a) Heat given out = mass of water (lb) × temperature fall (deg C)

$$= 1 \times 10(85 - 15)$$
$$= 10 \times 70$$
$$= \mathbf{700\ chu.}$$

(b) $1 \text{ Chu} = \frac{9}{5} \text{ Btu},$

$$\therefore \quad 700 \times \frac{9}{5} = \mathbf{1260\ Btu.}$$

(c) Heat given out = mass of water × temperature fall (deg C)

$$= 1 \times 10 \times 454(85 - 15)$$
$$= 10 \times 454 \times 70$$
$$= \mathbf{317{,}800\ cal.}$$

7.11. Specific heat

It has been found by experiment that different materials require different amounts of heat to raise their temperature through the same range. We found by definition that a mass of 1 lb of water required 1 Chu to raise the temperature 1 deg C, but if we heat a mass of 1 lb of copper through 1 deg C it only requires approximately 0·1 Chu; similarly, aluminium requires approximately 0·2 Chu to heat a mass of 1 lb through 1 deg C. We say therefore the SPECIFIC HEAT of copper and aluminium is less than that of water.

"The specific heat of a substance is the quantity of heat required to raise the temperature of unit mass of the substance by one degree compared with the heat required to raise the same quantity of water through the same temperature range."

TABLE 7.2

TYPICAL SPECIFIC HEAT VALUES

Substance	Specific heat
Water	1·00
Aluminium	0·211
Brass	0·09
Copper	0·093
Asbestos	0·2
Iron	0·119
Glass	0·16
Tin	0·056
Solder	0·042

EXAMPLE. Find the heat required to raise the temperature of a mass of 1 ton of iron from 65°F to 750°F. Specific heat of iron 0·119.

$$\text{Heat required} = \text{mass of substance} \times \text{specific heat} \times \text{temperature rise}$$
$$= 1 \times 2240 \times 0{\cdot}119(750-65) \text{ Btu}$$
$$= 2240 \times 0{\cdot}119 \times 685 \text{ Btu}$$
$$= \mathbf{182{,}600\ Btu.}$$

EXAMPLE. An object has a mass of 200 g. It is found that 800 cal changes its temperature 20 deg C. What is the specific heat of the material?

$$\text{Heat required} = \text{mass} \times \text{specific heat} \times \text{temperature rise},$$
$$\therefore \quad \text{specific heat} = \frac{800}{200 \times 20} = \mathbf{0{\cdot}2.}$$

7.12. Water equivalent

To heat a liquid it is necessary to provide a container and in the process of applying heat the container will absorb some of the heat supplied.

"The water equivalent of a body is the mass of water which would take up as much heat as the body when the temperature rise was the same for each."

Water equivalent = mass of body × specific heat.

"The thermal capacity of a body is the heat necessary to raise the temperature of the body by one degree."

Thermal capacity = mass of body × specific heat.

EXAMPLE. Find the water equivalent of a copper container having a mass of 7·2 lb. The specific heat of copper is 0·091.

Heat absorbed by copper

= heat absorbed by water.

7·2 × 0·091 × temperature difference

= mass of water × temperature difference.

Mass of water = 0·6552 lb.

Water equivalent of container

= **0·6552 lb.**

EXAMPLE. A copper vessel contains 80 g of water at 32°C. If the vessel has a mass of 65 g and its specific heat is 0·091, find the amount of heat required to raise the temperature of the water and calorimeter from 32°C to 60°C, assuming no heat loss to atmosphere, etc.

Water equivalent of calorimeter = mass × specific heat

= 65 × 0·091 g

= 5·915 g.

Heat required = (mass of water + water equivalent of calorimeter) × temperature rise

= (80 + 5·915) × (60 − 32) cal

= 85·915 × 28 cal

= **405·6 cal.**

EXAMPLE. In an experiment to determine the water equivalent of a calorimeter or vessel, the following readings were obtained.

Mass of calorimeter empty	47·7 g
Mass of calorimeter with cold water added	156·6 g
Initial temperature of calorimeter and cold water	43°F
Initial temperature of hot water added	178°F
Final temperature of the mixture	56°F
Final mass of calorimeter and water	168·6 g

Find the water equivalent (W.E.) and the specific heat of the calorimeter.

Mass of cold water $= 156{\cdot}6 - 47{\cdot}7 = 108{\cdot}9$ g,

Mass of hot water $= 168{\cdot}6 - 156{\cdot}6 = 12{\cdot}0$ g.

Heat lost by water in cooling

= heat gained by cold water and calorimeter

$$12(178-56) = (108{\cdot}9 + \text{W.E. of calorimeter}) \times (56-43),$$

$$\frac{12 \times 122}{13} = 108{\cdot}9 + \text{W.E. of calorimeter}.$$

$\therefore$ W.E. of calorimeter $= 112{\cdot}56 - 108{\cdot}9 =$ **3·66 g.**

W.E. of calorimeter = mass × specific heat,

$$\therefore \quad \text{specific heat} = \frac{3{\cdot}66}{47{\cdot}7} = \mathbf{0{\cdot}0767.}$$

EXPERIMENT. *To find the specific heat of a metal.*

Equipment required. Thick walled copper calorimeter, heat-insulating jacket, beaker, bunsen, tripod, gauze, stand, thermometer, piece of aluminium.

Determine the mass of the calorimeter and stirrer and then half fill the calorimeter with water and weigh again. Weigh the piece of aluminium and tie a piece of thread around it to enable it to be carried. Place the copper calorimeter in the heat-insulating jacket and take the temperature of the cool water. Place the aluminium in

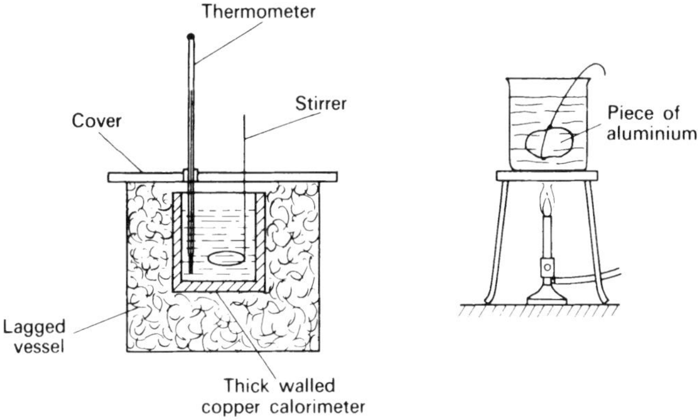

FIG. 7.15

the beaker containing water and boil the water so that the aluminium reaches a uniform temperature throughout, i.e. the same temperature as the boiling water. Quickly transfer the aluminium to the calorimeter and gently stir the water taking note of the highest temperature recorded on the thermometer. Typical results are given, and from them the specific heat of aluminium can be found.

Take the specific heat of copper $= 0{\cdot}1$

Mass of calorimeter and stirrer $= 70$ g

Water equivalent of calorimeter and stirrer $= 70 \times 0{\cdot}1 = 7{\cdot}0$ g

Mass of water and calorimeter $= 190$ g

Mass of water $= 190 - 70 = 120$ g

Initial temperature of water $= 14°C$

Final temperature of water $= 18°C$

Temperature rise of water $= 18 - 14 = 4$ deg C

Initial temperature of aluminium $= 100°C$

Mass of aluminium $= 29{\cdot}5$ g

Heat gained by water $=$ heat lost by aluminium

(Weight of water + W.E. of calorimeter) × temperature rise
= weight of aluminium × specific heat × temperature fall.

Let c = specific heat of aluminium

$$(120+7)4 = 29{\cdot}5c(100-18),$$

$$\therefore \quad c = \frac{127 \times 4}{29{\cdot}5 \times 82} = \mathbf{0{\cdot}21}.$$

Care must be taken to transfer the sample quickly from the beaker of boiling water to the calorimeter otherwise heat will be lost in the transfer.

Problems

7.1. 1000 g of water are placed in a copper calorimeter weighing 50 g and heated by an electric heater from 10°C to 80°C. The specific heat of copper is 0·1. Calculate the amount of heat required (a) for the water, (b) for the calorimeter.

(70,000 cal, 350 cal.) (N.C.T.E.C.)

7.2. Describe how to determine the water boiling point and freezing point marks on a thermometer. State the two values on (a) the Centigrade, (b) the Fahrenheit scale.

Convert the following readings: 50°C to °F, 140°F to °C, −20° C to °F.

(122°F, 60°C, −4°F.)

7.3. Explain what is meant by the specific heat of a metal. A piece of iron weighing 80 lb falls through a vertical height and the energy is all used in raising the temperature of the iron. Find the height from which the weight falls if its temperature rise is 2°F.

(Specific heat of iron = 0·2; 1 Btu = 780 ft-lbf.)

(312 ft.) (N.C.T.E.C.)

7.4. A 1500-W immersion heater is used to heat 45 gallons of oil. If the efficiency of the operation is 75% find the time needed to raise the temperature of the oil from 55°F to 105°F. (One gallon of oil weighs 9 lb; specific heat of oil 0·49; 1 kWh = 3410 Btu/h.)

(2 h 35 min.) (C. & G.)

7.5. (a) Define specific heat. (b) A metal cylinder which contained 6 gallons of water was heated until the temperature of both cylinder and water reached 212°F. The whole was then plunged into 3 ft^3 of a liquid which had a density of 50 lb/ft^3, a temperature of 50°F and a specific heat of 0·84. The final temperature of the whole mixture was 104°F.

Calculate (a) the water equivalent of the cylinder, and (b) its weight.

Take the specific heat of the metal as 0·12.

Neglect any heat losses. 1 gallon of water weighs 10 lb.

(3 lb, 25 lb.) (N.C.T.E.C.)

7.6. Define the coefficient of linear expansion of a solid.

A copper wire is 120 yards long. How much longer will it be if the increase in temperature is 45 deg C?

Coefficient of linear expansion of copper = 0·000017 per deg C.

(3·305 in.)

7.7. A block of steel measures $3 \times 3 \times 4\frac{1}{2}$ in. at a temperature of 15°C and weighs 10·53 lb. It is heated to a temperature of 600°C. Find the volume and density of the block at this temperature.

Coefficient of linear expansion = 0·0000125 per deg C.

(41·3884 in^3; 0·254 lb/in^3.)

7.8. How much heat is required to raise the temperature of 0·2 lb of copper through 189 deg F? Take the specific heat of copper to be 0·1. If this heated copper is plunged into 212 lb of water at 59°F, what will be the final temperature of the water assuming no losses by radiation?

(37·8 Btu; 76·2°F.) (U.E.I.)

No answers are given to the following problems:

7.9. The metal of a boiler weighs 12 tonf and it contains 9 tonf of water. How much heat is required to raise the temperature of the boiler and water from 68°F to 180°F? Assume the specific heat of the metal is 0·11.

7.10. (a) Define "coefficient of linear expansion" and explain why it is important that its value for each engineering material should be known.

(b) A brass rod is 100 in. long at 0°C. What will be its length at 80°C if the coefficient of linear expansion of brass is = 0·000019/deg C? What would be the area of a square sheet of brass at 80°C if at 0°C the length of a side were 10 in.?

(E.M.E.U.)

7.11. In a certain cooling apparatus cold water enters the cooling tank at the bottom and leaves by an outlet near the top of the tank. The hot liquid passes through a coil, entering near the top and leaving at the bottom of the tank; 3000 lb of liquid of specific heat 0·42 enters the cooler per hour, the liquid entering at a temperature of 329°F and leaving at 59°F. How much heat is liberated per hour? The cooling water enters at 53°F and leaves at 143°F. Calculate the quantity of water in gallons per hour passing through the cooling apparatus.

1 gallon of water weighs 10 lb. (U.E.I.)

CHAPTER 8

States of Matter and Changes of State

8.1. States of matter

There are three states of matter. These can be described as solid, liquid, and gaseous. If we assume that the molecules of a substance always have motion and that there is an attractive force acting between them which quickly disappears when the molecules separate, then we can explain the three states of matter.

The solid state occurs when the molecules are very close to each other and the forces acting on each molecule prevent any dispersion. The molecules form a definite ordered pattern and this creates mechanical strength. The body in the solid state has a fixed volume and shape and only by breaking down the ordered arrangement by force or the application of heat can the shape of the body be changed.

The liquid state allows more freedom of movement of the molecules and the restraining forces on each molecule are not as great as in the solid state. For this reason a liquid has a definite volume but not a shape of its own; it takes up the shape of the container holding the liquid. If heat is applied to the liquid the molecules speed up their motion until the liquid changes to gas. If the liquid is cooled sufficiently the molecules slow down and form the regular pattern of the solid.

When the gaseous state occurs the molecules have sufficient freedom to fill the container into which the gas is introduced. In this state a pressure is formed inside the container which can be increased either by introducing more gas into the container or heating the gas. The collisions of the particles inside the container and consequent hammering on the walls of the container cause the pressure to increase.

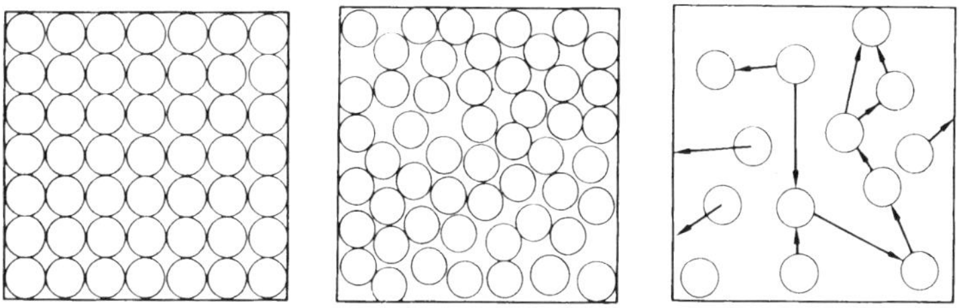

FIG. 8.1. (a) In the solid state the molecules form an ordered pattern; (b) In the liquid state the molecules have no particular arrangement and can flow easily; (c) In the gaseous state the molecules travel at speed and collide with each other and with the sides of the container, building up pressure. The faster the molecules move the higher the pressure becomes.

8.2. Changes of state

The state of a substance at a particular time depends on its pressure and temperature. In general if heat is added to a solid it changes into a liquid and if further heat is added the liquid changes into a gas. Conversely if a gas is cooled it changes into a liquid and then into a solid.

Figure 8.2 is a diagram showing the relationship between the temperature of a substance, its state and the amount of heat energy added or extracted from the substance at a constant pressure.

Starting with 1 lb of a solid at an initial temperature shown by point (1), heat is added represented by *AB* and the temperature rises to point (2). The heat added to increase the temperature of a substance is called Sensible Heat. The heat added to a substance which causes no increase in temperature is called Latent Heat. Point (2) represents the melting point of the solid and the temperature remains constant whilst the solid is completely changed to liquid. The amount of heat added, represented by *BC*, is called the LATENT HEAT OF FUSION. A further application of heat increases the temperature of the liquid until the boiling point is reached at point (3). Until all the liquid has been turned into gas the temperature remains constant. The amount of heat added represented by *DE* is called the LATENT HEAT OF VAPORISATION. Any further addition of heat to the gas superheats the gas.

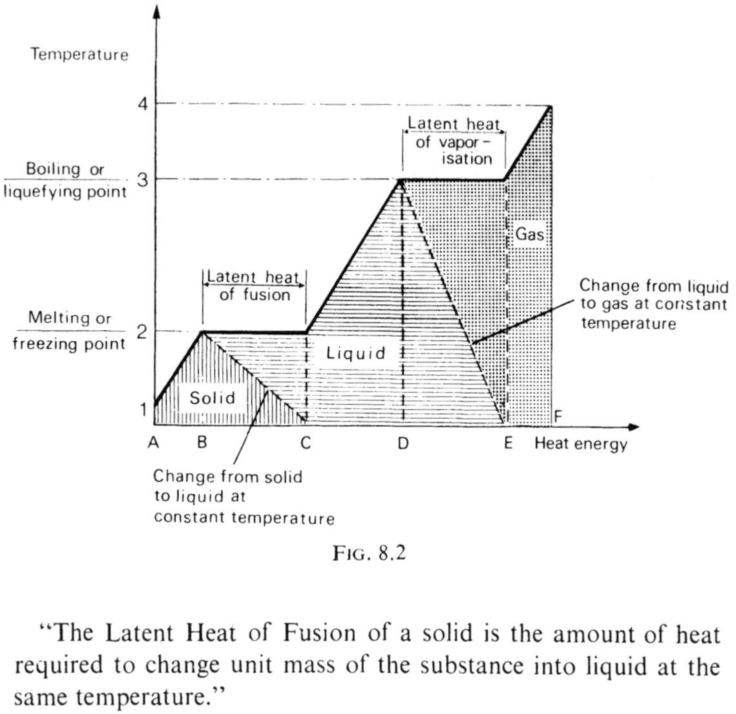

FIG. 8.2

"The Latent Heat of Fusion of a solid is the amount of heat required to change unit mass of the substance into liquid at the same temperature."

"The Latent Heat of Vaporisation of a liquid is the amount of heat required to change unit mass of the liquid into gas at the same temperature."

The melting points and boiling points and latent heats of a variety of substances are given in Table 8.1. The pressure of the substance in each case is atmospheric, i.e. 14·7 lbf/in^2.

TABLE 8.1

Substance	Melting point		Boiling point		Latent heat of fusion, per lb		Latent heat of vaporisation, per lb	
	°F	°C	°F	°C	Btu	Chu	Btu	Chu
Aluminium	1215	657	3272	1800	138·6	77		
Copper	1944	1062			77·4	43		
Lead	621	327	2777	1525	9	5		
Iron	2741	1505	4442	2450				
Steel	2642	1450						
Alcohol	−170	−112	172·8	78·2			369	205
Mercury	−38·0	−38·9	675	357	5·4	3	122·4	68
Water	32	0	212	100	144	80	970·6	539·2

EXAMPLE. If 10 g of ice at −15°C is added to 60 g of boiling water contained in a copper calorimeter of water equivalent 7 g, what will be the resulting temperature of the calorimeter and contents, neglecting losses? Take the specific heat of ice as 0·5 and the latent heat of fusion of ice as 80 cal/g.

Let t°C represent the final temperature.

Heat lost by boiling water $= 60(100-t)$ cal,

Heat lost by the calorimeter $= 7\times(100-t)$ cal,

Heat to raise temp. of ice to 0°C $= 10\times 0{\cdot}5(15-0)$ cal,

Heat to melt ice at 0°C $= 10\times 80$ cal,

Heat to raise the temp. from 0°C to t°C $= 10(t-0)$ cal,

Heat lost by boiling water = heat gained by ice,

$$60(100-t)+7(100-t)=(10\times 0{\cdot}5\times 15)+(10\times 80)+10t,$$

$$6000-60t+700-7t=75+800+10t,$$

$$77t=5825,$$

$$t=\mathbf{75{\cdot}6°C}.$$

8.3. Change of boiling point of water with pressure

When water boils we take the temperature as 100°C or 212°F, but we should also state that the pressure on the surface of the water is atmospheric pressure, accepted as 14·7 lbf/in². If the pressure on the surface of the water is altered then the temperature at which the water boils will alter. An increase in pressure will lift the temperature of the boiling water above 100°C or 212°F. A reduction in pressure below atmospheric will cause the water to boil at a temperature less than 100°C or 212°F.

8.4. Absolute pressure

The pressure of the atmosphere varies slightly from day to day but an average figure is 14·7 lbf/in² at sea level. The pressure gauges seen on boilers measure the difference in pressure between the steam in the boiler and the pressure of the atmosphere and the

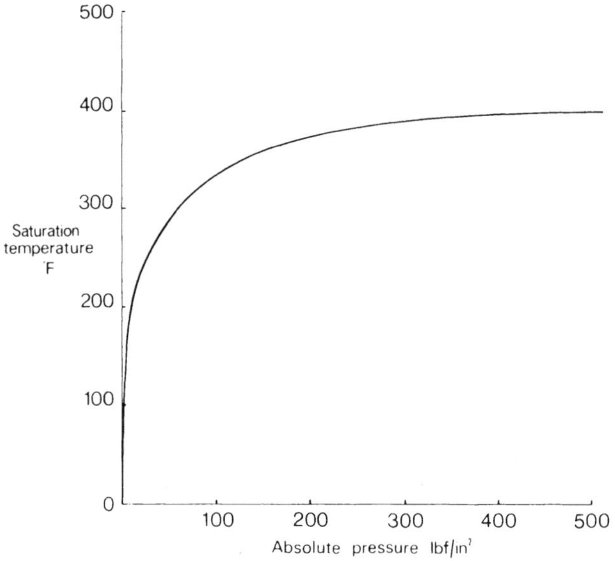

FIG. 8.3

gauge reading is called the "Gauge Pressure". The actual pressure of the steam is called the "Absolute Pressure".

Absolute pressure = gauge pressure + atmospheric pressure.

Experiments have been performed very carefully to measure the temperature of steam in contact with water at various pressures and these figures are compiled into steam tables. The temperature at which steam is formed from the boiling water is called the "Saturation Temperature" for that particular pressure.

Figure 8.3 shows a graph giving the saturation temperatures at various absolute pressures.

8.5. Sensible or liquid heat (h)

It has been stated previously that to raise the temperature of 1 lb of water through 1 deg C requires 1 Chu.

"The sensible heat of a liquid is the amount of heat contained in 1 lb of the liquid at saturation temperature."

Therefore the sensible heat in 1 lb of water at atmospheric pressure is 100 Chu. If we look at Table 8.2 we find a value of 100·06 Chu. This slight difference is due to variation in the specific heat of the water at different temperatures.

8.6. Latent heat of steam (L)

"The latent heat of steam is the amount of heat required to change one pound of water into steam at the same temperature and pressure."

Table 8.2 gives values of latent heats of steam at various pressures. It is seen that the latent heat decreases as the saturation temperature increases.

8.7. Total heat (H_s)

This is the total heat content of one pound of dry saturated steam; that is

$$\text{Total heat} = \text{sensible heat} + \text{latent heat},$$

$$H_s = h + L.$$

TABLE 8.2

Properties of dry saturated steam in foot-pound Fahrenheit units

Absolute pressure (lb/in²)	Temp. (°F)	Vol. (ft³/lb)	Heat Content (Btu/lb)		Absolute pressure (lb/in²)	Temp. (°F)	Vol. (ft³/lb)	Heat Content (Btu/lb)	
			Liquid	Latent				Liquid	Latent
0·5	79·6	643·0	47·6	1048·5	62	294·9	6·957	264·4	914·6
0·6	85·3	540·6	53·1	1045·4	64	296·9	6·752	266·5	913·1
0·7	90·1	466·6	58·1	1042·7	66	299·0	6·560	268·6	911·6
0·8	94·4	411·7	62·4	1040·3	68	301·0	6·378	270·7	910·1
0·9	98·2	368·7	66·2	1038·1	70	302·9	6·206	272·7	908·7
1·0	101·7	334·0	69·7	1036·1	72	304·8	6·044	274·6	907·4
1·2	107·9	281·1	75·9	1032·5	74	306·7	5·890	276·5	906·0
1·4	113·3	243·0	81·3	1029·5	76	308·5	5·743	278·4	904·6
1·6	118·0	214·3	86·0	1026·8	78	310·3	5·604	280·3	903·2
1·8	122·2	191·8	90·2	1024·4	80	312·0	5·472	282·1	901·9
2·0	126·1	173·7	94·0	1022·2	82	313·7	5·346	283·9	900·6
2·4	132·9	146·4	100·8	1018·3	84	315·4	5·226	285·6	899·4
2·8	138·8	126·7	106·7	1014·8	86	317·1	5·110	287·3	898·1
3·2	144·0	111·7	112·0	1011·7	88	318·7	5·000	289·0	896·8
3·6	148·7	100·0	116·7	1009·0	90	320·3	4·896	290·7	895·5
4·0	153·0	90·63	121·0	1006·7	92	321·9	4·796	292·3	894·3
5·0	162·3	73·52	130·2	1001·6	94	323·3	4·699	293·9	893·1
6·0	170·1	61·98	138·1	996·6	96	324·8	4·607	295·5	891·9
7·0	176·9	53·64	144·9	992·2	98	326·3	4·519	297·0	890·8
8·0	182·9	47·35	151·0	988·5	100	327·8	4·434	298·5	889·7
9·0	188·3	42·40	156·5	985·2	110	334·8	4·046	305·7	884·2
10·0	193·2	38·42	161·3	982·5	120	341·3	3·729	312·5	878·9
12·0	202·0	32·40	170·1	976·9	130	347·3	3·456	318·8	874·0
14·0	209·6	28·03	177·7	972·2	140	353·0	3·222	324·9	869·1
14·69	212·0	26·80	180·1	970·6	150	258·4	3·015	330·6	864·5
15·0	213·0	26·28	181·2	970·0	160	363·6	2·835	336·0	860·1
16·0	216·3	24·74	184·5	967·9	170	368·4	2·677	341·2	855·9
18·0	222·4	22·17	190·6	964·0	180	373·1	2·534	346·1	851·9
20·0	228·0	20·09	196·3	960·4	190	377·5	2·407	350·9	847·9
22	233·1	18·38	201·4	957·2	200	381·8	2·290	355·5	844·0
24	237·8	16·94	206·3	954·0	220	389·9	2·089	364·2	836·5
26	242·2	15·72	210·8	951·1	240	397·4	1·918	372·3	829·4
28	246·4	14·67	215·0	948·3	260	404·4	1·775	379·9	822·6
30	250·3	13·73	219·0	945·6	280	411·1	1·651	387·1	816·1
32	254·0	12·93	222·7	943·1	300	417·3	1·543	394·0	809·8
34	257·6	12·21	226·3	940·7	320	423·3	1·448	400·6	803·7
36	260·9	11·58	229·7	938·5	340	429·0	1·364	406·9	797·8
38	264·1	11·02	233·0	936·4	360	434·4	1·289	412·8	792·3
40	267·2	10·50	236·1	934·4	380	439·6	1·222	418·6	786·7
42	270·3	10·03	239·1	932·3	400	444·6	1·161	424·2	781·3
44	273·1	9·600	242·0	930·3	420	449·4	1·106	429·6	775·9
46	275·8	9·209	244·9	928·3	440	454·0	1·056	434·8	770·8
48	278·5	8·848	247·6	926·4	460	458·5	1·009	439·9	765·7
50	281·0	8·516	250·2	924·6	480	462·8	0·967	444·8	760·7
52	283·5	8·208	252·7	922·9	500	467·0	0·928	449·6	755·8
54	258·9	7·992	255·2	921·1	600	486·2	0·770	471·8	732·4
56	288·3	7·656	257·6	919·4	700	503·1	0·655	491·7	710·5
58	290·5	7·407	259·9	917·8	800	518·2	0·569	509·9	689·7
60	292·7	7·175	262·2	916·2	1000	544·6	0·446	542·6	650·2

Note that the sensible heat on the Centigrade scale starts from 0°C but on the Fahrenheit scale it is measured from 32°F so that water at 212°F contains 212 − 32 = 180 Btu per lb.

EXAMPLE. What is the total heat in Btu of 80 lb of steam at 100 lbf/in² absolute pressure?

From table of properties of steam for 1 lb of steam:

$$\begin{aligned} \text{Sensible or liquid heat } h &= 298{\cdot}5 \text{ Btu} \\ \text{Latent heat } L &= 889{\cdot}7 \text{ Btu} \\ \text{Total heat } H_s &= 1188{\cdot}2 \text{ Btu} \end{aligned}$$

∴ for 80 lb of steam the total heat = 1188·2 × 80 = **95,056 Btu.**

These values can be converted from Btu to Chu by using the fact that 1 Btu $= \frac{5}{9}$ Chu.

Problems

8.1. (a) Complete the following extract from the steam tables:

Pressure lbf/in²	Temperature (°F)	Heat in Btu/lb		
		h	L	H
50	281		924·6	1174·8
100	327·8	298·5	889·7	
150	358·4	330·6		1195·1

(b) A boiler is required to produce 500 lb of dry saturated steam at 250 lbf/in² from feed water at 98°F. Calculate the total heat required.

Extract from steam tables

Pressure (lbf/in²)	Temperature (°F)	h	L
250	401	376·1	826

(*continued*)

(c) Explain what is meant by
(i) Sensible heat.
(ii) Latent heat.
(iii) Total heat of saturated steam.

((a) 250·2, 1188·2, 864·5; (b) 568,050 Btu.) (U.E.I.)

8.2. A boiler contains 20 tons of water at 60°F. The boiler weighs 16 ton f and its specific heat is 0·1. Heat is added until 2 tons of the water is turned into dry saturated steam at 250 lbf/in^2. How much heat is required? At 250 lbf/in^2 steam is produced at 401°F, sensible heat is 376 Btu/lb and the latent heat is 833 Btu/lb.

($20{\cdot}54 \times 10^6$ Btu.)

8.3. Calculate the amount of heat required to convert 10 lb of water at 68°F into steam at 212°F. Express your answer in Chu and in Btu.
Latent heat of steam at 212°F = 961 Btu/lb.

(11,052 Btu; 6140 Chu.) (N.C.T.E.C.)

8.4. A copper calorimeter contains 7 lb of water and the temperature is raised from 12°C to 18°C when 12 oz of dry saturated steam at 110°C are passed into the water and condensed. Calculate the water equivalent of the calorimeter.

For steam at 110°C, L = 533 Chu/lb.

(2·12 lb.) (N.C.T.E.C.)

No answers are given to the following problems:

8.5 A piece of ice weighing 200 gf at −6°C is heated until it completely melts. The heating is continued until the water is at a temperature of 70°C. Sketch a graph showing these changes using heat units as the horizontal scale and temperature as the vertical scale. Determine the total heat required to effect this change.

Specific heat of ice = 0·5; latent heat of fusion of ice = 80 cal/g.

8.6. A steel ball weighs 10 lb and has a specific heat of 0·116. The ball is placed at a temperature of 550°F in 1 lb of water at 62°F and at atmospheric pressure. What weight of water will be evaporated?

Latent heat of steam = 970 Btu/lb.

8.7. Explain the meaning of the terms "sensible heat" and "latent heat". 5 lb of dry steam at a pressure of 190 lbf/in^2 absolute are passed into a tank containing 60 lb of water at 50°F. Assuming no loss of heat and neglecting the water equivalent of the tank, what will be the final temperature of the water?

Steam at a pressure of 190 lbf/in^2 absolute has a temperature of formation of 377·5°F, h = 350·9 Btu/lb, L = 847·9 Btu/lb. (U.E.I.)

CHAPTER 9

Electrical Properties of Materials

9.1. Protons and electrons

Before beginning to study the effects of an electric current, it is desirable to have some knowledge of the constitution of matter.

An element is a substance from which only one kind of matter has ever been obtained. The smallest part of an element having the same properties as the element is called an ATOM (Fig. 9.1). Some atoms can exist alone in normal circumstances, e.g. those of copper, aluminium, neon, etc., whereas some others such as oxygen and hydrogen can only normally exist in pairs. The smallest group of atoms that can exist alone are called MOLECULES.

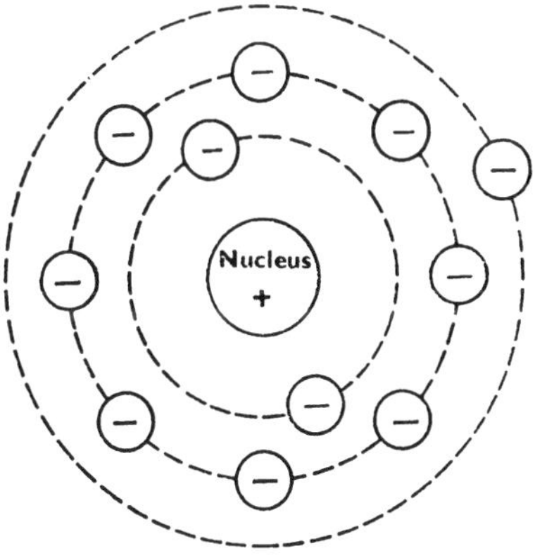

FIG. 9.1. Simplified diagram of a sodium atom showing the central nucleus and eleven orbital atoms.

Molecules of elements contain only one type of atom. The molecules of COMPOUNDS, such as water, contain atoms of different types. Each molecule of water contains two atoms of hydrogen and one atom of oxygen (using symbols, H_2O).

An atom consists of a NUCLEUS containing electrical charges called PROTONS and spinning around the nucleus in various paths or orbits are other electrical charges called ELECTRONS. The size of the charge on an electron is the same as that on a proton but they produce different effects. A proton is said to possess a positive (+) charge and an electron a negative (−) charge.

An analogy to the structure of the atom can be made by considering the satellites (the electrons) orbitting the earth (the nucleus). Just as there is a vast amount of space between the earth and the satellites so there is a relatively large space in the atom between the nucleus and its electrons. In fact, the atom is nearly all space.

There may be other particles in the nucleus of the atom but it is not necessary to consider them here.

The particles mentioned previously are very small. One cubic inch of copper contains about $1{\cdot}385 \times 10^{24}$ atoms. The proton has a mass of $1{\cdot}672 \times 10^{-27}$ kg and the electron mass is $9{\cdot}109 \times 10^{-31}$ kg. The mass of the proton is about 1840 times as great as that of an electron.

Positive and negative charges can be produced in small quantities on insulated rods and their presence demonstrated by the use of a gold-leaf electroscope (Fig. 9.2). When the gold-leaf is deflected, a stable position is reached when the deflecting force is balanced by the effect of gravity on the deflected leaf.

When a glass rod is rubbed with silk and brought near to the plate of an uncharged electroscope (Fig. 9.2a), the gold-leaf will be displaced (Fig. 9.2b). If the plate is now connected to earth (c), the positive charges will flow to earth whilst the negative charges are held by the mutual attraction between them and the positive charges on the glass. If the earth connection and then the glass rod are removed (d), then the negative charges distribute themselves over the whole metal surface of the electroscope and the leaves again diverge because of the force of repulsion between like charges.

If, now, a positively-charged body, such as an ebonite rod that has been rubbed with fur, were brought near to the plate of the electroscope, the divergence of the leaves would be reduced (e). A larger divergence would occur if a negatively charged body were used.

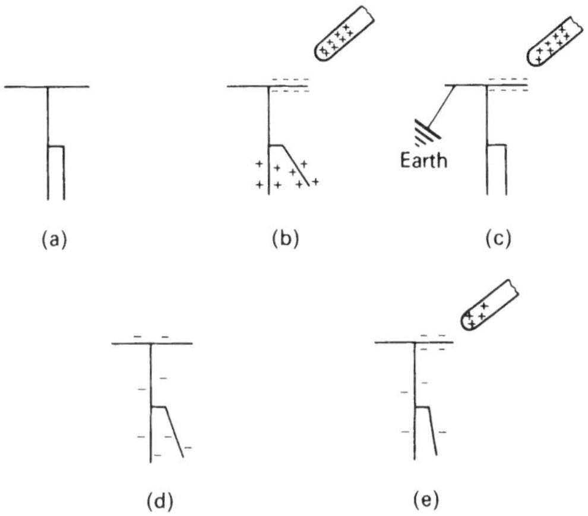

FIG. 9.2

It can be shown, also, that like charges repel each other and unlike charges attract each other. A charged body will exert a force of attraction on an uncharged body.

Brief mention may be made here of the meaning of an electric field. An electric field can be considered as existing where there is a force exerted on an electric charge. Electric charges are often obtained by rubbing two materials together. For example, the teeth of a comb may attract small pieces of paper after rubbing the comb on one's sleeve. Rubbing a balloon may enable it to "stick" to the ceiling or wall. If two unlike charges are large enough a spark discharge may occur between them. On a very large scale this effect is observed as lightning.

9.2. Conductors, insulators, and resistors

It has been stated that each atom consists of a nucleus containing protons (positively charged) and around the nucleus in various orbits are electrons (negatively charged). The electrons are in continuous motion around the nucleus and normally they are unable to leave the atom because of the force of attraction between the positive charges in the nucleus and the negative charge on the electron.

All atoms in a material are not, at any instant, in this normal state. If it were possible to see the electrons of, for example, a piece of copper, some electrons from the outer orbits would be drifting in a random manner in the spaces between the atoms. Generally, the atom which has lost an electron gains one almost immediately from another atom. These drifting electrons are known as free electrons and are least tightly held to the nucleus because they are furthest away from it. Individually, these free electrons are moving rapidly in a random manner.

When the right kind of energy is supplied to the electrons in the copper, they drift in one direction and this flow of electrons is known as an electric current. The source of energy required to produce this electron drift is known as an electromotive force (e.m.f.). (This is perhaps an unfortunate choice of name since it is not a force but a measure of the energy available from a source.) This e.m.f. is most commonly available from a battery or an electric generator.

The electron drift is usually at the rate of very much less than an inch per second but the number of electrons passing is very large. Now the speed at which the *effects* of electricity are observed to travel is approximately 188,000 miles per second. In view of what has been said this appears to be impossible. A similar effect is noticed with sound. If one person shouts to another the message travels at about 760 miles per hour, but a hurricane is not set up. The particles of air vibrate rapidly but do not travel any appreciable distance. An electric current is similar in effect but a complete explanation must be left to advanced studies of travelling waves.

Copper was used as an example of a material in which there is a large number of free electrons. Experience has shown that all metals have a large number of free electrons and a fairly small e.m.f. will make the electrons in these materials drift in one direction. Materials which possess this property are referred to as CONDUCTORS. Copper and aluminium are the two metals most frequently used in electrical engineering for conducting electrical energy to where it is to be usefully employed. Silver, which is very expensive, is an excellent conductor used for special applications.

A conductor often used but which is non-metallic is carbon. It is used in the form of graphite and one use is for conducting an electric current into and out of the windings on the rotating part (rotor or armature) of electrical machines.

There is another group of materials known as INSULATORS which do not have a large number of free electrons and these materials are used to restrict the electrons to paths or circuits along which they are required to flow. A very large e.m.f. applied to these materials would cause only a very small electron drift. Examples of these are air, oil, paper, mica, rubber, enamel, polyvinyl chloride (P.V.C.), varnish, as well as many other natural and man-made materials. There is, however, no such thing as a perfect insulator, although in most practical cases the current flowing through it is usually negligible.

A third group of materials could be described as rather poor conductors. They have quite a number of free electrons but less than the materials called conductors. These materials are known as RESISTORS and examples are eureka, constantan, manganin and many other alloys. They are frequently used in the part of a circuit where it is desired to produce heat. Alternatively, they are used in a circuit to control the current and are then called variable resistors or rheostats.

A further group of materials called semi-conductors will be discussed in Chapter 20.

9.3. The direction of current

It was stated that a source of e.m.f. was necessary in order to obtain an electron flow. Now it becomes necessary to identify, in some way, the source of e.m.f. with the direction of current.

A source of e.m.f. has two terminals called respectively the positive and the negative terminal. When the effects of an electric current were first studied scientists had no knowledge of electron drift, and from other considerations a direction of current flow from the positive terminal of the source of e.m.f. through the external circuit to the negative terminal was assumed. So by long-established convention this has become the assumed direction of direct current (d.c.) flow. For certain branches of electrical engineering, it is more convenient to use actual electron flow, which is in the opposite direction to "conventional flow". Figure 9.3 indicates these points.

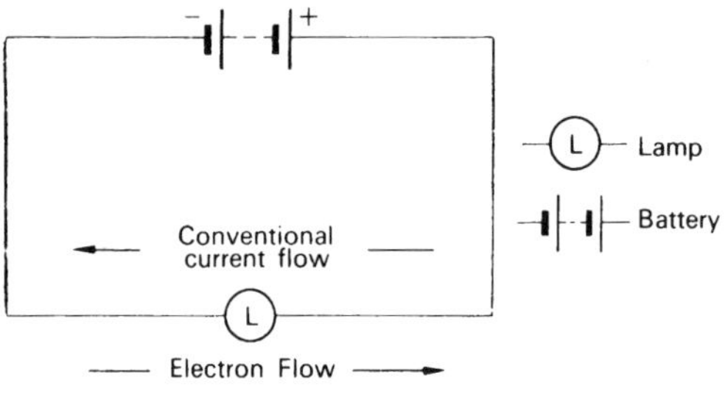

FIG. 9.3

9.4. Insulating materials

An insulator, or an insulating material, may be regarded as being a very poor conductor. A little more will now be said about these materials and this necessitates the use of some units which are explained later.

The effectiveness of insulating materials can be compared by measuring their resistivities (see Section 10.14). The resistivity of a material is the resistance between opposite faces of a unit cube of the material and the figures given refer to a cube having sides of 1 centimetre length. Typical values are:

Mica	2×10^{17} ohm-cm
Porcelain	3×10^{14} ohm-cm
Rubber	1×10^{18} ohm-cm
Air	Much greater than these.

For comparison, a common conductor, copper, has a resistivity of $1{\cdot}7 \times 10^{-6}$ ohm-cm.

Another important property of an insulating material is the voltage that it will withstand before a sudden large increase of current occurs. The voltage across a flat sheet of an insulating material of uniform thickness divided by the thickness is called the voltage gradient across the material. When the voltage gradient exceeds a certain value, the atomic structure of the material changes and its resistance suddenly decreases, allowing a much larger current flow across a comparatively small area. This is known as an electrical breakdown and the material, if it is solid, is permanently damaged.

Should this critical voltage be exceeded with a liquid or a gas, there is an electric arc, but permanent damage does not necessarily result after removal of the voltage.

Typical breakdown voltages for common materials are:

Air	3×10^4 volts per centimetre (as for any gas, this value varies with temperature and pressure).
Mica	$3{\cdot}4 \times 10^6$ up to 7×10^6 V/cm.
Porcelain	$2{\cdot}9 \times 10^{14}$ V/cm.
Oil	1×10^{12} V/cm.
Paper	$1{\cdot}1 \times 10^5$ up to 4×10^5 V/cm.

Insulating materials have their usefulness reduced by chemical changes or by the absorption of moisture. Chemical changes occur very slowly at room temperatures but high temperatures accelerate these changes. For this reason, it is normal practice to specify the maximum temperature at which a material may be permitted to operate. There are several classes of insulation specified in the

British Standards Institution's publication and more information is given in B.S.S. 2757. In some cases, ultraviolet light can cause chemical changes resulting in a reduction in the usefulness of the material as an insulator.

The absorption of moisture is prevented in porous materials such as paper and cotton by impregnating them with insulating varnishes or oil. More recently, silicone varnishes have been used and they have an added advantage that they can withstand moderately high temperatures.

The following are examples of the uses of insulating materials:

Paper (dry). Telephone cables.

Paper (oil immersed). High-voltage power cables.

Asbestos. Conductors connected to heating apparatus such as fires, immersion heaters, etc.

Polyvinyl chloride (P.V.C.). Covering copper conductors used for house wiring, etc.

Rubber. Similar to P.V.C.

Silicone Varnishes. As a covering on the conductors used for winding the armatures of electrical machines.

Oil. Switchgear, transformers, and oil-impregnated cables.

Cotton. Used for covering the conductors used for winding low-voltage coils. Double-cotton-covered (d.c.c.) wire has two layers of cotton thread.

Air. An insulator between overhead lines, busbars, etc. It prevents a current from flowing when a switch is open.

9.5. The heating effect of an electric current

An electric current can be detected by the effects it produces. These can be summarised as (1) heating, (2) chemical and (3) magnetic. The chemical effect is dealt with in Chapter 12 and the magnetic effect in Chapter 13. Only the heating effect will be discussed here.

It is believed that the molecules of any substance are in a state of continuous vibration and so, like any body in motion, they possess kinetic energy. When a current in a circuit is increased, there is a greater electron movement and so there is an increased number of collisions between the free electrons and atoms, this resulting in increased vibration of the molecules. Heat, as was shown by Dr. Joule in the early 1840's, is a form of energy and it is as kinetic energy associated with the vibrating molecules that heat is stored in a body.

Whenever a current flows, heat is produced. If the current is varying in magnitude and alternating in its direction (i.e. an alternating current), heat is still produced but the amount of heat is, in general, different at every instant. More is said about heat produced by an alternating current in Chapter 19.

Heat produced by a current may be useful or wasteful. It is useful in electric fires, soldering irons, immersion heaters, etc., and in fuses (see Chapter 22). In metal filament lamps, only a small amount of the electrical energy supplied produces useful light. Most of the energy is wasted as heat. Fluorescent lamps are much more efficient as light producers and are much cooler in operation.

Heat energy is useless in the windings of electrical machines and often energy has to be used driving a fan to blow air through the machine to keep it cool. In very large alternators, hydrogen is often used instead of air, as windage power losses are reduced, and a given volume of hydrogen removes more heat than the same volume of air. In some cases, hollow conductors are used through which cooling water is circulated. Heat is wasted in the conductors carrying the current to where it is required to perform a useful function. These losses are called copper losses.

In most cases, it is heat which limits the rating of a piece of equipment. As stated in Section 9.4, if the temperature of an insulating material exceeds a certain value, its properties deteriorate more rapidly and an electrical breakdown may result.

Electrical equipment is rated so that it will carry a certain current indefinitely without overheating. This is called its continuous rating. It will also have a short-time rating. For example, a piece of equipment may be able to carry 50% overload for 30 minutes from cold without excessive temperature rise. This is its half-hour rating. It should be remembered that the hottest part of a machine is in some inaccessible position within the machine. If it is hot externally, it is very hot internally!

CHAPTER 10

Introduction to Electrical Units and Circuits

10.1. Coulombs and amperes

As previously stated, an electron flow in a conductor constitutes an electric current. A unit of quantity of electricity is the charge of ONE COULOMB, being $6{\cdot}242 \times 10^{18}$ times as great as the charge of a single electron.

If a charge of one coulomb of electricity passes a point in an electrical circuit in one second, the current is said to be ONE AMPERE. Thus

$$\frac{\text{coulombs}}{\text{seconds}} = \text{amperes}.$$

There are standard symbols for all electrical quantities. For example, if a quantity or charge (Q) of twelve coulombs (12 C) passes a point in a circuit in 4 seconds (4 sec), the current (I) flowing in amperes (A) is

$$I = \frac{12}{4} \text{ C/sec} = 3 \text{ A}.$$

Smaller units of current occur in various branches of electrical engineering. Examples of these are:

$$1 \text{ milliampere } (1 \text{ mA}) = \frac{1}{1000} \text{ ampere } (10^{-3} \text{ A}),$$

$$1 \text{ microampere } (1\ \mu\text{A}) = \frac{1}{1{,}000{,}000} \text{ ampere } (10^{-6} \text{ A}).$$

10.2. Electromotive force

Electromotive force, or, as it is often written, e.m.f., may be described as that source of energy which enables electrons to move around an electric circuit. It is now necessary to define this quantity.

For anything to move from rest, there must be some energy change. To enable electrons to move round an electrical circuit, they must receive energy from the source of e.m.f. which is usually a battery or a generator.

For each coulomb of electricity to move completely round an electrical circuit, a certain amount of electrical energy is required, and this depends on the particular circuit. The e.m.f. is expressed in VOLTS and is numerically the number of joules (see Section 10.7) of energy given by the source of e.m.f. to each coulomb to enable it to move round the circuit. The symbol for volt is the capital letter V.

Thus

$$\frac{\text{joules}}{\text{coulombs}} = \text{volts}.$$

It follows that:

$$\begin{aligned}\text{joules} &= \text{volts} \times \text{coulombs}\\ &= \text{volts} \times \text{amperes} \times \text{seconds}.\end{aligned}$$

A 12-volt (12-V) battery is capable of giving 12 joules (12 J) of energy to each coulomb to enable it to move round an electrical circuit.

The symbol for e.m.f. is the capital letter E.

EXAMPLE. Calculate the energy supplied by a 12-V battery when a current of 4 A flows for 10 minutes (10 min).

$$\begin{aligned}\text{Energy supplied} &= \text{volts} \times \text{amperes} \times \text{seconds},\\ &= 12 \times 4 \times (10 \times 60)\ \text{joules}\\ &= \mathbf{28{,}800\ J.}\end{aligned}$$

10.3. Potential difference

If each coulomb of electricity receives energy to move it completely around a circuit, then part of the energy must be used in moving it between two points in the circuit. This amount of energy per coulomb is known as the potential difference (p.d.) between the two points.

The unit of potential difference is also the volt and it follows that the sum of the p.d.s must equal the e.m.f. The symbol for potential difference is V, and this symbol must not be confused with V when it is used to represent volts.

10.4. Measurement of current and voltage

An instrument that will indicate the current in a circuit is called an ammeter. An e.m.f. or a p.d. can be measured by a voltmeter. These instruments are classified as indicating instruments.

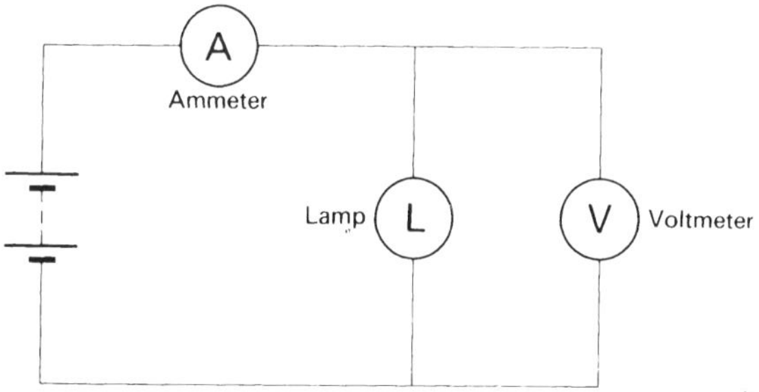

FIG. 10.1

As will be seen later, the principle of operation of the ammeter and the voltmeter is the same. At this stage, it will be assumed that when an ammeter is in a circuit there is no p.d. across its terminals. It is also assumed that a voltmeter takes no current when it is connected across a circuit. In most cases, negligible errors are made by these assumptions but in some cases, as shown in Section 15.6, care has to be taken in the choice of instrument if accurate results are required.

Figure 10.1 shows an ammeter and a voltmeter connected in a circuit. An ammeter is connected in series with the part of the circuit in which it is desired to know the current, whilst a voltmeter is connected in parallel with the part of the circuit across which the value of the p.d. is required.

10.5. Electrical units of power

Power is defined as the rate of doing work or the rate of expending energy. Units of energy divided by time give units of power. The energy used in part of a circuit is given by

$$\text{joules} = \text{volts} \times \text{amperes} \times \text{seconds}.$$

Using symbols, $W = VIt$ joules.

Hence Power, $P = \dfrac{VIt}{t} = VI$ joules per second (J/s).

Joules per second are known as watts (symbol W). The power consumed in a part of a circuit in which a current I is flowing and across which a potential difference V exists is VI watts.

Larger units of power include

$$1 \text{ kilowatt } (1 \text{ kW}) = 1000 \text{ W}$$

and

$$1 \text{ megawatt } (1 \text{ MW}) = 1{,}000{,}000 \text{ W}.$$

An electric fire may have a rating of 3 kW whereas a modern alternator in a power station may have a power output of 500 MW.

In electronics, a smaller unit of power may be used,

$$1 \text{ milliwatt } (1\text{mW}) = \frac{1}{1000} \text{ watt } (10^{-3}\text{W}).$$

Note how important the m or M is in the abbreviation.

10.6. The commercial unit of electrical energy

The joule or watt-second as a unit of energy is too small for everyday use and a larger unit, the kilowatt-hour, is used. This is the "unit" of electricity referred to on electricity bills. The electricity meter in the house registers kilowatt-hours.

One kilowatt-hour (1 kWh) is equivalent to a power of 1 kW being used for 1 hour, or 0·5 kW for 2 hours, 2 kW for 0·5 hour, etc.

$$\begin{aligned} 1 \text{ kWh} &= 1 \text{ kilowatt} \times 1 \text{ hour} \\ &= 1000 \text{ watts} \times 3600 \text{ seconds} \\ &= 3{,}600{,}000 \text{ watt-seconds or joules.} \end{aligned}$$

EXAMPLE. Calculate the energy consumed (a) in joules, (b) in kilowatt-hours, when a 2-kW fire is on for 1·5 hours.

$$\begin{aligned} \text{(a) Energy} &= \text{watts} \times \text{seconds} \\ &= 2000 \times 1{\cdot}5 \times 60 \times 60 \text{ joules} \\ &= \mathbf{10{,}800{,}000\ J.} \end{aligned}$$

$$\begin{aligned} \text{(b) Energy} &= \text{kilowatts} \times \text{hours} \\ &= 2 \times 1{\cdot}5 \text{ kilowatt-hours} \\ &= \mathbf{3\ kWh,} \\ \text{or} \qquad &= \frac{10{,}800{,}000}{3600} \text{ kilowatt-hours} \\ &= \mathbf{3\ kWh.} \end{aligned}$$

10.7. The joule or newton-metre

The system of units used in electrical engineering is the rationalised metre-kilogramme-second (m.k.s.) system. Prior to the general use of this system, two other systems were in common use for studying electrical engineering. These were the electromagnetic (e.m.c.g.s.) and the electrostatic (e.s.c.g.s.) centimetre-gram-second systems. The practical units were related to these by a number of constants.

When the rationalised m.k.s. system of units was established, all existing practical units were incorporated in it. One of these units was the joule which, in the electromagnetic c.g.s. system, was equivalent to 10^7 ergs or dyne-centimetres, the dyne being the unit of force in the e.m.c.g.s. system. In the rationalised m.k.s. system, the unit of length is the metre = 100 cm. To maintain the joule as the unit of energy,

$$1 \text{ joule} = (\text{a unit of force in dynes}) \times (100 \text{ cm})$$
$$= 10^7 \text{ dyne-cm or ergs.}$$

Thus the unit of force must be equal to 10^5 dynes.

This force of 10^5 dynes was called the newton and so

$$1 \text{ newton (1 N)} = 10^5 \text{ dynes}$$

and

$$1 \text{ joule} = 1 \text{ newton-metre.}$$

10.8. Unit of mass in the rationalised m.k.s. system

A force of 1 dyne would accelerate at 1 cm/sec^2 a mass of 1 gramme, if it were free to move.

A force of 10^5 dynes would accelerate at 10^5 cm/sec^2 a mass of 1 gramme.

A force of 1 newton would accelerate at 10^2 cm/sec^2 a mass of $\frac{10^5}{10^2} = 10^3$ grammes, i.e. a force of 1 newton would accelerate at 1 m/sec a mass of 1 kilogramme.

As the newton, metre and second are all units of the m.k.s. system, the kilogramme must be the unit of mass in this system.

Reference may be made to Section 5.8 for the relationship between the units of force, energy, and power in the various systems of units. Note particularly the similarity of the SI and m.k.s. sysems.

10.9. Heat energy

Heat is a form of energy and when a current passes through a conductor, the electrical energy is converted into heat. The amount of heat produced by a given amount of electrical energy has been found by experiment and the relationships between these quantities are

$$1 \text{ calorie} = 4{\cdot}19 \text{ joules,}$$
$$1 \text{ kilowatt-hour} = 3410 \text{ British Thermal Units (Btu).}$$

Further information about heat energy is given in Section 7.9.

EXAMPLE. An electric kettle raises the temperature of 3 pints of water from 10°C to 100°C in 5 min. If the kettle element is rated at 2500 W when operating from a 240-V supply, calculate the efficiency of the kettle.

[1 gallon (8 pints) of water weighs 10 lb, 1 lb = 0·454 kg.]

$$\text{Energy supplied} = 2500 \times 5 \times 60 \text{ joules}$$
$$= 750{,}000 \text{ J.}$$
$$\text{Heat utilised} = 10 \times 454 \times (100 - 10) \times \tfrac{3}{8} \text{ calories}$$
$$= 153{,}225 \text{ cal.}$$

Converting this to joules,

$$\text{Heat utilised} = 153{,}225 \times 4{\cdot}19 \text{ joules}$$
$$= 642{,}013 \text{ J.}$$
$$\text{Kettle efficiency} = \frac{642{,}013}{750{,}000} \times 100\,\%$$
$$= \mathbf{85{\cdot}6\,\%}.$$

10.10. Ohm's law

Tests can be carried out in a laboratory which help students to understand electrical theory. In many cases the reader will carry out some of the tests described in this book. Some results are, however, given.

Let us first investigate the relationship between the current through a length of wire and the p.d. across it.

The piece of wire can be of copper, aluminium, eureka, constantan, etc. A wire made of eureka or constantan is more convenient for use in this test as only a short length will be necessary. The wire sample may be called a resistor.

The current flowing in the circuit may be varied in magnitude by the use of a variable resistor (or rheostat). Figure 10.2 shows the circuit required.

The results obtained should be tabulated and before the equipment is disconnected, a graph should be drawn to verify that the results obtained are satisfactory. Any that are obviously incorrect can then be checked by repetition. Typical results were obtained as follows:

Potential difference across the wire:	0	1·48	2·90	4·80	6·10	7·4	volts
Current through the wire:	0	0·50	0·98	1·52	2·00	2·45	amperes

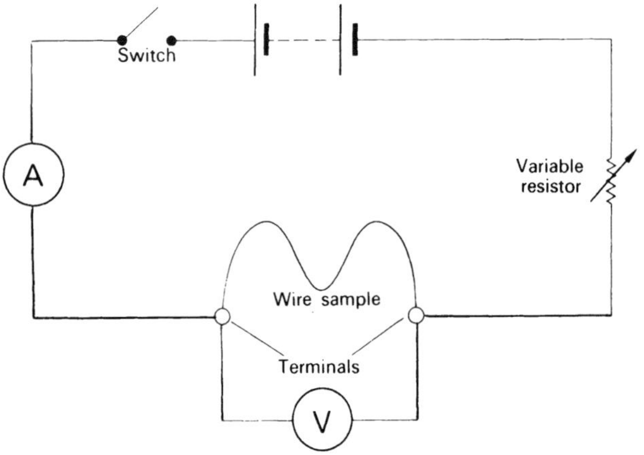

FIG. 10.2

Only one of these results is absolutely accurate. If there is no current, there is no potential difference across the wire. All other results are subject to experimental error, the commonest of these being due to inaccurate reading of the instruments. In a well-run laboratory, errors due to faulty instruments should be rare. This excuse is too frequently offered in laboratory reports as a reason for incorrect results.

In this test, the current was varied and as a result, the p.d. across the wire changed. It is usual to plot the first quantity, i.e. the one that was intentionally varied, on the horizontal axis of the graph and the other quantity on the vertical axis. Figure 10.3 shows this.

When the results are plotted, it will be apparent that they lie approximately in a straight line. A straight line drawn so that the plotted points are distributed as nearly as possible along the line, must pass through the only correct point, that is, zero current, zero voltage. If a point is obviously incorrect, as one is in this case, the point must be ignored or checked if the equipment is still available. Plotting the results given will show this.

When two quantities are so related that, when plotted, a straight line passes through the origin, then one quantity is proportional to the other. Figure 10.3 shows that in this case, the p.d. is proportional to the current. Using symbols, this can be written as $V \propto I$.

$$\text{Hence } V = (\text{a constant}) \times I.$$

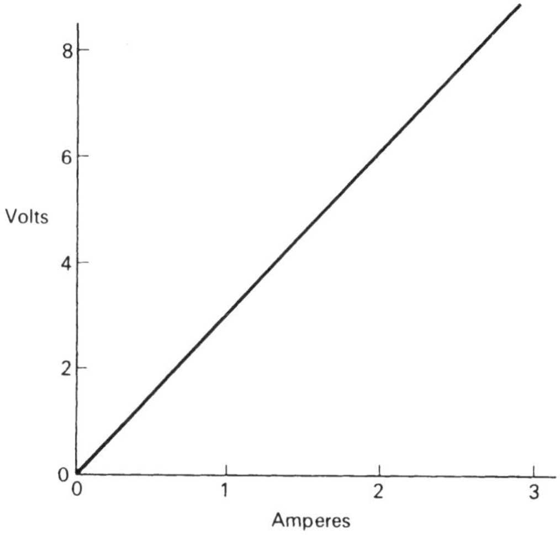

FIG. 10.3

This constant is known as the resistance of the wire. When V is measured in volts and I is in amperes the resistance is expressed in ohms (symbol Ω). Ω is the Greek letter omega. Note that the resistance R is equal to the gradient of the graph.

Hence

potential difference (volts) = resistance (ohms) × current (amperes).

In symbols, $V = RI$ volts.

A smaller unit of resistance is the microhm.

$$1 \text{ microhm } (1\ \mu\Omega) = \frac{1}{1{,}000{,}000} \text{ ohm } (10^{-6}\ \Omega).$$

These observations were summarised by Ohm as follows:

Provided that the physical conditions remain unchanged, then for certain materials the potential difference across them is proportional to the current through them.

This important law applies to all common metals and many alloys but there are many materials to which it does not apply. For example, if the voltage across a piece of metrosil is doubled, the current through it may increase several times. Typical characteristics for this material are given in Section 10.18.

10.11. Further expressions for power and energy

It has been shown that energy $= VIt$ joules.

From Ohm's law,

$$V = RI,$$

$$\therefore \quad \text{energy} = (RI)It = I^2Rt \text{ joules.}$$

Alternatively,

$$I = \frac{V}{R},$$

$$\therefore \quad \text{energy} = V\frac{V}{R}t = \frac{V^2t}{R} \text{ joules.}$$

Likewise, power is given by the expressions VI, I^2R and V^2/R watts.

As 1 N $= 0{\cdot}225$ lbf (Section 5.9),

then

$$1 \text{ J} = 1 \text{ Nm} = 1 \times 0{\cdot}225 \text{ (lbf)} \times \frac{100}{2{\cdot}54 \times 12} \text{ (ft)}$$

$$= 0{\cdot}739 \text{ ft-lbf.}$$

$$\therefore \ 1 \text{ lbf} = 1{\cdot}356 \text{ J or Nm.}$$

Also

$$1 \text{ hp} = 550 \text{ ft-lbf/s},$$

$$= 550 \times 1{\cdot}356 \text{ J/s},$$

$$= 746 \text{ J/s or W.}$$

EXAMPLE. If a current of 5 A flows through a resistance of 12 Ω for 3 minutes, how much energy is expended?

$$\begin{aligned}\text{Energy} &= I^2Rt\\ &= 5^2 \times 12 \times (3 \times 60)\ \text{J}\\ &= \mathbf{54{,}000\ J.}\end{aligned}$$

Remember that all this energy is converted into heat.

10.12. Effect of length on the resistance of a piece of wire

It has been shown that a piece of wire could be used in order to find the relationship between the current through it and the potential difference across it. It is now necessary to find how the resistance of a wire having a uniform cross-sectional area varies with its length.

If a suitable piece of wire is connected between two terminals, the circuit shown in Fig. 10.4 can be used to determine the resistance of any length of it. By altering the length of the wire across which the p.d. is measured, a set of readings can be obtained such as those in Table 10.1. Remember that these are actual results and may contain some errors.

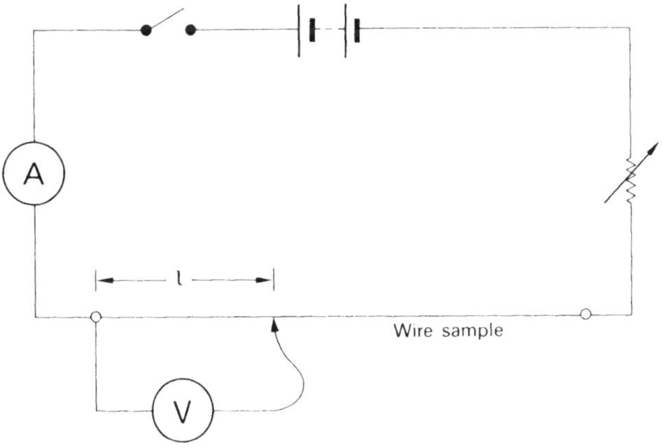

FIG. 10.4

TABLE 10.1

Length of wire	0	20	40	60	80	100	120	cm
P.D. across wire	0	0·25	0·65	0·98	1·29	1·58	1·90	volts
Current through wire	0	0·5	0·5	0·5	0·5	0·5	0·5	amperes
Resistance of wire = V/I	0		1·30				3·80	ohms

In the last line, two values of the resistance have been calculated. The reader should complete the table. A graph should then be plotted of the resistance of the wire against its length and this will be found to be a straight line passing through the origin (corresponding to resistance = 0 Ω when the length = 0 cm). Experimental error may occur if the wire is not exactly uniform or if the wire under test becomes warm.

From the graph obtained it may be stated that the resistance of a uniform wire is proportional to its length. Using symbols, $\boldsymbol{R} \propto \boldsymbol{l}$.

10.13. Effect of the cross-sectional area of a wire on its resistance

If equal lengths of wire made from the same material are connected between pairs of terminals, then if each sample of wire has a different cross-sectional area, the relationship between the resistance of the wire and its area can be investigated. To obtain the results shown in Table 10.2, pieces of wire each having a length of 100 cm were used.

TABLE 10.2

Diameter of wire	0·1129	0·1598	0·1955	0·2257	0·2521	cm
Resistance of wire	1·09	0·545	0·363	0·272	0·218	Ω
Area of cross-section	0·01					cm^2
$\frac{1}{\text{Area of cross-section}}$	100					—

The area of cross-section of the wire may be calculated from the expression

$$\frac{\pi \times (\text{diameter})^2}{4}.$$

The resistance can be measured by the same method as was used for finding the resistance of wire of various lengths. More accurate results can be obtained using a commercial instrument designed for resistance measurement (Section 15.7).

If a graph of resistance is plotted against the area of the wire, no relationship between these quantities is immediately obvious. When resistance is plotted against 1/area, it will be found that there is a linear relationship. The reader should calculate the required quantities and verify this for himself. One set of results has been worked out.

As the graph of resistance against the reciprocal of the area is a straight line passing through the origin, it can be said that the resistance of each piece of wire is proportional to 1/area. Another way of saying the same thing is to state that the resistance is inversely proportional to the area. Using symbols, $\boldsymbol{R \propto (1/a)}$.

10.14. Resistivity

The results of the last two tests show that the resistance of a uniform wire of a given material is proportional to its length (l) and inversely proportional to its area of cross section ($1/a$). Combining these results,

$$\text{Resistance, } R \propto \frac{l}{a}.$$

Thus for a given material, the resistance

$$R = (\text{a constant}) \times \frac{l}{a} \text{ ohms.}$$

This constant is denoted by the symbol ρ (pronounced rho) and the constant is called the resistivity of the material.

Hence $$R = \frac{\rho l}{a} \text{ ohms.}$$

If the length and the area are in inches and square inches, respectively, ρ is expressed in ohm-inches (Ω-in.). If centimetre measure is used, the unit is in ohm-cm (Ω-cm) whilst with metre measure, the unit is the ohm-metre (Ω-m).

Rearranging the expression,

$$\rho = \frac{Ra}{l}.$$

The dimensions of these quantities are

$$\frac{\text{ohms} \times (\text{length})^2}{\text{length}} = \text{ohms} \times \text{length}.$$

The numerical value of ρ is the resistance between opposite faces of a cube of the material having sides of unit length. For example, the resistivity of copper may be taken as $\frac{2}{3} \times 10^{-6}$ Ω-in. In other words, the resistance between the opposite sides of a cube of copper having sides 1 in. in length is $\frac{2}{3} \times 10^{-6}$ Ω. For the results given in Section 10.13, $\rho = 109$ μΩ-cm.

The resistivity of other conductors is given below, some values for insulating materials having been given in Section 9.4.

Iron	8·71 μΩ-cm.
Copper	1·55 μΩ-cm.
Silver	1·5 μΩ-cm.
Aluminium	2·5 μΩ-cm.

The reader should, as an exercise, convert these to μΩ-in. units.

The relationship for resistance can be written

$$\frac{V}{I} = R = \frac{\rho l}{a}, \qquad \therefore \frac{I}{a} = \frac{1}{\rho}\frac{V}{l}$$

or (current density) = (circuit constant) × (voltage gradient). This can be compared later with relationships used in magnetic and electrostatic circuit calculations.

EXAMPLE. A copper conductor is 150 yd in length and has a cross-sectional area of 0·04 in^2. Calculate its resistance if the resistivity of the copper is $\frac{2}{3} \times 10^{-6}$ Ω-in.

$$\text{Resistance} = \frac{\frac{2}{3} \times 10^{-6} \times 150 \times 36}{0.04}\ \Omega$$
$$= \mathbf{0{\cdot}09\ \Omega.}$$

10.15. Effect of temperature on resistance

In order to find out the effect of temperature on resistance, special pieces of equipment have been manufactured. One of these essentially consists of coils of wire made from different materials which are immersed in oil. The oil is heated by a small immersion heater and if the oil is stirred, the thermometer indicates the temperature of the oil and, therefore, the temperature of the wire. The resistance of the wire can be measured as its temperature changes by using a commercial instrument.

The following results were obtained for a copper coil:

Resistance	5·3	5·6	5·8	6·13	6·43	6·63	7·02	Ω
Temperature	20·5	34·5	44·5	59·5	72·0	82·5	98·5	°C

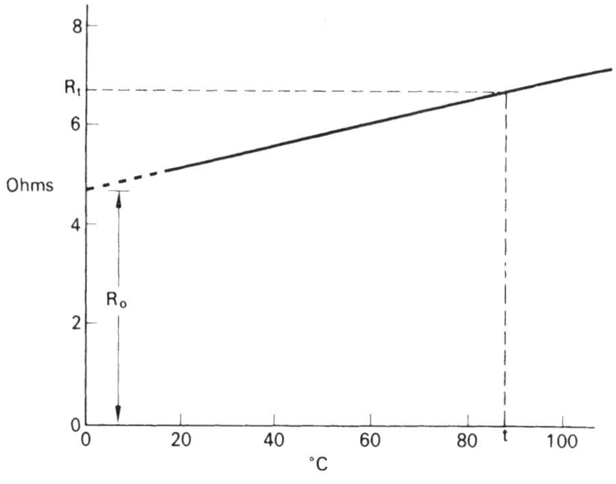

FIG. 10.5

In plotting these results on a graph, resistance is plotted vertically using a scale from 0 to 8 ohms and temperature horizontally using a scale from 0 to 100°C. It will be found that over this range the graph is linear. By producing the graph backwards, the resistance at 0°C can be found. The resistance at 0°C is represented by R_0 (see Fig. 10.5).

Now suppose that a piece of copper wire had a value of 1 ohm at 0°C and every time the temperature rose by 1 deg C, the resistance increased by α_0 ohms (α is the Greek letter alpha). At 2°C the wire would have a resistance of $(1+2\ \alpha_0)$ ohms and at t°C it would have a resistance of $(1+\alpha_0 t)$ ohms. If the resistance at 0°C had been R_0 ohms, then its final resistance would have been $R_t = R_0(1+\alpha_0 t)$ ohms.

The constant α_0 is known as the Temperature Coefficient of Resistance at 0°C. It can be found by substituting values in the above formula.

The following results were obtained for iron and for constantan respectively and the reader can plot these resistance–temperature graphs.

Iron

Resistance	5·13	5·35	5·63	5·97	6·27	6·50	6·84	Ω
Temperature	20·5	34·5	44·5	59·5	72·0	82·5	98·5	°C

Constantan

Resistance	5·3	5·3	5·3	5·31	5·31	5·31	5·31	Ω
Temperature	20·5	34·5	44·5	59·5	72·0	82·5	98·5	°C

It can be seen that the change in temperature has little effect on the resistance of the constantan, i.e. its temperature coefficient of resistance is very small.

Carbon is a material whose resistance decreases with increase in temperature, i.e. it has a negative temperature coefficient of resistance.

The temperature -273°C is known as absolute zero. It is interesting to note that at this temperature, materials lose their resistance and experiments are taking place to try to operate equipment near this temperature as there would be very small power losses.

In many cases the temperature t_1 and the resistance R_{t_1} are known under one set of conditions and it is required to find, say, the temperature t_2 when the resistance is R_{t_2}. The average temperature of a winding can be found using these results.

EXAMPLE. The field coil on a d.c. machine takes a current of 0·8 A from a 110-V supply when it has been in a room all night at a temperature of 18°C. (In other words, the ambient temperature is 18°C.) When the machine has been operating for some time, the current falls to 0·7 A, the voltage remaining the same. What is the average temperature of the copper coil?

Assume $\alpha_0 = 0{\cdot}004\ \Omega/\Omega/\text{deg C}$ at 0°C.

Under the first condition $R_{t_1} = R_0(1+\alpha_0 t_1)$,

$$\frac{110}{0{\cdot}8} = R_0(1+0{\cdot}004\times 18),$$

$$137{\cdot}5 = 1{\cdot}072R_0. \qquad (1)$$

Under the new operating conditions

$$\frac{110}{0{\cdot}7} = R_0(1+0{\cdot}004\times t_2),$$

$$157{\cdot}1 = R_0(1+0{\cdot}004t_2). \qquad (2)$$

Hence,

$$\frac{137{\cdot}5}{157{\cdot}1} = \frac{1{\cdot}072R_0}{R_0(1+0{\cdot}004t_2)}.$$

The R_0 cancels. Solving this equation gives

$$t_2 = \mathbf{56{\cdot}2°C}.$$

10.16. Variation of the resistance of a metal filament lamp

Calculate the resistance of the lamp for each pair of values given in the table on the following page and plot graphs of resistance against p.d. and also current against p.d. across the lamp.

Voltage across lamp (V)	0	4	12	27	45	66	77	90	102	107
Current through lamp (A)	0	0·15	0·25	0·35	0·45	0·55	0·6	0·65	0·7	0·73

Has the lamp a positive or a negative temperature coefficient of resistance? Does Ohm's law apply? If not, why not?

10.17. Variation of the resistance of a carbon filament lamp

For the values in the following table, carry out calculations as in the preceding section and also plot graphs as before. Answer the same questions as before in respect of the carbon lamp.

Voltage across lamp (V)	0	20	30	42	52	61	71	80	88	97	107
Current through lamp (A)	0	0·08	0·15	0·25	0·35	0·45	0·55	0·65	0·75	0·85	0·98

10.18. Tests on a piece of metrosil

Metrosil is an example of what is known as a non-linear resistor. Its resistance varies as the voltage across it changes. The following results were obtained when a sample was tested:

Voltage across metrosil (V)	0	10	20	30	40	50	60	70	75
Current through metrosil (mA)	0	0·7	2·8	7·7	24·5	38	75	135	210

Graphs may be plotted of resistance and current against voltage. If it is assumed that

$$\text{Current} = (\text{a constant}) \times (\text{voltage})^N,$$

or

$$I = KV^N,$$

where K and N are constants, then

$$\log I = \log K + N \log V.$$

If $\log I$ is plotted against $\log V$, then a straight-line graph should be obtained if the results are experimentally correct. The constants K and N can then be found.

10.19. Use of a variable resistor as a potential divider

A potential divider is used to obtain a variable voltage output from a constant-voltage supply. This device was used to vary the voltage applied to the lamps in the results given earlier. It could equally well be used to vary the current through the field windings of a d.c. machine. In this test, the output of the potential divider, *AB*, was connected to a fixed resistor *R* and instruments were connected as shown in Fig. 10.6.

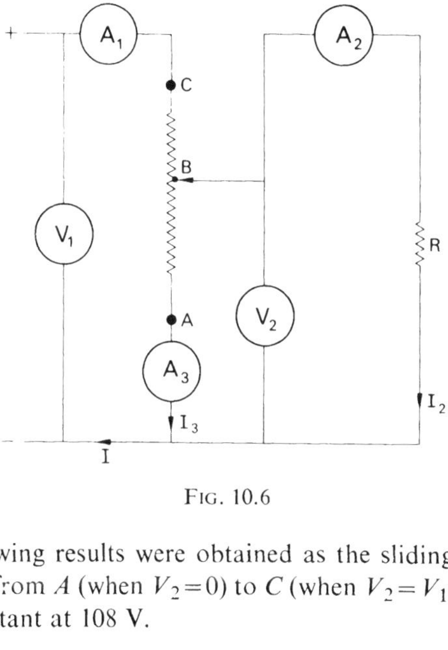

FIG. 10.6

The following results were obtained as the sliding contact (*B*) was moved from *A* (when $V_2=0$) to *C* (when $V_2=V_1$). In the test, V_1 was constant at 108 V.

Output voltage V_2	0	20	30	40	50	60	70	80	90	108	(V)
Input current I_1	0·88	0·92	0·98	1·07	1·17	1·28	1·42	1·56	1·72	1·97	(A)
Output current I_2	0	0·22	0·33	0·44	0·54	0·66	0·76	0·86	0·98	1·17	(A)
Common current I_3	0·88	0·70	0·65	0·64	0·63	0·62	0·66	0·70	0·74	0·80	(A)

Plot graphs of each of the currents against the output voltage V_2 and comment on their shape. Show that the resistance R_{CA} is about 120 Ω and that the value of the fixed resistor is about 91 Ω. State how I_1, I_2 and I_3 are related.

Problems

(For copper, $\alpha_1 = 0{\cdot}004$ Ω/Ω/deg C at 0°C)

10.1. Calculate the energy required to move 10 coulombs of electricity from one point in a circuit to another if the potential difference existing between the points is 12 V.

(120 J.)

10.2. Calculate the energy supplied by a 12-V battery to a circuit taking a current of 2 A for a period of 5 min.

(7200 J.)

10.3. The cost of operating twenty 150-W lamps for 12 h is 4*s*. 6*d*. What is the cost of the energy per kilowatt-hour?

($1\frac{1}{2}d$.)

10.4. Calculate the resistivity of a material if 1000 ft of wire made from it has a resistance of 4 Ω when the cross-sectional area is 0·025 in^2.

(8·33 μΩ-in.)

10.5. A 20-hp motor operates on full-load when driving a hoist which lifts a load 200 ft in 2 min. If the motor efficiency is 85% and the hoist efficiency 80%, calculate (a) the load being lifted in lbf, (b) the energy used in kWh to make one lift.

(5280 lbf; 0·584 kWh.)

10.6. A field coil wound with copper wire has a resistance of 200 Ω at 0°C. What would be its resistance if the temperature were raised to 70°C?

(259·4 Ω.)

10.7. A copper coil has a resistance of 240 Ω at a temperature of 10°C. What would be the resistance of the coil if its temperature were raised to 70°C?

(298 Ω.)

10.8. A calorimeter contains 2400 g of water. A coil whose resistance is 7 Ω was inserted in the calorimeter and a current of 12 A passed through it for 10 min. The calorimeter and the coil weigh 250 g and their specific heat is 0·1. If the initial water temperature was 20°C, what would be the final temperature of the water? Neglect all losses. (1 calorie = 4·2 J.)

(79·3°C.) (N.C.T.E.C.)

10.9. An electric kettle contains 2 pints of water and it is required to raise the temperature of the water from 20°C to 100°C in 5 min. If 80% of the heat which is produced in the heating element is usefully employed in heating the water, calculate the resistance of the element, the supply being at 240 V.

(1 lb = 454 g; take 1 gallon of water as weighing 10 lb; 1 calorie = 4·2 J.)

(36·25 Ω.) (N.C.T.E.C.)

Answers are not given to the following questions:

10.10. Calculate the energy expended as heat when a current of 10 A flows through a resistor of 3 Ω for 5 min.

10.11. An electric fire is rated at 1 kW at 240 V. What power is consumed if the supply voltage is reduced to 230 V?

10.12. A 5-hp motor takes 26 A from a 200-V d.c. supply. What is its efficiency under this operating condition?

10.13. A 240-V, 100-W lamp is switched on for 30 min. Calculate (a) the lamp resistance, (b) the quantity of electricity used, (c) the energy consumed.

10.14. An electric kettle has a rating of 2500 W at 240 V. What is the resistance of the element?

The kettle heats 1600 cm^3 of water from 59°F to 203°F. What range of temperature is this on the Centigrade scale?

Assuming an overall efficiency of 85%, calculate the time taken to heat the water. (1 calorie = 4·2 J.) (N.C.T.E.C.)

10.15. A 240-V d.c. electric furnace is used to heat the temperature of 8 lb of brass from 60°F to the annealing temperature of 1100°F in a time of 25 min at an overall efficiency of 78%.

Calculate: (a) the resistance of the heating element, (b) the current taken from the supply, (c) the power in the circuit and (d) the energy used.

The specific heat of brass may be taken as 0·09.

1 Btu = 1055 J (or watt-seconds). (C. & G.)

10.16. A coil of copper wire has a resistance of 100 Ω at 0°C. Calculate its resistance at 5°F.

10.17. What is the total resistance of a two-core copper cable 200 yd long if each core has a cross-sectional area of 0·06 in^2?

Resistivity of copper = $\frac{2}{3} \times 10^{-6}$ Ω-in.

CHAPTER 11

Resistors in Series and Parallel

11.1. Resistors in series

When resistors are connected to each other end to end in such a way that they carry the same current, they are said to be connected in series (Fig. 11.1).

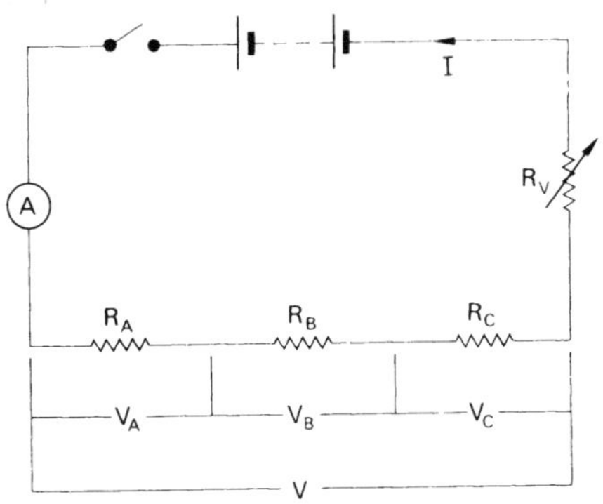

FIG. 11.1

In Section 10.10 it has been shown that if readings were taken of the voltage across a resistor and the current through it, the gradient of a graph of voltage (V) and current (I) would give the resistance in ohms if V and I were in volts and amperes respectively.

Three such resistors, A, B, and C, were found to have resistances of:

$$R_A = 3\ \Omega, \quad R_B = 2\ \Omega \quad \text{and} \quad R_C = 6\ \Omega.$$

These three were then connected in series. The current (I) through them was varied by altering the variable resistor R_V and the p.d. (V) across them was noted. The results obtained were as follows:

P.D. (V)	0	2·15	4·42	6·55	8·83	10·95
Current (A)	0	0·2	0·4	0·6	0·8	1·0

If the reader plots these results with current on the horizontal axis, he can determine the total resistance from the gradient of the graph.

The total resistance can also be found by calculation.

From Fig. 11.1,

$$V = V_A + V_B + V_C \text{ volts.}$$

Now $\quad V_A = IR_A, \quad V_B = IR_B \quad \text{and} \quad V_C = IR_C.$

If the three resistors were replaced by a single resistor R ohms such that it will take the same current from the same supply, then

$$V = IR \text{ volts,}$$

$$\therefore \quad IR = IR_A + IR_B + IR_C.$$

Dividing both sides by I gives

$$R = R_A + R_B + R_C.$$

Using the values of resistance used in the tests,

$$R = 3+2+6$$

$$= \mathbf{11\ ohms.}$$

11.2. Resistors in parallel

When resistors are connected to each other such that the same voltage is across each of them, they are said to be in parallel (Fig. 11.2).

The same three resistors (3, 2 and 6 ohms) were connected in parallel and the voltage across them noted as the current I to the resistors was varied by the same means as previously.

Results obtained in a test were:

P.D. across the resistors	(V)	0	0·18	0·42	0·63	0·79	1·02
Current to the resistors	(A)	0	0·2	0·4	0·6	0·8	1·0

It should be noted that when the current reaches the junction to which all the resistors are connected, the current splits in such a way that

$$I = I_A + I_B + I_C.$$

This is an application of Kirchhoff's first law, which is given in Section 16.1.

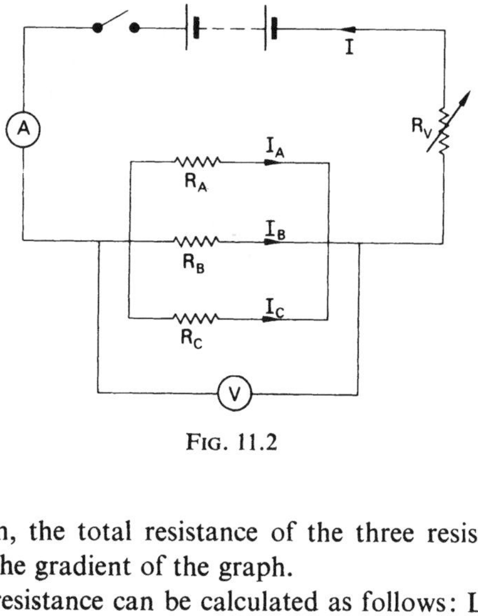

FIG. 11.2

Once again, the total resistance of the three resistors can be found from the gradient of the graph.

The total resistance can be calculated as follows: Let the total resistance of the three resistors R_A, R_B, and R_C be R ohms. Then the resistor R will have a p.d. of V volts across it when a current I is flowing through it and therefore $I = V/R$ amperes.

Using Ohm's law:

$$I_A = \frac{V}{R_A}, \quad I_B = \frac{V}{R_B} \quad \text{and} \quad I_C = \frac{V}{R_C}.$$

As $$I = I_A + I_B + I_C,$$

then $$\frac{V}{R} = \frac{V}{R_A} + \frac{V}{R_B} + \frac{V}{R_C}.$$

Dividing both sides of this equation by V,

$$\frac{1}{R} = \frac{1}{R_A} + \frac{1}{R_B} + \frac{1}{R_C}.$$

Using the values

$$R_A = 3\ \Omega, \quad R_B = 2\ \Omega \quad \text{and} \quad R_C = 6\ \Omega,$$

then if the total resistance is R ohms,

$$\frac{1}{R} = \frac{1}{3} + \frac{1}{2} + \frac{1}{6}.$$

The common denominator of the right-hand side of the expression is 6,

$$\therefore \quad \frac{1}{R} = \frac{2+3+1}{6} = \frac{6}{6}.$$

Inverting both sides of the equation

$$\frac{R}{1} = R = \frac{6}{6} = 1\ \Omega.$$

11.3. Mixed series and parallel circuits

Generally problems of this kind can be solved by replacing resistors connected in series or in parallel by a single equivalent resistor. An example is the best way of showing the method.

EXAMPLE. In Fig. 11.3(a), calculate the current in each part of the circuit and the p.d. across each resistor.

If the circuit is not lettered, it is convenient to insert letters to enable easier identification of any part of the circuit. For example, R_{AB} is the total resistance between A and B, consisting of a 2-, 3-, and 6-Ω resistor in parallel.

First find a single resistor R_{AB} equivalent to the three resistors (2, 3, and 6 Ω). As previously shown (Section 11.2):

$$\frac{1}{R_{AB}} = \frac{1}{2}+\frac{1}{3}+\frac{1}{6},$$

$$R_{AB} = 1\ \Omega.$$

For the parallel circuit R_{CD},

$$\frac{1}{R_{CD}} = \frac{1}{6}+\frac{1}{12} = \frac{2+1}{12} = \frac{3}{12},$$

$$\therefore\quad R_{CD} = \frac{12}{3}\ \Omega = 4\ \Omega.$$

The circuit can now be drawn as in Fig. 11.3(b). The three resistors R_{AB}, R_{BC}, and R_{CD} in series can be replaced by a single resistor (Fig. 11.3(c)) such that

$$R_{AD} = 1+3+4 = 8\ \Omega.$$

The circuit can be further simplified to Fig. 11.3(d).

$$\frac{1}{R_{AD}} = \frac{1}{8}+\frac{1}{4} = \frac{1+2}{8} = \frac{3}{8},$$

$$R_{AD} = \frac{8}{3}\ \Omega = 2\tfrac{2}{3}\ \Omega.$$

Current from the battery $I = \dfrac{24}{2\frac{2}{3}}\ \text{A} = \mathbf{9\ A}.$

Current in the 4-Ω resistor (Fig. 11.3(b)) $I_1 = \dfrac{24}{4}\ \text{A} = \mathbf{6\ A}.$

This could, of course, have been obtained much earlier as this resistor is connected straight across the battery.

$$I_2 = \frac{24}{8}\ \text{A} = \mathbf{3\ A}.$$

Or this could be found from

$$I_2 = I - I_1 = 9-6 = \mathbf{3\ A}.$$

The next step could be to find the p.d. across the other resistors. Using Fig. 11.3(b),

$$V_{AB} = I_2 R_{AB} = 3 \times 1 \text{ V} = \mathbf{3\ V},$$
$$V_{BC} = I_2 R_{BC} = 3 \times 3 \text{ V} = \mathbf{9\ V},$$
$$V_{CD} = I_2 R_{CD} = 3 \times 4 \text{ V} = \mathbf{12\ V}.$$

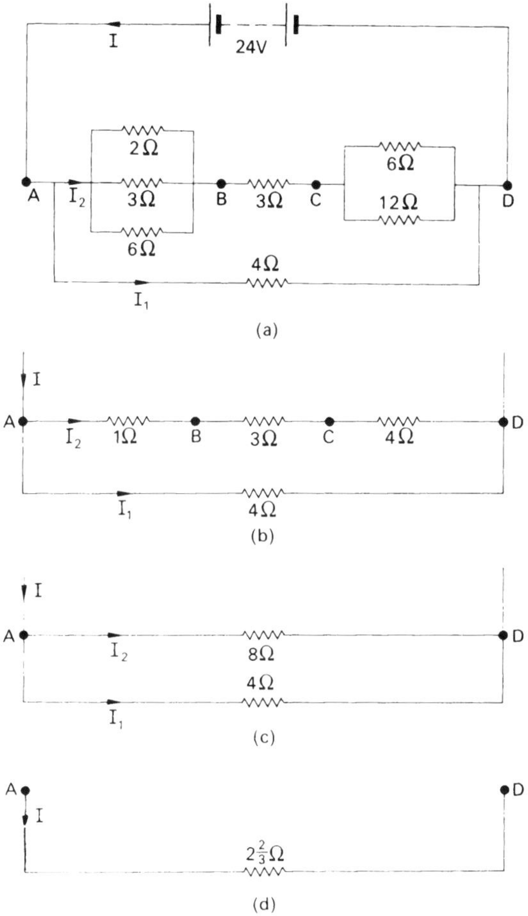

FIG. 11.3

Note that if no errors have been made, the sum of these three voltages must equal the battery voltage. Thus if any two of these p.d.s had been calculated, the third could have been found by subtracting their sum from the battery voltage.

Referring now to the original circuit (Fig. 11.3(a)) and considering the resistors between A and B, then as $V_{AB}=3$ V,

$$\text{Current in 2-}\Omega\text{ resistor}=\frac{3}{2}\text{ A}=\mathbf{1{\cdot}5\ A},$$

$$\text{Current in 3-}\Omega\text{ resistor}=\frac{3}{3}\text{ A}=\mathbf{1\ A},$$

$$\text{Current in 6-}\Omega\text{ resistor}=\frac{3}{6}\text{ A}=\mathbf{0{\cdot}5\ A}.$$

Note that if these values are correct, their total must be the current flowing in the 3-Ω resistor R_{BC} (i.e. $I_2=3$ A).

For the resistors between C and D,

$$\text{Current in the 6-}\Omega\text{ resistor}=\frac{12}{6}\text{ A}=\mathbf{2\ A},$$

$$\text{Current in the 12-}\Omega\text{ resistor}=\frac{12}{12}\text{ A}=\mathbf{1\ A}.$$

These, too, total to I_2.

The total power taken by the circuit is given by multiplying the battery terminal voltage (24 V) by the battery current (9 A).

$$\text{Total power}=24\times 9=\mathbf{216\ W}.$$

If the power in a single resistor is required, e.g. the 12-Ω resistor, where the voltage across this resistor is $V_{CD}=12$ V and the current through it is 1 A,

$$\text{Power}=VI=12\times 1\text{ W}=12\text{ W},$$

or

$$\text{Power}=I^2R=1^2\times 12\text{ W}=12\text{ W},$$

or

$$\text{Power}=\frac{V^2}{R}=\frac{12^2}{12}\text{ W}=12\text{ W}.$$

As an exercise, the power in every resistor could be found. If the calculations made are correct, their total will be 216 W.

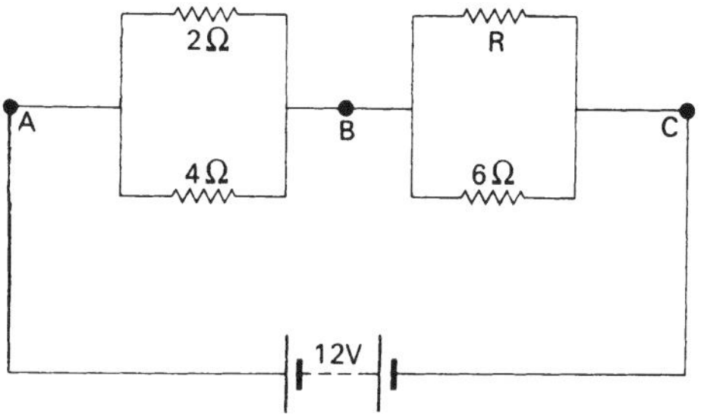

FIG. 11.4

EXAMPLE. If the total power consumed in the circuit shown in Fig. 11.4 is 36 W, what is the value of the resistor R?

$$\text{Current } I=\frac{W}{V}=\frac{36}{12}\text{ A}=3\text{ A}.$$

Total circuit resistance

$$R_{AC}=\frac{12}{3}=4\ \Omega,$$

$$\frac{1}{R_{AB}}=\frac{1}{2}+\frac{1}{4}=\frac{2+1}{4}=\frac{3}{4},$$

$$R_{AB}=\frac{4}{3}\ \Omega=1\tfrac{1}{3}\ \Omega,$$

$$R_{BC}=R_{AC}-R_{AB}=4-1\tfrac{1}{3}=2\tfrac{2}{3}\ \Omega,$$

$$\therefore\quad \frac{1}{R_{BC}}=\frac{1}{2\frac{2}{3}}=\frac{3}{8}=\frac{1}{R}+\frac{1}{6},$$

$$\frac{1}{R}=\frac{3}{8}-\frac{1}{6}=\frac{9-4}{24}=\frac{5}{24},$$

$$R=\frac{24}{5}=4\tfrac{4}{5}\ \Omega\text{ or }\mathbf{4{\cdot}8\ \Omega}.$$

EXAMPLE. A 6-Ω and a 12-Ω resistor are connected in parallel and the combination connected in series with three resistors connected in parallel, two of these having values of 2 Ω and 6 Ω respectively. When connected to a source of supply whose e.m.f. is 20 V, the supply current is 4 A. Calculate (a) the value of the unknown resistor, (b) the energy dissipated in it in 9 min.

A conductor of negligible resistance is connected in parallel with the 2-Ω resistor. What current is now taken from the supply? (N.C.T.E.C.)

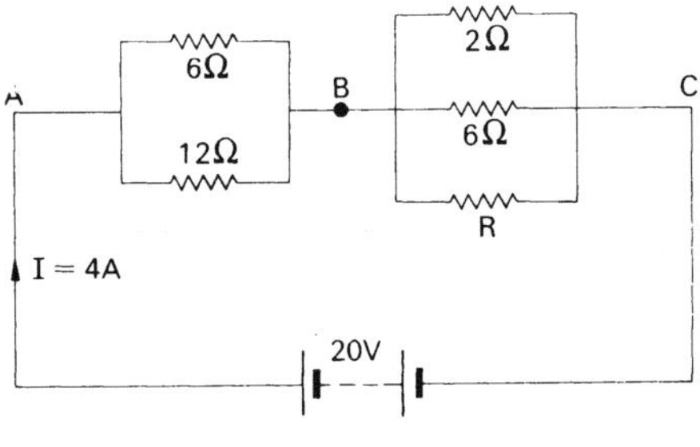

FIG. 11.5

The circuit is given in Fig. 11.5.

(a) Total resistance $=\frac{20}{4}\ \Omega=5\ \Omega$,

Resistance $R_{AB}=\frac{6\times 12}{6+12}\ \Omega=4\ \Omega$,

Resistance $R_{BC}=5-4=1\ \Omega$,

$\therefore \frac{1}{R_{BC}}=\frac{1}{2}+\frac{1}{6}+\frac{1}{R}$, where R is the unknown resistor,

$R=3\ \Omega$.

(b) $V_{BC}=4\times 1\ \text{V}=4\ \text{V}$,

Current (in the resistor R) $=\frac{4}{3}\ \text{A}=1\frac{1}{3}\ \text{A}$,

Energy expended $=(1\frac{1}{3})^2\times 3\times 9\times 60\ \text{J}=$ **2880 J.**

When a conductor of negligible resistance is connected in parallel with the 2-Ω resistor, the total resistance of the branch *BC* is reduced to zero. The only resistance is that due to the part *AB* of the circuit. Hence

$$I=\frac{20}{4}=\mathbf{5\ A.}$$

Problems

11.1. A 100-V, 100-W lamp is to operate normally from a 240-V supply by connecting a resistor in series with the lamp. Calculate: (a) the resistance of the resistor; (b) the total power taken from the supply; (c) the current from the supply if a similar lamp is connected in parallel with the first.
(140 Ω, 240 W, $1\frac{5}{19}$ A.) (N.C.T.E.C.)

11.2. A circuit *AB*, consisting of a 4-Ω resistor in parallel with a resistor R_1, is connected in series with a resistor R_2 to a 4-V supply. The energy consumed in 15 min is 1800 J and the potential difference across *AB* is 1·2 V. Calculate the values of R_1 and R_2 and the total power consumed in each of the three resistors.
($R_1=6\,\Omega$, $R_2=5{\cdot}6\,\Omega$, 0·36 W, 0·24 W and 1·4 W.) (U.L.C.I.)

11.3. A 6-Ω and a 12-Ω resistor are connected in parallel and the combination connected in series with a resistor *R* to a battery which has an e.m.f. of 24 V. The current from the battery is 4 A. Calculate: (a) the total energy dissipated in 5 min; (b) the value of the resistor *R*; (c) the current in the 6-Ω resistor.
(28,800 J, 2 Ω, 2·67 A.)

11.4. A battery consists of six cells connected in series, each cell having an e.m.f. of 1·45 V and internal resistance of 4 Ω. A resistor of 50 Ω is connected across the terminals of the battery. Determine: (a) the current; (b) the terminal p.d. of the battery; (c) the power dissipation in the external load.
(0·118 A, 5·87 V, 0·69 W.) (N.C.T.E.C.)

11.5. A 30,000-Ω and a 50,000-Ω resistor are connected in series across a 240-V supply. What is the voltage across the 50,000-Ω resistor? If 100,000 Ω is connected across the 50,000-Ω resistor, what is the voltage across it?
(150 V, 126 V.)

11.6. A 1-Ω and a 5-Ω resistor are connected in parallel and these are connected in series with two resistors in parallel having values of 4 Ω and 8 Ω. The combination is supplied by a battery and the power dissipated in the 8-Ω resistor is 32 W. Determine the p.d. at the terminals of the battery.
(21 V.)

11.7. A 6-Ω and 12-Ω resistor are connected in parallel and the combination connected in series with a 2-Ω resistor. A battery of e.m.f. 28 V and internal resistance 1 Ω is connected to the resistors. Calculate: (a) the current from the battery; (b) the power consumed in the 6-Ω resistor.
(4 A, $42\frac{2}{3}$ W.)

Answers are not given to the following questions:

11.8. Calculate the total resistance between the terminals *AB* (Fig. 11.6). If a 20-V supply were connected to the terminals *AB*, calculate: (a) the current in each part of the circuit; (b) the total power consumed.

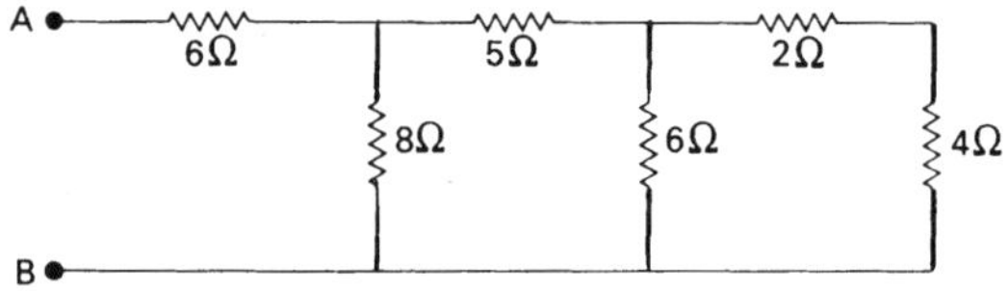

FIG. 11.6

11.9. A field coil consumes 400 W when connected to a 110-V supply. It is desired to operate this coil at its normal voltage from a 200-V supply. State how this may be done and calculate the total power taken from the supply.

11.10. A metal filament lamp, whose resistance may be assumed constant at 150-Ω, is connected in series with a 50-Ω resistor and the circuit connected to a 200-V supply. Determine the supply current when the lamp is shunted (i.e. has connected in parallel with it), by (a) a 70-Ω resistor, (b) a 300-Ω resistor. State in which case the lamp will be brighter.

11.11. Explain why a 100-V, 40-W lamp should not be connected in series with a 100-V, 100-W lamp across a 200-V supply. What would happen if they were connected as described? (U.L.C.I.)

11.12. Show three different methods of connecting three 10-Ω resistors together. Calculate the current taken in each case from a 20-V battery.

CHAPTER 12

Primary and Secondary Cells

12.1. Primary cells

In general, if pieces of different materials (called electrodes) are immersed in a liquid (known as an electrolyte), an electromotive force (e.m.f.) will exist between the electrodes. Some current can be taken from this arrangement which is known as a primary cell. Primary cells cannot be recharged by passing a current through them in the opposite direction to that in which current is taken from them. During use, one of the electrodes is used up by chemical action and eventually the cell becomes useless.

This action can occur in places where it is not wanted. A steel ship's rudder and a bronze propeller in sea water constitute a cell and this combination would rapidly cause the rudder to corrode.

The earliest cell was due to Volta about the year 1800 and consisted of alternate sheets of zinc and copper separated by pieces of cloth soaked in brine. The later cell had a copper electrode and a zinc electrode, the electrolyte being dilute sulphuric acid. The Daniel cell had a copper vessel containing copper sulphate solution and into this was placed a porous earthenware pot containing dilute sulphuric acid. A zinc electrode was placed in this acid.

In each case, the copper electrode is known as the anode and the zinc electrode as the cathode. The anode is the positive of the cell and current conventionally flows from the anode, through the external load, to the cathode or negative of the cell.

In about 1865 Georges Leclanché produced a cell using the same materials as are used today.

12.2. The wet Leclanché cell

The wet Leclanché cell has carbon and zinc rods for its anode and cathode respectively. Ordinary commercial zinc has many impurities. These form tiny cells when the zinc is immersed in an electrolyte, rapidly dissolving the zinc. If mercury is rubbed over the zinc, forming an amalgam, the uniform surface exposed to the electrolyte prevents this local action. As the zinc is still in contact with the electrolyte, the action of the cell is not affected.

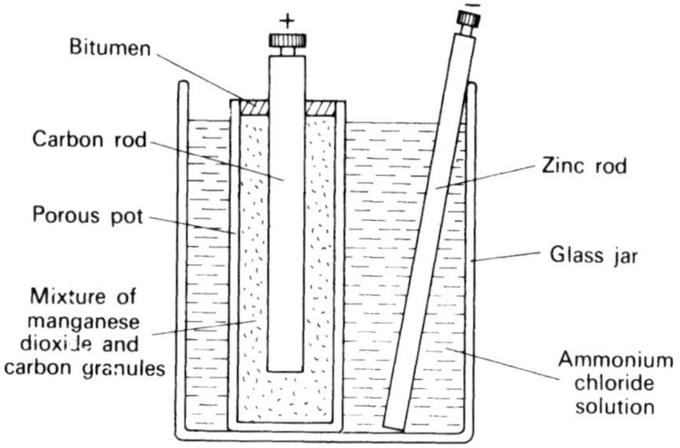

FIG. 12.1. Wet Lechlanché cell.

The electrolyte is a strong solution of ammonium chloride (sal-ammoniac), this often being available in tablet form for dissolving in water. A glass jar holds the electrolyte.

The carbon rod is placed in a porous earthenware pot and the rod is surrounded by manganese dioxide. When current is taken from the cell, hydrogen is formed at the anode, and if allowed to remain, the hydrogen would affect the operation of the cell as explained below. The oxygen from the manganese dioxide, called the depolarising agent, combines with the hydrogen to form water, thus removing the hydrogen from the anode. This is known as depolarisation.

If the current taken from the cell is excessive, the hydrogen is produced faster than it can be removed. The hydrogen alters the character of the electrode and results in an e.m.f. being produced, between the electrodes, which is in opposition to the e.m.f. causing the current flow. Thus the effective voltage at the terminals is reduced. In effect, it is as if the internal resistance of the cell had been increased.

This effect can be demonstrated by connecting a cell to two platinum electrodes immersed in dilute sulphuric acid. Hydrogen forms at the cathode and oxygen at the anode. When there are plenty of gas bubbles on the electrodes, the cell should be disconnected and the electrodes connected to an instrument such as a galvanometer that can indicate small currents. It will be observed that a current flows until the bubbles on the electrodes have disappeared. The presence of this back e.m.f. is known as polarisation.

These wet cells are suitable for operating bells and have an e.m.f. of about 1·5 V.

12.3. The dry Leclanché cell

CONSTRUCTION

These primary cells are manufactured in many different sizes and there are two types of construction:

1. Round cells which are available as a single unit or forming multi-cell units.
2. Layer cells which are available in multi-cell units only. Sectional views of these are shown in Figs. 12.2(a) and (b).

The chemical components of both types of cell are the same. Carbon forms the positive electrode and zinc the negative electrode of the cell. The cell is not truly dry, the electrolyte consisting of a paste containing ammonium chloride as the active ingredient, some zinc chloride having been added to assist in retaining the moisture. The depolarising agent is manganese dioxide, mixed with crushed carbon so as to provide a good electrical path.

Recent improvements have replaced the paste electrolyte with a special type of thin paper separator which permits a larger volume of the material forming the depolarising agent. This results in the life of the cell being about double that of a paste cell of comparable physical size.

Advantages of the layer cell are that it has no air space or carbon rod. This permits a larger amount of active material within a given cell volume than in a round type. When layer cells form a multi-

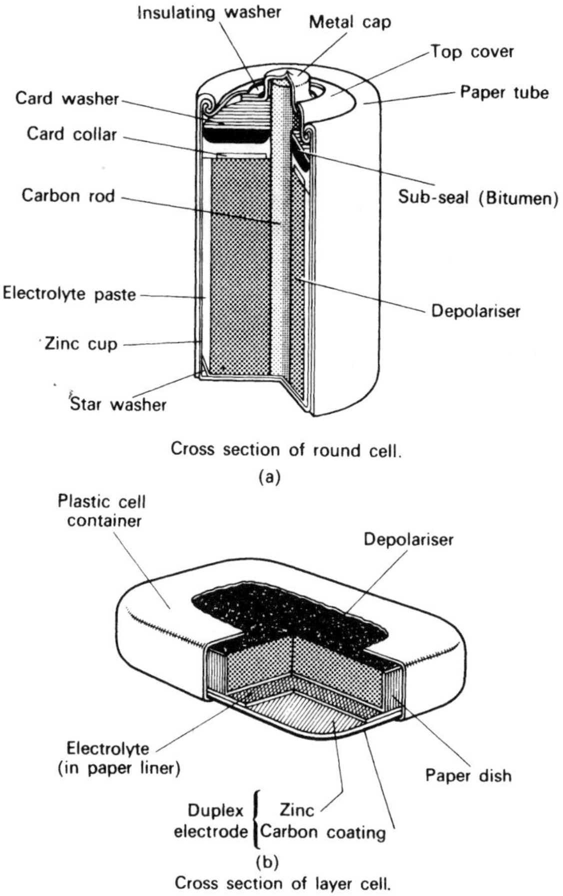

FIG. 12.2. (a) Cross-section of round cell; (b) Cross-section of layer cell. (Ever Ready Co. (G.B.) Ltd.)

cell unit there is much less waste space than when the round type is used. This results in nearly a two to one increase in energy for a given volume of cell.

Applications

Several cells connected in series are known as a battery. The life of a battery depends on the load it supplies and different batteries are manufactured to supply different types of load. It is, therefore, important to use a battery for the type of load for which it was designed. They have many uses, including transistor hearing aids, photoflashes, portable radios, etc. Their voltages cover the range of 1·5 V to 300 V whilst their designed current output may be as low as 0·1 mA and as high as 400 mA. Use of the wrong type of battery in a particular application may result in premature failure of the equipment.

Battery storage

Batteries in storage should not be subjected to excessive temperature. Above about 120°F, rapid deterioration will almost certainly occur, loss of moisture from the cell being one reason for loss of performance. Temperatures of 40–50°F are ideal for battery storage.

There is no simple way of finding out the state of discharge of a dry cell. A common fallacy is that by short-circuiting the cell and noting the short-circuit current, some indication is given. This is, however, not reliable. If its full life history is known, it may be compared with a similar cell working under similar conditions whose length of useful life is known, possibly, from the manufacturer's data.

12.4. Connection of cells

Cells are frequently connected in series to form a battery in order to obtain a higher value of e.m.f. If four Leclanché cells were connected in series, each cell having an e.m.f. of 1·5 V, then the total e.m.f. would be $4 \times 1 \cdot 5 = 6$ V provided that the positive of one cell was connected to the negative of the next cell. This is what is implied when it is stated that cells are connected in series (Fig. 12.3(a)).

If one of the cells had not been connected correctly as is illustrated in Fig. 12.3(b), then its e.m.f. would in effect cancel the e.m.f. of one of the correctly connected cells and so the output voltage would be decreased from 6 V to 3 V.

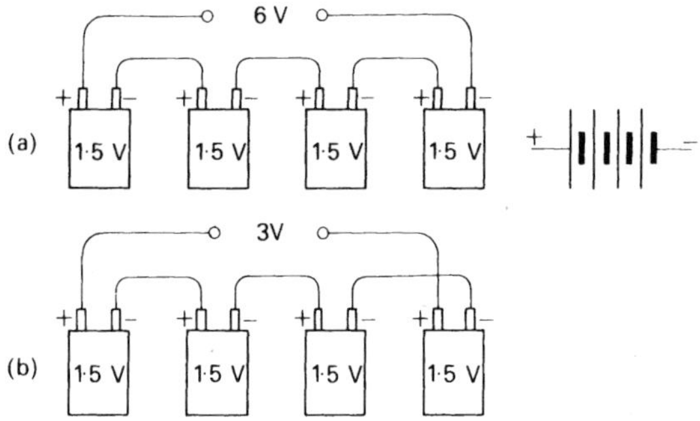

FIG. 12.3

Each cell has its own internal resistance and as these are effectively in series, the total resistance is the sum of the resistance of each cell, irrespective of whether they are connected correctly or incorrectly.

Parallel connection of these cells is comparatively rare. Should this connection be necessary in order to obtain a larger current, only cells having equal e.m.f.s must be used otherwise there will be circulating currents which will result in cells being discharged. If five cells each having an e.m.f. of 1·2 V and internal resistance 0·5 Ω were connected in parallel, the combination is equivalent to a single cell whose e.m.f. is 1·2 V and internal resistance $\frac{0 \cdot 5}{5} = 0 \cdot 1\ \Omega$.

EXAMPLE. A circuit has six identical cells, each with an e.m.f. of 1·5 V and an internal resistance of 0·8 Ω. What is the e.m.f.

available if the cells are correctly connected in series? If one cell has its polarity reversed, what is now the available e.m.f.? What current would flow in each case if a connection of negligible resistance were connected to the battery terminals?

$$\text{With correct connections } E = 6 \times 1{\cdot}5 = \mathbf{9\ V},$$

$$\text{Current flowing } I = \frac{9}{6 \times 0{\cdot}8} = \mathbf{1{\cdot}87\ A},$$

$$\text{With incorrect connections } E = 4 \times 1{\cdot}5 = \mathbf{6\ V},$$

$$\text{Current flowing } I = \frac{6}{6 \times 0{\cdot}8} = \mathbf{1{\cdot}25\ A}.$$

12.5. The standard cell

A standard cell is one whose e.m.f. is very accurately known. It is used in conjunction with a precision measuring device called a potentiometer for calibrating instruments.

As early as 1836 an attempt was made by Daniel to produce a standard cell, but the cell he produced had only a short life and was not a very reliable source of constant e.m.f. The Weston cell, first made about 1892, is still the standard cell used today. One electrode is made of mercury and the other is of cadmium sulphate. The cell has an e.m.f. of about 1·0196 V at 20°C and will, when properly used, maintain this value for many years. When a standard cell is purchased, a certificate is included which gives the exact e.m.f. at various temperatures. Its internal resistance may exceed 500 Ω.

A modified version of the above cell is used for general measurement work. Its useful life is only about 10 years and its e.m.f. is very little affected by variations of temperature.

A standard cell must never be allowed to deliver much current. When correctly used with a potentiometer, this requirement is fulfilled. If an ordinary voltmeter were connected to the terminals of such a cell, the current taken by the instrument would be much more than that which the cell can give without impairing its accuracy.

12.6. The Mallory mercury cell

This type of cell was developed during World War II for use in portable communications equipment in which maximum energy within minimum space was of prime importance.

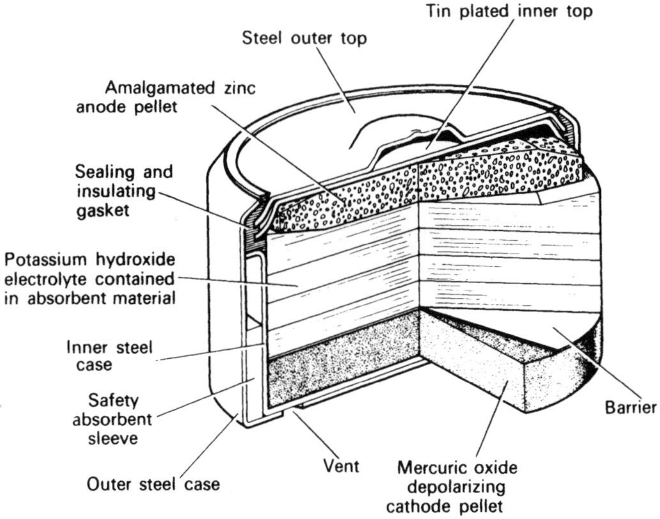

FIG. 12.4. Mallory mercury cell—flat pellet structure. (Mallory Batteries Ltd.)

CONSTRUCTION

Three types of structure are available, one of which is illustrated in Fig. 12.4. The flat pellet structure has a pressed powder pellet anode, whilst the cylindrical structure has a pressed powder cylindrical anode. The wound anode flat has an amalgamated wound zinc anode and is designed specifically for low-temperature performance. This cell has a large surface area of anode. The choice of structure is decided by the physical and electrical requirements of the application.

Venting occurs automatically if operating abnormalities, reverse currents or short-circuits of any magnitude or duration, produce excessive gas within the cell. Should an abnormally high internal pressure be set up within the cell, the cell top is forced upwards against the external crimped edge of the outer case, so tightening that portion of the seal and relieving the portion between the inner container and the outer case. Venting then occurs between the inner cell container and the outer case. Any electrolyte carried into this space is absorbed by the safety sleeving, so that corrosive materials are not carried with the escaping gas through the vent at the bottom of the case. After venting, the internal pressure is reduced, the seal reseats, and normal operation continues.

Chemical principle

The anode is formed from pure amalgamated zinc. The depolarising cathode consists of high purity mercuric oxide to which a small percentage of micronised graphite and, in certain types of cell, manganese dioxide is added. The electrolyte consists of a concentrated aqueous solution of potassium hydroxide containing zincate ions, these ions limiting the possibility of hydrogen evolution during storage or normal discharge and giving these cells their high stability.

During operation, the conductivity of the cathode increases under load due to the production of metallic mercury from the mercuric oxide and this in turn results in a flat voltage discharge curve. The e.m.f. of these cells is 1·35 V if no manganese dioxide is used in the cathode.

Applications

These cells are frequently used where their extremely low deterioration rate under adverse conditions is advantageous. Examples include rescue radio receivers, emergency alarm devices, etc. If stored at between 50°F and 70°F in dry conditions, there will be negligible deterioration after 2 years.

In medical electronics, rocketry and miniaturised communications equipment, these cells find common use. They will withstand mechanical shock and corrosion. Effects of high pressure and vacuum have negligible effect on their operation and unlike most other primary cells they will deliver their designed output continuously without requiring recuperation periods.

Their stability is such that Mallory mercury cells have gained wide acceptance of their use as secondary standards of voltage reference.

12.7. Secondary or storage cells (accumulators)

There are two common secondary cells, the lead–acid and the nickel–iron. These differ from the Leclanché cell in that they can be charged and discharged many times.

12.8. The lead–acid cell

Planté's discovery of the lead–acid cell was in 1859, and it was not until about 20 years later that the electric generator was developed as a means of charging it. Until the development of the generator, lead–acid cells were recharged from Leclanché cells and so the Leclanché cell was referred to as a primary and the lead–acid battery as the secondary cell.

If a current were passed through an electrolyte of dilute sulphuric acid by means of two platinum electrodes, it was possible to get a small current from the "cell" when the supply was disconnected. By using lead electrodes instead of platinum, it was found that a greater current could be obtained. Planté utilised this idea and found that by repeatedly passing current into this arrangement (charging) and then taking current from it (discharging), the available current was increased to a useful value.

In 1881 Fauré introduced an idea which is still used in principle in car batteries. A lattice grid of lead was used and to it was applied a lead oxide paste. By the use of an electric current, an active plate was produced of lower weight than previously.

Cells using this construction had the disadvantage that after repeated charge and discharge, active material from the positive plate fell to the bottom of the cell and these small particles built up. Eventually the cell could no longer be used. Modern construction reduces the possibility of this occurring.

MANUFACTURE

The lattice grids are first cast and usually consist of lead alloyed with antimony (10%) to make the grid hard. A paste is made of lead oxide and dilute sulphuric acid and this is put on the grids. The grids are placed side by side and are prevented from touching each other by separators of wood, plastic, or glass. Alternate grids are joined to form one electrode and the remainder are joined to form the other electrode.

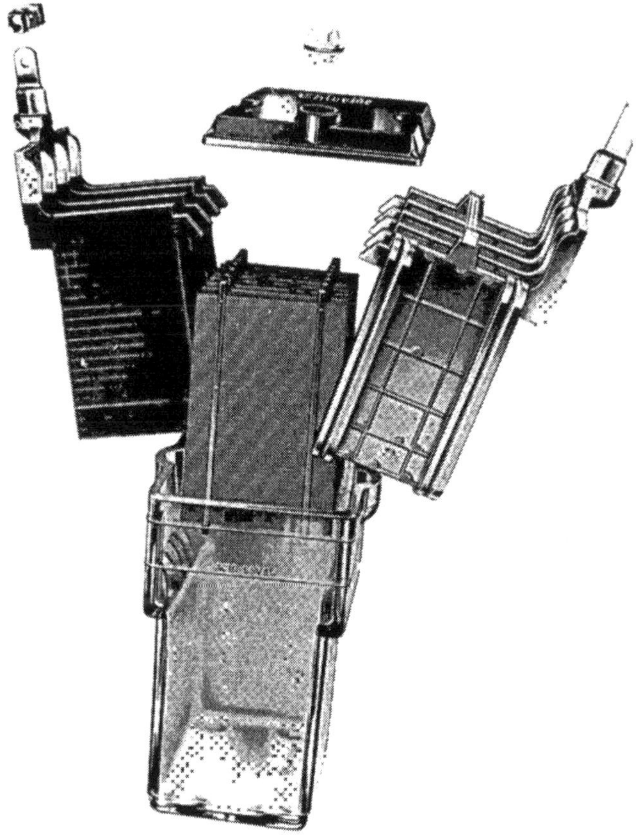

FIG. 12.5. Lead-acid battery construction. (Chloride Batteries Ltd.)

The grids are then immersed in the dilute sulphuric acid and a current is passed through the cell. The material in the grid connected to the positive of the supply is converted into lead peroxide and the other is changed into a spongy lead.

When the plates are formed, the electrolyte is removed and the cell sealed until it is ready to be put into use. This is found to result in a longer life of the cell.

Battery discharging

At both electrodes, lead sulphate is produced. Sulphuric acid is consumed during the process and so a measure of the amount of discharge is indicated by the specific gravity of the acid. For example, a fully-charged battery would have a specific gravity of about 1·28, and this would fall to about 1·12 when the battery became fully discharged.

An unused battery will lose about 25% of its charge in a month so it is advisable to charge a battery periodically even if it has not been used.

It should be appreciated that a lead–acid battery may be discharged at any rate of current that it is capable of giving without damage to the plates. However, if it is regularly over-discharged as is indicated by the specific gravity falling below 1·12, permanent damage will result.

Low temperature will temporarily reduce the ampere-hour capacity which can be taken from a battery, but it will return to its normal state even without charging when the temperature returns to normal. Freezing of the electrolyte is very unlikely. If water has been added without subsequent mixing, the water could freeze.

Battery charging

The process is the reverse of discharging. The sulphate is removed from the plate and the specific gravity of the acid rises. The manufacturer's instructions should always be followed when charging batteries. The initial charging current for a fully discharged battery may be high but as the sulphate is removed from

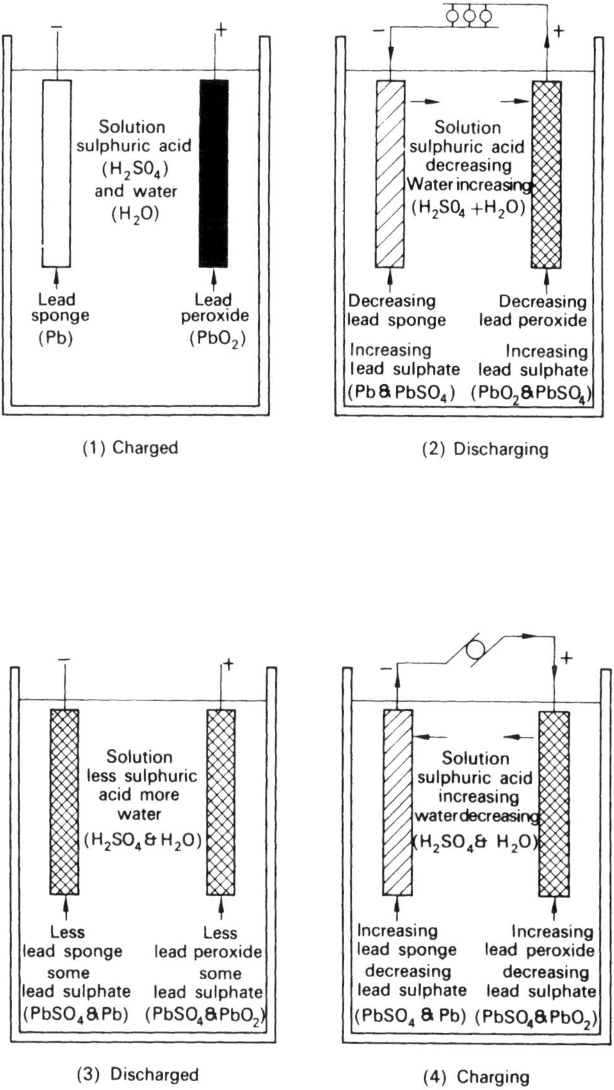

FIG. 12.6. Lead-acid battery—chemical changes.
(Chloride Batteries Ltd.

the plates, the charging current is wasted in decomposing the water in the electrolyte, so producing excessive gassing and temperature rise. The electrolyte temperature must not be allowed to exceed 110°F. The excessive gassing causes shedding of material from the plates and so shortens the battery life. This mixture of gases is explosive and many accidents have been caused by disconnecting a lead from a battery when current is still flowing, a spark causing an explosion. Batteries should only be charged where there is adequate ventilation and smoking or other form of naked light must be forbidden.

When a battery is being charged, the terminal voltage will remain at about 2·5 V per cell for a considerable time and then will increase to about 2·7 V. When this value is reached and the specific gravity ceases to rise, the cell is considered to be charged.

If the level of the electrolyte falls—it should cover the plates by about $\frac{1}{4}$ in.—preferably only distilled water should be added to raise the level. The total amount of water added each year may be more than the electrolyte volume of the cell. If impurities are introduced, the operation and condition of the cell may be affected by (a) abnormal loss of charge due to local action, (b) deterioration of the separators, (c) premature corrosion of the positive grids.

Batteries are rated to have a capacity of so many ampere-hours. When charging they may be charged, if the manufacturer's instructions are not known, at the 10-hour rate. This means that a 40-ampere-hour (40-Ah) battery would be charged at 40 Ah/10 h =4 A and if fully discharged initially (its specific gravity down to about 1·12), it would take about 10 hours to become fully charged.

The open-circuit terminal voltage of a cell is about 2·2 V and when fully charged it would have a terminal voltage of about 2 V at the normal discharge rate. A 12-V car battery has six cells connected in series.

Applications

Apart from car batteries, lead-acid cells are used in aircraft, traction, emergency lighting, submarines, stationary batteries for operating switchgear, guided missiles, portable T.V., etc. Improvements are still taking place in battery manufacture and cell capacity within a given cell volume is considerably greater than a few years ago.

The lead–acid battery has available 8–10 watt-hours per pound of chemical energy content. Other types of high energy batteries are being developed in this country and the Ford Motor Company in America claim that a new battery that they have developed has a value of about 150 watt-hours per pound of energy content.

Such a battery would greatly increase the distance that an electrically-driven vehicle could travel before battery recharging became necessary. This battery has liquid electrodes of metallic sodium and sulphur and these are separated by a solid electrolyte composed mainly of alumina. The battery operates at a temperature of 250°–300°C and as the whole system is sealed, no topping up is necessary. The battery is still a laboratory project but it should be noted that the materials from which the battery is made are freely available.

As a contrast in size, a range of rechargeable silver–cadmium button batteries has been designed for microelectronics, etc., the smallest of the range having a capacity of 250 mAh.

12.9. Measurement of the internal resistance of a battery

The following results were obtained using the circuit shown in Fig. 12.7 when a 6-V lead–acid car battery was connected to a variable resistance load:

Terminal voltage	6·13	6·10	6·07	6·03	6·01	6·00	V
Current	0	3·0	7·5	12·5	17·0	20·0	A

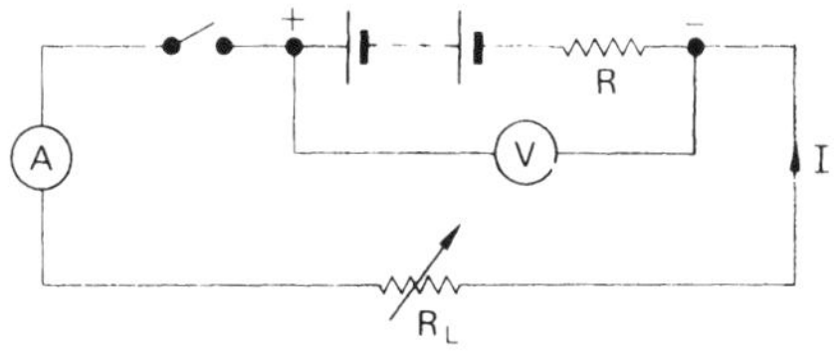

FIG. 12.7

In each case, the current was only allowed to flow for a short period of time. A graph of voltage plotted against current will be found to be almost linear. When the battery is on open circuit (i.e. no current is being taken from the battery), the voltmeter indicates the e.m.f. of the battery (E volts). For any current I amperes,

$$E - IR = V,$$

where R is the internal resistance of the battery. The above results give an internal resistance of about 0·006 Ω and the reader should verify this.

12.10. Power output of a source of e.m.f.

If a battery or other source of e.m.f. has a value E volts and an internal resistance R ohms, a current I amperes will flow through a load resistor R_L ohms connected to it (Fig. 12.8). The power dissipated in R_L is I^2R_L watts and this value will be a maximum when the external resistance R_L is equal to the internal resistance R of the source of e.m.f.

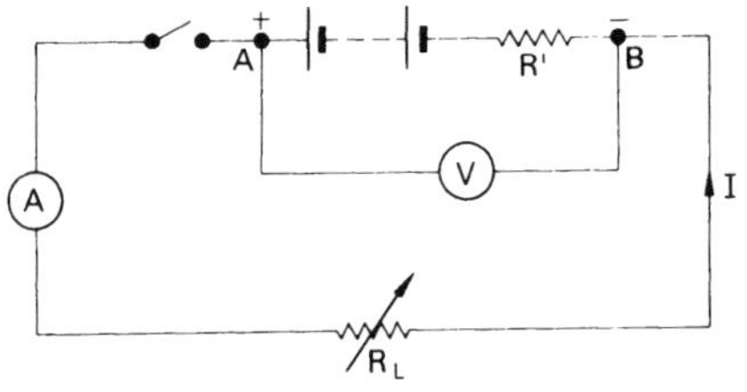

FIG. 12.8

As shown above, the internal resistance of a lead–acid cell is quite small and so for the purpose of the following test, a resistor R' is connected in series with the battery, R' being so much greater than R that for all practical purposes, the total battery resistance is R' ohms.

The following results were obtained as the load resistor R_L was varied:

Terminal voltage (V_{AB})	13·0	8·65	8·10	7·4	6·5	5·2	3·2	0	V
Current (I)	0	0·72	0·81	0·93	1·08	1·3	1·62	2·16	A

The load resistance $R_L = V_{AB}/I$ ohms and the power in this resistance can be calculated. A graph of power dissipated in the resistor R_L can be plotted against the value of this resistor. In the experiment, the effective resistance of the battery $R' = 6\ \Omega$ and so the graph should show a maximum power in R_L when its value is also 6 Ω.

12.11. The nickel–cadmium alkaline cell

CONSTRUCTION

The first patent on an electric storage cell based on the nickel–cadmium principle was taken out in 1899. Today there are cells using this principle which have a widely varying range of characteristics so as to be able to satisfy the requirements of a wide variety of applications. These characteristics are obtained by differences in the mechanical construction of the cells but the chemical principle of operation is the same in every case.

Both plates are constructed in a similar manner. The active materials are formed into briquettes under high pressure and these are fed automatically between two finely perforated ribbons of high-grade steel. The ribbon edges are then double-folded in a longitudinal direction and a continuous flat strip filled with active material is produced.

In the double-folding process, each side of the strip is left with a recessed lip and the long strips are placed side by side so that the lip on one strip keys into the recess in the next one. As many strips as are necessary to produce the required length of plate are keyed together and then passed through pressure rollers which join them into a uniform plaque. This is then cut into the width of the required plates.

The plate frame consists of two channelled steel side pieces welded to a top bar and the cut plaque is placed in the channelled frame. The complete plate is subjected to high pressure which corrugates the surface, indents the grooves which will hold the plate separators in position and seals the ends of the plates into the side frames.

Figure 12.9 shows the plates in the cell container. The cell container is of welded construction, being made from high-grade steel sheet. It will be realised that these cells are very robust.

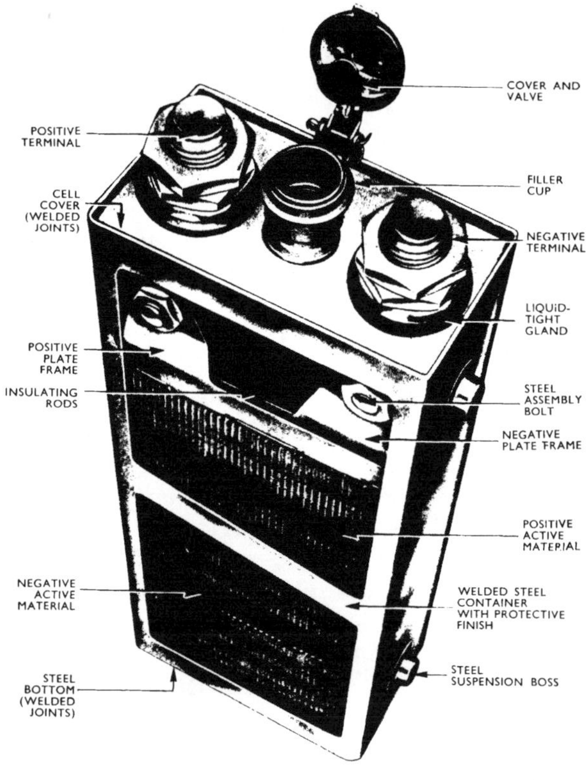

FIG. 12.9. Nickel-cadmium alkaline battery. (Nife Batteries Ltd.)

Chemical principles

The active material in the positive plate consists of nickel hydrate with a conducting mixture of pure graphite. The negative plate has cadmium oxide as its active material with an admixture of oxide of iron. The electrolyte is usually a 21% solution of potassium hydroxide which has a normal specific gravity of about 1·19.

In a fully charged battery the nickel hydrate is at a high degree of oxidation and the negative material is reduced to pure cadmium. On discharge the nickel hydrate is reduced to a lower degree of oxidation and the cadmium in the negative plate is oxidised.

Thus the action on charge or discharge consists of the transfer of oxygen from one plate to the other. The electrolyte acts as an ionised conductor and as it does not react with either plate, the specific gravity of the electrolyte remains constant. Should the specific gravity fall due to, for example, absorption of carbon dioxide from the atmosphere, then if the value is as low as 1·16, the electrolyte should be replaced. If it is not replaced, the capacity of the cell will be decreased. Fresh electrolyte will restore its capacity.

Battery charging

A suggested rate of charging for these batteries is $\frac{1}{5}$ of the ampere-hour capacity for 7 hours. Thus a fully discharged battery of 100 Ah capacity should be fully recharged in 7 h with a current of 20 A. However, depending on the type and use of the battery, other rates of charge may be desirable and the manufacturer's advice should always be obtained.

The batteries cannot be damaged by over-charging provided that the electrolyte level is maintained. Thus it is advisable to arrange the charge/discharge cycle so that the battery is "over-charged". The average discharge voltage is about 1·2 V per cell at normal currents and depending on the charging current, the voltage per cell may be as high as 1·5 V. Batteries can be discharged at high rates without damage and will withstand short-circuit conditions.

USES OF NICKEL–CADMIUM BATTERIES

Modern battery requirements can be subdivided into (1) applications where the full ampere-hour capacity of the battery is regularly required, (2) applications where high currents are required for short periods. The conditions required in (2) are met by increasing the plate area and decreasing the thickness of both positive and negative plates. The internal resistance of these batteries is lower than those used in (1).

These batteries have many uses including switchgear operation, train lighting, engine starting, railway signalling, electric trucks, hand lamps, etc.

12.12. Efficiency of batteries

When a battery is in a state of discharge, energy must be supplied in order to charge it. If a load is connected to the now charged battery, it can become discharged to the same state as previously. The ratio of energy output to energy input may be considered as the efficiency of the battery which may be expressed as follows:

$$\text{Watt-hour efficiency} = \frac{\text{Watt-hours given out during discharge}}{\text{Watt-hours supplied during charging}}$$

As there is relatively little variation in the terminal voltage during the charging and discharging periods, the ampere-hour efficiency is often used commercially.

$$\text{Ampere-hour efficiency} = \frac{\text{Ampere-hours given out during discharge}}{\text{Ampere-hours supplied during charging}}$$

EXAMPLE. A battery was charged at 6 A for 24 h after being discharged in 6 h by a current of 20 A. The average terminal voltage during charging was 2·15 V and during discharge 1·9 V. Calculate the watt-hour and ampere-hour efficiencies.

$$\text{Watt-hour efficiency} = \frac{1{\cdot}9 \times 20 \times 6}{2{\cdot}15 \times 6 \times 24} = \mathbf{0{\cdot}739} \text{ or } \mathbf{73{\cdot}9\%}.$$

$$\text{Ampere-hour efficiency} = \frac{20 \times 6}{6 \times 24} = \mathbf{0{\cdot}833} \text{ or } \mathbf{83{\cdot}3\%}.$$

12.13. The chemical effect of an electric current

The chemical effect of an electric current may be demonstrated by immersing two pieces of metal, known as electrodes, into a solution, known as an electrolyte, and passing a direct current through it. The electrode connected to the positive supply terminal is called the anode and the other which is connected to the negative supply terminal is called the cathode. When a current is passed through this arrangement, a process takes place which is known as electrochemical decomposition or electrolysis.

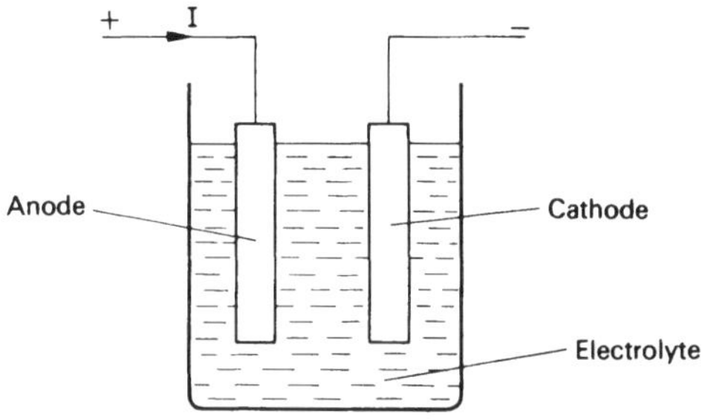

FIG. 12.10

12.14. Faraday's laws of electrolysis

Experiment has shown that a certain mass of substance is always deposited on the cathode, this being extracted from the electrolyte. For example, if both electrodes were of copper and the electrolyte was copper sulphate solution, metallic copper would be deposited on the cathode. It would also be found that the mass of copper deposited would equal the loss in mass of copper at the anode. A similar effect would occur with silver electrodes and a solution of silver nitrate. In electroplating, the object to be plated is made the cathode and for silver plating there would be a silver anode and silver nitrate solution as the electrolyte.

The mass of material deposited can be found using Faraday's laws. These may be expressed as follows:

1. The mass of substance deposited on the cathode is directly proportional to the quantity of electricity which has passed through the electrolyte.
2. The mass of substance deposited for a given quantity of electricity is directly proportional to the electrochemical equivalent (E.C.E.) of the substance.

The electrochemical equivalent is numerically equal to the mass of substance deposited when 1 coulomb of electricity passes. It may be expressed in grams per coulomb (g/C) or in kilograms per coulomb (kg/C).

Expressing this in the form of an equation,

$$\underset{\text{(grams)}}{\text{Mass deposited}} = \underset{\text{(g/C)}}{\text{E.C.E.}} \times \underset{\text{(A)}}{\text{current}} \times \underset{\text{(sec)}}{\text{time.}}$$

Typical values of electrochemical equivalent are:

Copper	0·00033 g/C
Silver	0·001118 g/C
Nickel	0·000304 g/C.

EXAMPLE. Calculate the weight of copper deposited in an electrolytic tank having copper electrodes and containing a solution of copper sulphate when a current of 1·5 A flows for a period of 200 min. What mass of silver would be deposited in the same time by the same current using silver electrodes in a solution of silver nitrate?

$$\text{Mass of copper deposited} = 0{\cdot}00033 \times 1{\cdot}5 \times (200 \times 60) \text{ g}$$
$$= \mathbf{5{\cdot}94\ g.}$$

$$\text{Mass of silver deposited} = 0{\cdot}001118 \times 1{\cdot}5 \times (200 \times 60) \text{ g}$$
$$= \mathbf{20{\cdot}08\ g.}$$

Alternatively, mass of silver deposited

$$= 5{\cdot}94 \times \frac{0{\cdot}001118}{0{\cdot}00033} \text{ g}$$
$$= \mathbf{20{\cdot}08\ g.}$$

The electrochemical effect of a current is used commercially in electroplating and in extracting some metals from their ores. It is also used for refining metals. Direct currents in the region of 10,000 A are used in producing aluminium from a molten electrolyte. (The magnetic field adjacent to busbars carrying currents of this magnitude is so strong that nails will stand on the hand and watches stop because the hair-spring becomes magnetised.)

Problems

12.1. Twelve cells, each having an e.m.f. of 1·5 V, are connected in series with a 1·6-Ω resistor. Each cell has an internal resistance of 0·7 Ω. Calculate (a) the current flowing, (b) the power in the resistor, (c) the current if one cell were incorrectly connected.

((a) 1·8 A; (b) 5·18 W, (c) 1·5 A.)

12.2. Four cells, each having an e.m.f. of 1·5 V and internal resistance of 1 Ω, are connected in series to a resistor and a current of 0·75 A flows. What is the value of the resistor?

(4 Ω.)

12.3. A battery whose e.m.f. is 12 V has an internal resistance of 1 Ω. It is connected to a variable resistor. Calculate the maximum power that can be taken by the resistor.

(36 W.)

12.4. A battery consists of six cells connected in series, each cell having an e.m.f. of 1·45 V and internal resistance 4 Ω. A resistor of 50 Ω is connected across the terminals of the battery. Determine: (a) the current; (b) the terminal p.d. of the battery; (c) the power dissipation in the external load.

((a) 0·117 A; (b) 5·87 V, (c) 0·69 W.) (N.C.T.E.C.)

12.5. A battery whose e.m.f. is 12 V and internal resistance 0·5 Ω is connected to resistors of 1, 1·5, and 3 Ω which are connected in parallel. Calculate (a) the total current, (b) the current in each resistor, (c) the energy expended in joules in the 3-Ω resistor in 0·25 h.

((a) 12 A; (b) 6 A, 4 A, and 2 A respectively; (c) 10,800 J.)

12.6. A battery has an e.m.f. of 6 V and an internal resistance of 1 Ω. Two 10-Ω resistors are connected in parallel and are connected to the battery. Draw a circuit diagram of this arrangement and calculate: (a) the current from the battery, (b) the voltage across the resistors, (c) the power in the resistors.

((a) 1 A; (b) 5 V; (c) 5 W.) (N.C.T.E.C.)

Answers are not given to the following questions:

12.7. Determine the number of cells each having an e.m.f. of 1·5 V and internal resistance of 0·8 Ω necessary to pass a current of 1·5 A through a resistor of 2·4 Ω.

12.8. A 9-Ω and an 18-Ω resistor are connected in parallel and the combination connected in series with a 4-Ω resistor. A battery having an e.m.f. of 30 V and internal resistance of 2 Ω is connected to the resistors. Calculate (a) the current from the battery, (b) the power consumed in the 9-Ω resistor, (c) the energy consumed in all the load resistors in 50 min.

12.9. A battery whose internal resistance is 1 Ω is connected to a circuit consisting of a 3-Ω and a 6-Ω resistor connected in parallel. If the energy expended in the 3-Ω resistor in 1 min is 5120 J, calculate the e.m.f. of the battery.

12.10. A battery whose e.m.f. is 12 V is connected to a 2·5-Ω resistor and the energy expended in it in 2 h is 0·08 kWh. Calculate (a) the internal resistance of the battery, (b) the current which would flow from the battery if it were short-circuited.

12.11. Calculate the weight of (a) copper, (b) silver deposited on the cathode of an appropriate electrolytic cell when a current of 2 A flows for $1\frac{1}{2}$ h.

CHAPTER 13

Magnetism

13.1. Introduction

The natural magnet, called lodestone or magnetite (an oxide of iron), was mentioned by Thales of Miletus who lived from 640 to 546 B.C. At this time it was believed that there were hooks on the surface of the magnet and little rings on any iron that was attracted to it. The first use as a compass was by the Chinese at the end of the eleventh century. William Gilbert in about 1600 carried out experiments on magnetism and mentioned magnetic dip and inclination of a compass. He stated that the earth itself was a giant magnet and also showed that a magnet lost its magnetism when heated to a bright red colour. Coulomb and Weber did further work at the end of the eighteenth century, as also did Maxwell, and it was in 1820 that Oersted found that an electric current would affect a magnet. Ampere followed up this discovery and in 1825 found by experiment and mathematics the mutual forces that exist between currents. Many electrical units were named in honour of these physicists.

13.2. The bar magnet

If a straight piece of magnetised steel, usually referred to as a bar magnet, is suspended by a thread, then provided that it is at some considerable distance from any magnetic material, it will always come to rest in the same direction, this direction being approximately from the north to the south pole of the earth. The end of the magnet pointing north is called the north-seeking pole (or, more simply, the north pole), and the other end is called the south (seeking) pole. The magnetic effect is strongest at the poles of a magnet. If a magnet is dipped into iron filings, many are attracted at the ends but very few at the centre.

A magnet will attract various materials. Some of them, such as iron, are strongly attracted and they are said to be ferromagnetic. If the substance is weakly attracted, it is said to be paramagnetic. Some, such as bismuth and antimony, are repelled from the poles and these are said to be diamagnetic.

13.3. Force between magnetic poles

This can be summarised by the statement that like poles repel and unlike poles attract. Thus two north poles repel each other whilst a north and a south pole will attract each other.

13.4. The magnetic field

If the bar magnet is in the shape of an arrow and is mounted on a pivot, it can be used as a compass. A small compass, known as a plotting compass, can be used for obtaining the shape of a magnetic field, this being the space in which magnetic effects can be detected.

To plot the shape of the magnetic field of a bar magnet, the magnet should be placed on a sheet of paper lying on a flat surface. If the plotting compass is placed near to the north pole of the magnet, the compass needle will point away from the magnet. A dot should be made near to the part of the compass where the arrow head points and the centre of the plotting compass should then be placed over this dot. This process is repeated and it will be found that if a line is drawn through all the points, a line joining the north pole of the magnet to its south pole will be obtained.

If a number of different starting points are made from near to the north pole and the lines plotted, the resultant diagram is referred to as the shape of the magnetic field of the magnet. If this is done very carefully, it will be found that these "lines of magnetic force" never cross and that they always enter and leave a magnetic surface at right-angles. Sometimes it is convenient to regard these lines of force as being like elastic bands in that they are always trying to shorten themselves.

It should be appreciated that this magnetic field exists not only in the horizontal plane in which it has been drawn, but also in every other possible plane. It should also be realised that these lines of force do not actually exist as lines. The lines are drawn so as to indicate the direction of the force exerted by the magnetic field on a very small single magnetic pole. The conventional direction of a line of force is such that it indicates the path that a unit north pole would take if free to move in a magnetic field.

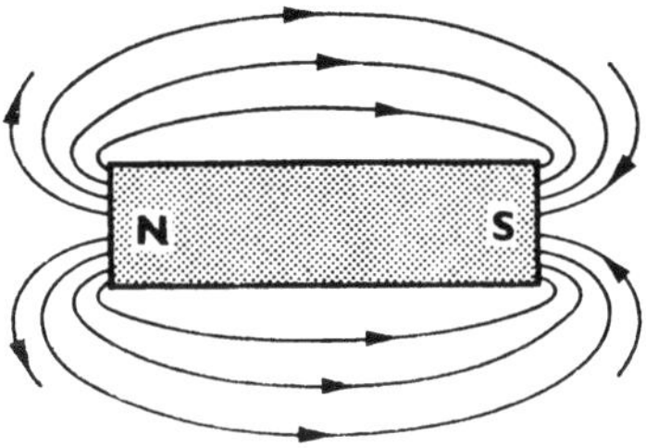

FIG. 13.1. Magnetic field of a permanent bar magnet.

Another method of obtaining the shape of the magnetic field of a bar magnet is to lay the magnet on a flat surface and to place a piece of smooth cardboard over it. Iron filings should then be scattered on to the cardboard and these will form themselves into "lines". The direction of the field may again be determined by a compass needle.

13.5. The magnetic effect of an electric current

The magnetic field of a permanent magnet is often used in engineering, for example, in instruments, magnetic chucks, etc. For many applications, the permanent magnet field is not strong enough or a field of variable strength is required. In these cases, an electric current is used to set up the field. The shape of the field and its direction of action can be found by the methods described previously. Figure 13.2(a) shows the field due to a straight conductor carrying current and Fig. 13.2(b) shows the field due to current flowing in a long coil (or solenoid).

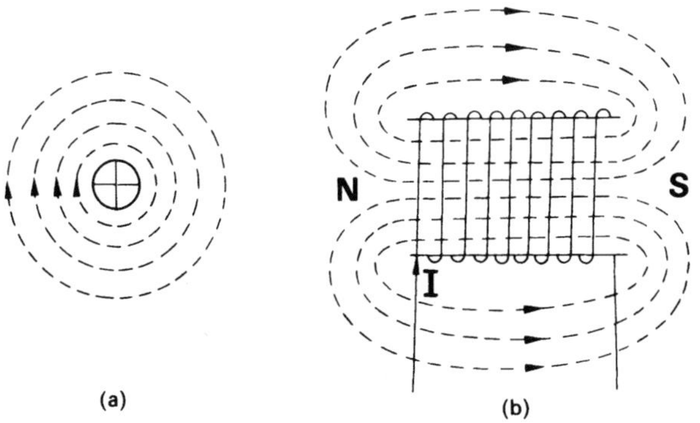

FIG. 13.2

Considering Fig. 13.2(a), the magnetic lines of force lie in concentric circles about the conductor and at right angles to it. The ⊕ indicates a cross-section of the conductor with the current flowing downwards: + is what the eye would see if a dart were travelling away from it. If ⊙ had been drawn, it would indicate current travelling upwards, what the eye would see if it stayed long enough to see the point of a dart coming towards it!

A simple rule enables the direction of the magnetic field to be found for any given direction of current. Grasp the conductor in the right hand with the thumb pointing in the direction of current flow. The fingers indicate the direction of the magnetic field.

The direction of the magnetic field of the solenoid shown in Fig. 13.2(b) can be determined using the above rule but perhaps the following rule may be easier. Grasp the solenoid in the right hand with the fingers pointing in the direction of the current flow. The thumb indicates the direction of the magnetic lines of force within the core.

13.6. Electromagnets

If an insulated wire is wound round a piece of soft iron and a current is passed through the wire, other magnetic materials will be attracted to the iron core. This constitutes a simple electromagnet. When the current is switched off, the pieces of magnetic material will fall away because the soft iron then loses nearly all its magnetism.

An electromagnet needs for its core a material in which a strong magnetic field can be set up with a relatively small magnetising force. It is also desirable that when the current is switched off, very little of the magnetic flux remains. Soft iron fulfils these two requirements.

Electromagnets have many applications in engineering and are used in electric bells, picking up magnetic materials in scrap yards, in electric motors and generators, telephone receivers, radio speakers, relays and circuit-breakers. Some magnets will lift a weight of about 80,000 lbf with a power input of about 40 kW. An electromagnet such as is used in a cyclotron may weigh 4000 tonf.

13.7. Electromagnetic induction

Suppose a coil having a number of turns is connected to a voltmeter of the centre-zero type. If a bar magnet is pushed into the coil, it will be seen that there is a deflection of the voltmeter pointer whilst the magnet is moving (Fig. 13.4(a)). Should the magnet be stationary at any time, the deflection will cease.

Although it was stated earlier that it would be assumed for the present that a voltmeter would take no current, it actually takes a small current to operate it. This small current flowing through the coil and instrument sets up a magnetic field around the coil and the direction of this field would be as shown in Fig. 13.4(a). In other words, the direction of the magnetic flux is such that there is a north pole at the left-hand end of the coil. Thus this magnetic field is trying to repel the north pole that is being pushed into it.

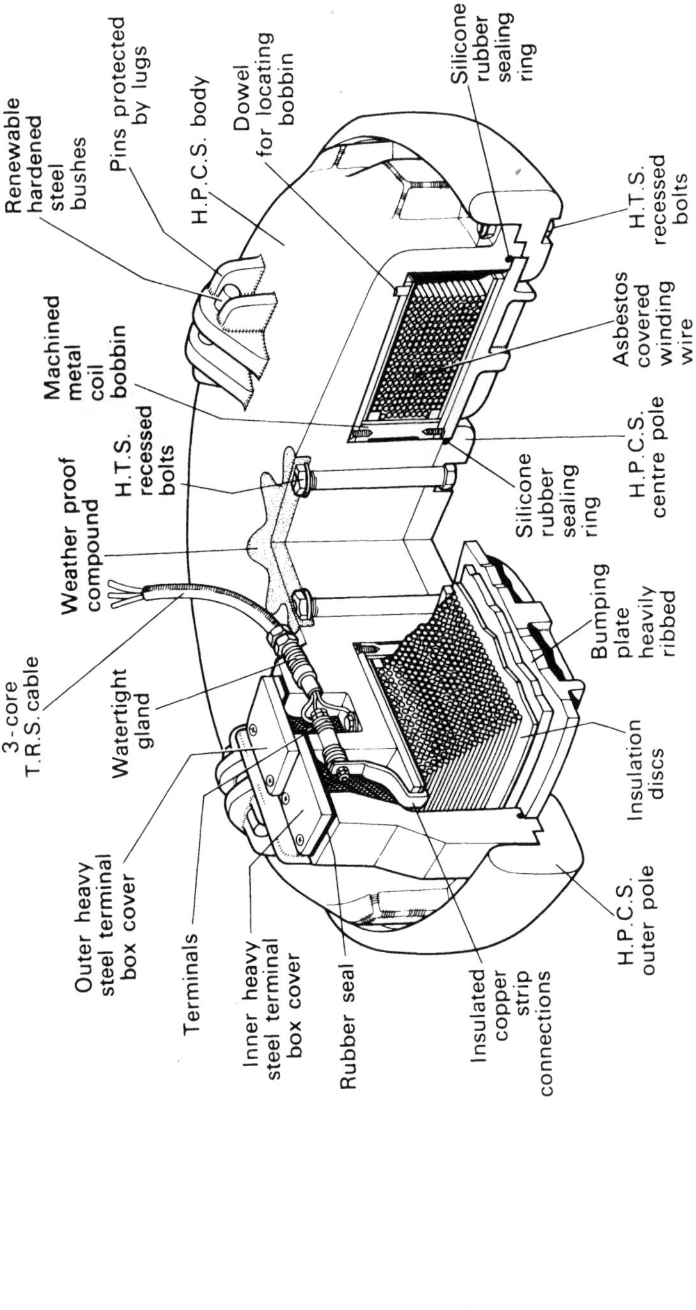

FIG. 13.3. Wire-wound lifting magnet. (Electromagnets Ltd.)

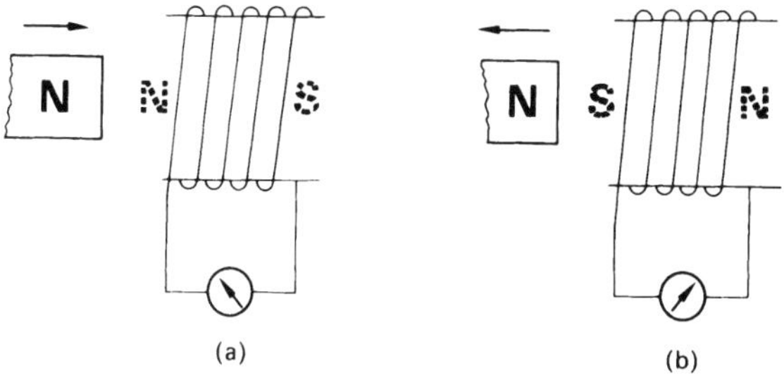

FIG. 13.4

In mechanics, action and reaction are equal and opposite. In electromagnetic circuits there is a similar effect which is summarised by Lenz's law.

13.8. Lenz's law

If in an electromagnet circuit there is a change in state, a force will be set up opposing the change.

Energy is expended in making the pointer of the instrument move. Some is expended as heat in the wires in which the current is flowing. This electrical energy is converted from the mechanical work done in pushing the magnet into the coil against the opposing force.

When the magnet is pulled out of the coil (Fig. 13.4(b)) the voltmeter will indicate a deflection in the opposite direction to previously and the reversed current flow will make the left-hand end of the coil a south pole. The force set up tries to prevent the magnet from leaving the coil.

It can be shown that the faster the magnet is moved, the bigger the deflection on the voltmeter, and for a given speed a stronger magnet will produce a bigger deflection than a weak one. In other words, a larger e.m.f. will be induced in the coil which in turn will cause a greater current.

13.9. E.M.F. induced in a coil

During the movement of the magnet, the magnetic flux is said to be changing its linkage with the coil.

Faraday's law of induction states that the average e.m.f. induced in the coil depends upon the average rate of change of flux linkages.

If N is the number of turns on the coil and Φ the flux linking the coil, $N\Phi$ is called the flux linkages. The average induced e.m.f. is

$$E_{av} = \frac{N\Phi}{t} \text{ volts}$$

where t is the time in seconds for the change of flux linkages and Φ is the flux in webers (pronounced "vayber").

13.10. Unit of magnetic flux

In the m.k.s. system of units, the weber (Wb) is the unit of flux. It is defined as follows: if a flux of 1 weber linking a coil of one turn were uniformly decreased to zero in one second, the e.m.f. induced in the coil would be 1 volt.

Smaller units of flux are:

$$1 \text{ milliweber } (1 \text{ mWb}) = \frac{1}{1000} \text{ weber } (10^{-3} \text{ Wb}) \text{ and}$$

$$1 \text{ microweber } (1\ \mu\text{Wb}) = \frac{1}{1{,}000{,}000} \text{ weber } (10^{-6} \text{ Wb})$$

EXAMPLE. A flux of 200 mWb, when linking a coil of 100 turns, is reduced uniformly to zero in 0·05 sec. What is the value of the e.m.f. induced in the coil during this period?

$$\text{Induced e.m.f.} = \frac{200 \times 100}{1000 \times 0{\cdot}05} \text{ weber-turns/second}$$

$$= \mathbf{400\ V.}$$

13.11. Unit of magnetic flux density

Suppose that a magnetic flux were uniformly distributed over an area and that it was at right angles to this area. The density of the flux would be expressed as the amount of flux per unit area. In the m.k.s. system of units, the unit of area is the square metre and so

$$\text{Magnetic flux density } (B) = \frac{\text{total flux (Wb)}}{\text{area (m}^2\text{)}} \text{ Wb/m}^2.$$

In the SI system the unit is the tesla (T).

EXAMPLE. If a flux of 1 mWb is at right angles to an area of 20 cm^2, what is the flux density?

$$\text{Flux density } (B) = \frac{10^{-3}\text{(Wb)}}{20 \times 10^{-4}\text{(m}^2\text{)}}$$
$$= \mathbf{0{\cdot}5\ Wb/m^2} \text{ (or T).}$$

13.12. Principle of the transformer

A simple transformer has two windings wound round a common magnetic core. If a current in one winding, called the primary, is continually changing its magnitude, there will be a continually changing magnetic flux. This flux will change its linkage with the other winding, called the secondary, and an e.m.f. will be induced in it. These conditions exist when the primary winding carries an alternating current. An alternating e.m.f. is induced in the secondary.

The alternating flux also induces an e.m.f. in the primary winding and this e.m.f. is in opposition to the applied voltage which causes the current to flow in the primary winding.

13.13. E.M.F. induced in a moving conductor

Suppose that a uniform magnetic field having a flux density of B Wb/m^2 is cut by a conductor of length l metres moving with a velocity of V metres/second (m/s) at right angles to this magnetic field. With suitable equipment, it can be shown that an e.m.f. is induced in the conductor. The effect would be the same if the conductor were stationary and the field was moving past it. The flux may be said to have been cut by the conductor or that there has been a change in the flux linkage.

The value of the induced e.m.f. is, in this case, given by

$$E = BlV \text{ volts}$$

as the conductor is moving at right angles to the field. That is, the angle between the direction of the conductor and the field is 90°. If this angle had been θ°, then the e.m.f. generated would have been proportional to $V \sin \theta$, that is, the component of the velocity at right angles to the field (Fig. 13.5).

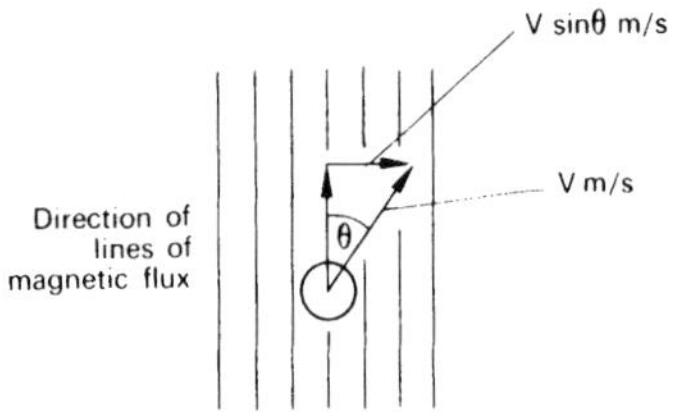

FIG. 13.5

EXAMPLE. A conductor 60 cm long is moving at a velocity of 3000 cm/s (a) at right angles to, (b) at 30° to the direction of a uniform magnetic field of density 0·2 Wb/m². Calculate the e.m.f. induced in each case.

(a) Generated e.m.f. $= BlV$ volts

$$= 0{\cdot}2 \times (60 \times 10^{-2}) \times (3000 \times 10^{-2}) \text{ volts}$$

$$= 0{\cdot}2 \times 0{\cdot}6 \times 30 \text{ volts}$$

$$= \mathbf{3{\cdot}6\ V.}$$

(b) Generated e.m.f. $= BlV \sin \theta$ volts

$= 3{\cdot}6 \sin 30°$ volts

$= 3{\cdot}6 \times 0{\cdot}5$ volts

$=$ **1·8 V.**

EXAMPLE. A rectangular coil wound with 50 turns of wire has a width of 50 cm and a length 40 cm. It is rotating at 1200 rev/min in a uniform magnetic field of density 0·8 Wb/m^2. Determine the e.m.f. induced in the coil when the plane of the coil is (a) parallel to the field, (b) at an angle of 60° to the direction of the field.

When the plane of the coil is parallel to the field, the conductors are moving at right angles to the magnetic flux. It should also be noted that each turn of the coil consists of two conductors in series so that their e.m.f.s are additive.

(a) In this case, the maximum e.m.f. is being generated as the sides of the coil are cutting the flux at right angles.

Radius of the coil $= \frac{50}{2} = 25$ cm or 0·25 m.

In 1 revolution the conductor moves $2\pi \times 0{\cdot}25 = 0{\cdot}5\pi$ m.

Velocity of conductor $= 0{\cdot}5\pi \times \frac{1200}{60} = 10\pi$ m/s.

The e.m.f. per conductor $= 0{\cdot}8 \times (40 \times 10^{-2}) \times 10\pi$ volts

$= 3{\cdot}2\pi$ volts

$= 10$ V.

As the coil has 50 turns, there are $2 \times 50 = 100$ conductors in series.

Hence e.m.f. induced in the coil

$E = 10 \times 100$ volts

$=$ **1000 V.**

(b) As the plane of the coil is at 60° to the direction of the field, the conductors are moving at 30° to the field. Therefore the e.m.f. induced at this instant is given by:

$$E = 1000 \sin 30° \text{ volts}$$
$$= 1000 \times 0{\cdot}5 \text{ volts}$$
$$= \mathbf{500\ V.}$$

13.14. Force on a conductor in a magnetic field

Figure 13.6 shows a conductor carrying a current and lying at right angles to a uniform magnetic field. Diagram (c) shows the resultant field and there is a force set up on the conductor in the direction shown. It should be noted that the original field has been distorted, being stronger in this case on the right-hand side of the conductor. The magnitude of this force is given by:

$$F = BlI \text{ newtons.}$$

In this case the angle between the conductor and the field is 90° but if this angle had been θ°, the force would have been given by:

$$F = BlI \sin \theta \text{ newtons.}$$

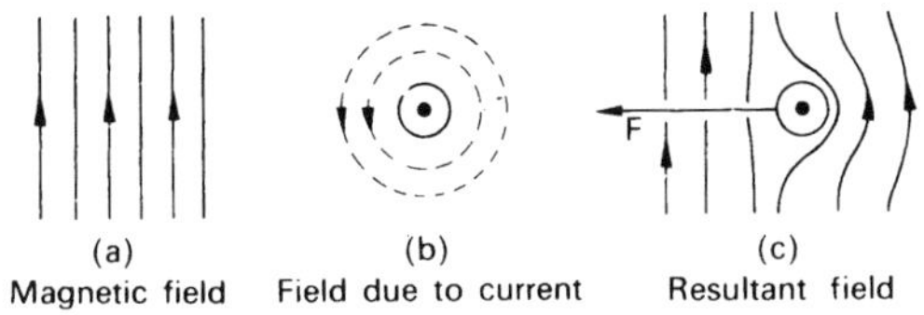

FIG. 13.6

13.15. Fleming's hand rules

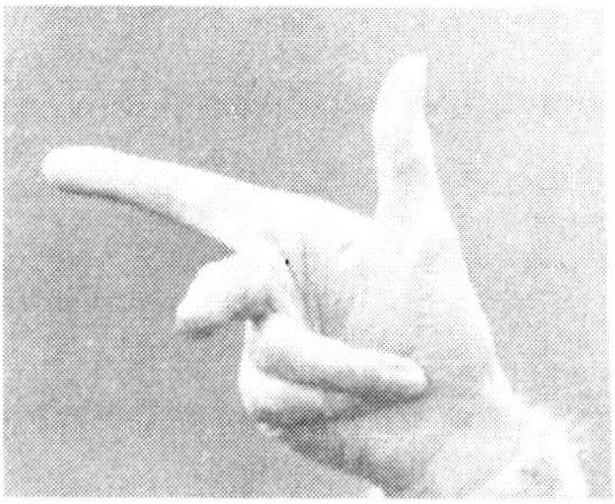

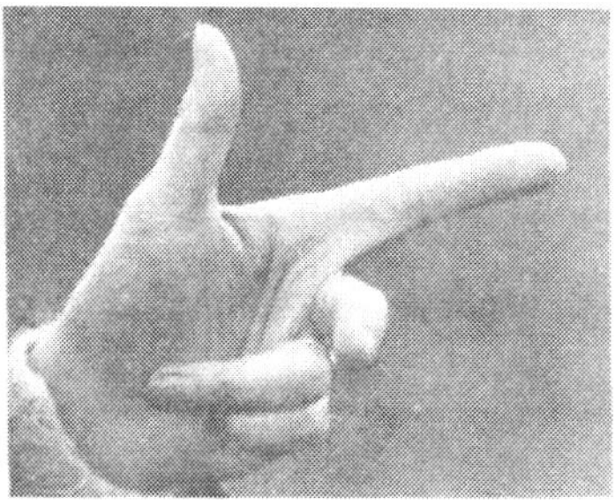

RIGHT HAND

Direction of e.m.f. in conductor.

First finger: direction of flux.

Thumb: direction of motion of the conductor **with respect to the field.**

Middle finger: direction of the e.m.f. in the conductor.

LEFT HAND

Direction of force on conductor.

First finger: direction of flux.

Middle finger: direction of current in the conductor.

Thumb: direction of the force on the conductor.

FIG. 13.7

Remember when using the right-hand rule that if the conductor were stationary and the field were moving from, say, left to right, the motion of the conductor relative to the field would be from right to left and this must be used in determining the direction of the induced e.m.f.

The direction of the induced e.m.f. and hence the direction of the resultant current which would flow if the conductor were connected to a resistor can be deduced as follows (Fig. 13.8):

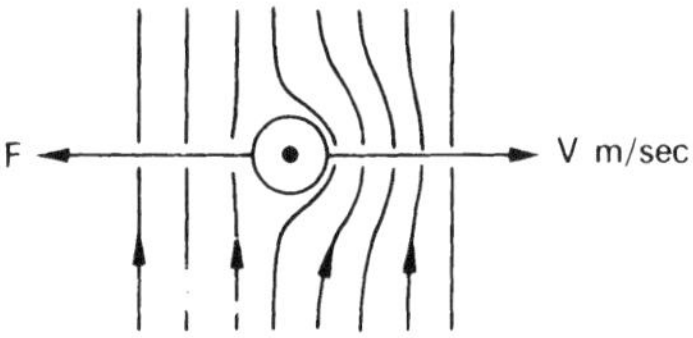

FIG. 13.8

If the conductor were moving from left to right, it would have an e.m.f. induced in it when it entered the magnetic field. By Lenz's law, the current flowing through the conductor must set up a field which would combine with the existing field in such a way as to set up a force in the opposite direction to that in which the conductor was being moved.

If the conductor were being moved at V metres/second by a force of F newtons, then

Work done per second $= F \times V$ newton-metres/second (Nm/s)

$= F \times V$ joules/second (J/s)

$= F \times V$ watts (W).

But $F = BlI$ newtons,

$\therefore$ work done per second $= BlIV$ newton-metres/second or watts.

But $BlV = E$ (the e.m.f. generated in the conductor when cutting the magnetic field),

$\therefore$ work done per second $= EI$ newton-metres/second or watts.

This shows that, neglecting any losses, the mechanical work done per second in moving the conductor is all converted into an electrical power output.

13.16. Magnetomotive force (m.m.f.)

The m.m.f. is that effect which sets up the magnetic flux in a magnetic circuit. In the m.k.s. system of units, it is equal to the product of the number of turns (N) and the current (I) in amperes. The symbol used for m.m.f. is F.

m.m.f. $= IN$ ampere-turns.

In some ways, the m.m.f. in the magnetic circuit is analogous to the magnetomotive e.m.f. of the electric circuit. Although this effect is called a magnetomotive force, it is not actually a force.

13.17. Magnetising force

Figure 13.9 shows a uniformly distributed coil wound on a ring-shaped former in which the radial thickness is small compared with the mean radius of the former. Practically all the magnetic flux is confined within the coil, i.e. there is negligible leakage flux,

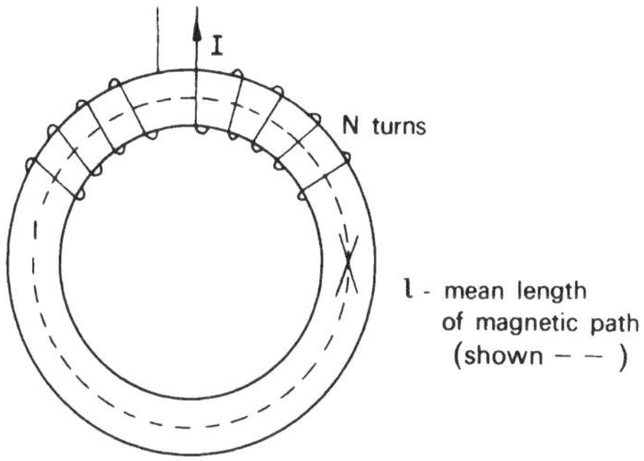

FIG. 13.9

so the mean length (l metres) of the magnetic flux path will be the mean circumference of the ring. The magnetising force is the m.m.f. per unit length of magnetic path.

Thus, magnetising force

$$H=\frac{F}{l}=\frac{IN}{l} \text{ ampere-turns per metre (AT/m).}$$

In the SI system the unit as the ampere per metre (A/m.)

Neglecting leakage flux, the value of the magnetising force is the same at all points within the coil provided that the cross-sectional area and the material in it are the same.

13.18. Permeability of free space

When a current is passed through the coil shown in Fig. 13.9, a flux is set up and, using suitable equipment, it is possible to measure the flux density. Suppose the coil were wound on a former having a non-magnetic core. If the value of the current were varied, both the flux density and the magnetising force would vary and if these quantities were plotted, a linear graph would result (Fig. 13.10),

$$\therefore\ B \propto H,$$

$$\therefore\ B = (\text{a constant}) \times H.$$

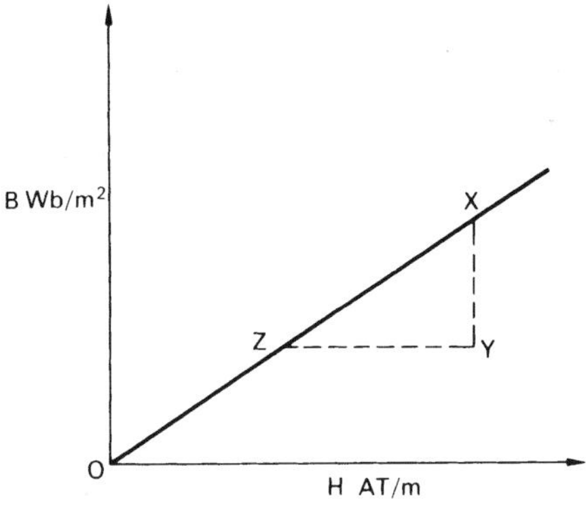

FIG. 13.10

The constant is equal to the gradient of the graph ($= XY/ZY$) and for a vacuum it is equal to $4\pi \times 10^{-7}$. Its unit is henry/metre (H/m). See Section 13.23 for the unit of inductance.

For all practical purposes, the constant would be the same for air or any other non-magnetic material.

This constant is called, for a vacuum, the permeability of free space, symbol μ_0.

EXAMPLE. What value of magnetising force is necessary to set up a flux density of 1 Wb/m^2 in air?

$$B = \mu_0 H \quad \text{or} \quad H = \frac{B}{\mu_0},$$

$$\therefore \quad H = \frac{1}{4\pi \times 10^{-7}} \text{ AT/m}$$

$$= \mathbf{796{,}000\ AT/m.}$$

If the coil described previously has the non-magnetic core replaced by a magnetic material, it will be found that the relationship between *B* and *H* is approximately linear at first but eventually magnetic saturation occurs, that is, there is no further increase in flux density for an increase in the magnetising force.

This is the region *XY* in Fig. 13.11. The relationship between the flux density and the magnetising force is usually referred to as a *B–H* curve.

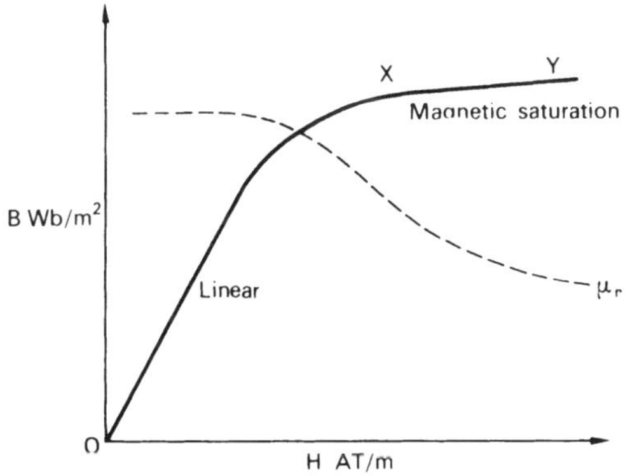

FIG. 13.11

13.19. Relative permeability

For a given magnetising force, the flux density when a magnetic material forms the core is very much greater (possibly a 1000 times or more) than with a vacuum core. The ratio of these two flux densities for a given magnetising force is called the relative permeability of the material for that particular value of magnetising force. Its symbol is μ_r and is a number—it has no physical dimensions.

Figure 13.11 also illustrates a typical curve of relative permeability for various values of the magnetising force.

EXAMPLE. What value of magnetising force is necessary to set up a flux density of 1 Wb/m^2 in a magnetic material if the relative permeability is known to be 1000?

$$B = \mu_0\mu_r H \quad \text{or} \quad H = \frac{B}{\mu_0\mu_r},$$

$$\therefore \quad H = \frac{1}{4\pi \times 10^{-7} \times 1000}$$

$$= \mathbf{796\ AT/m.}$$

13.20. Relationship between flux and m.m.f.

$$\text{Flux} = \text{Flux density} \times \text{area}.$$

$$\Phi = Ba \text{ webers},$$

$$\text{but } B = \mu_0\mu_r H \text{ webers}$$

$$\text{and } H = \frac{IN}{l} \text{ ampere-turns per metre},$$

$$\therefore \quad \Phi = \mu_0\mu_r\frac{IN}{l}a \text{ webers}$$

$$= \frac{IN}{\dfrac{l}{\mu_0\mu_r a}} \text{ webers.}$$

$\dfrac{l}{\mu_0\mu_r a}$ is called the reluctance of the magnetic circuit and as IN is the m.m.f. setting up the flux, then

$$\text{Flux} = \frac{\text{m.m.f.}}{\text{reluctance}}.$$

This is analogous in the electric circuit to

$$\text{Current} = \frac{\text{e.m.f.}}{\text{resistance}}.$$

The ratio $\frac{1}{\text{reluctance}}$ is called the permeance of the magnetic circuit.

EXAMPLE. An iron ring has a uniform cross-sectional area of 2 cm^2 and a mean diameter of 20 cm. It is wound with a coil of 250 turns which carries a current of 2·4 A. If the relative permeability of the iron is 460, calculate the flux set up in the ring.

$$\text{Mean length of magnetic path } l = \frac{20\pi}{100} = 0{\cdot}2\pi \text{ metres.}$$

(Note: lengths and areas must be in metre measure.)

$$\begin{aligned}
\text{Magnetising force } H &= \frac{250 \times 2{\cdot}4}{0{\cdot}2\pi} \text{ AT/m} \\
&= 955 \text{ AT/m.} \\
\text{Flux density } B &= \mu_0\mu_r H \\
&= 4\pi \times 10^{-7} \times 460 \times 955 \text{ Wb/m}^2 \\
&= 0{\cdot}552 \text{ Wb/m}^2. \\
\text{Total flux } \Phi &= 0{\cdot}552 \times 2 \times 10^{-4} \text{ Wb} \\
&= \mathbf{0{\cdot}0001104 \text{ Wb}} \\
\text{or } &= \mathbf{0{\cdot}1104 \text{ mWb}}.
\end{aligned}$$

13.21. Use of the magnetisation curve

In practice, problems involving magnetic circuit calculations are usually solved by using the magnetisation curve for the magnetic material.

EXAMPLE. An iron ring whose mean circumference is 60 cm has a cross-sectional area of 9 cm^2. It is cut into two equal parts which are separated by non-magnetic material 1 mm thick. The ring is uniformly wound with a coil having 1000 turns. Calculate the current required to produce a magnetic flux of 0·72 mWb in the ring. The magnetisation curve values for the iron are:

Magnetising force (H)	80	160	240	320	480	640	AT/m
Flux density (B)	0·4	0·63	0·8	0·9	0·98	1·02	Wb/m²

As the magnetic circuits are in series (they all carry the same flux Φ) the total m.m.f. required to set up the flux is the sum of the m.m.f.s in each part of the circuit.

$$\text{Total flux} = 0{\cdot}72 \text{ mWb} = 0{\cdot}72 \times 10^{-3} \text{ Wb}$$

$$\text{Flux density} = \frac{0{\cdot}72 \times 10^{-3}}{90 \times 10^{-4}} \text{ Wb/m}^2$$

$$= 0{\cdot}8 \text{ Wb/m}^2$$

$$\text{For the non-magnetic material, } H = \frac{B}{\mu_0}$$

$$= \frac{0{\cdot}8}{4\pi \times 10^{-7}}$$

$$= 636{,}800 \text{ AT/m.}$$

$$\text{m.m.F. required, } F = 636{,}800 \times (2 \times 0{\cdot}001)$$

$$= 1275 \text{ AT.}$$

(In this circuit, there were two non-magnetic materials each of length 0·001 m.)

For the iron part of the circuit, the value of magnetising force corresponding to a flux density of 0·8 Wb/m² is found from the characteristic given. If the curve is drawn, the value will be found to be 240 AT/m.

$$\text{m.m.F. required, } F = 240 \times 0{\cdot}6 \text{ AT}$$

$$= 144 \text{ AT.}$$

$$\text{Total m.m.f.} = 1275 + 144 \text{ AT}$$

$$= 1419 \text{ AT.}$$

$$\text{Current} = \frac{1419}{1000} \text{ A}$$

$$= \mathbf{1{\cdot}42 \text{ A.}}$$

13.22. Magnetic hysteresis

Suppose the magnetic specimen in a coil were initially unmagnetised when the magnetising force H was zero. If the current (and therefore H) is increased, the B–H curve OA could be plotted (Fig. 13.12). If the current is decreased to zero, the curve AC is obtained. OC is called the remanence or residual flux density, that is, the flux density remaining when the magnetising force has been

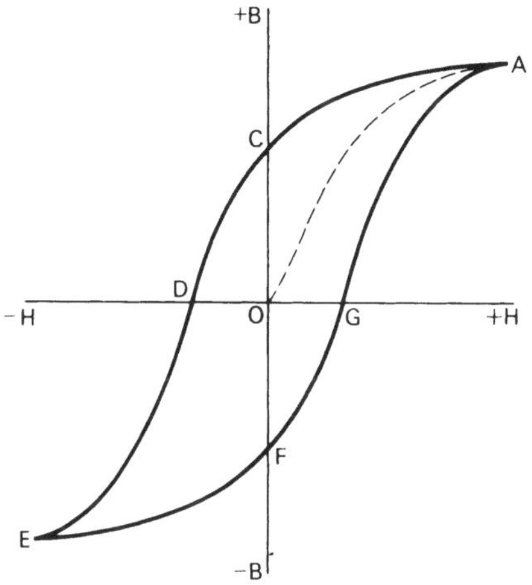

FIG. 13.12

reduced to zero. The current must be reversed and the magnetising force increased to OD in order to reduce the remanence to zero. OD is called the coercive force. Further variations of the magnetising force between the maximum values of magnetising force result in the relationship between corresponding values of H and B following the curve $DEFGA$. Note that $OC = OF$ and $OD = OG$.

The loop is called the hysteresis loop and this lagging of the flux density behind the magnetising force is known as magnetic hysteresis. It results in a power loss as heat in many electrical machines.

13.23. Inductance of a coil having a non-magnetic core

If a coil has a core of a non-magnetic material, the flux density is proportional to the magnetising force. Likewise, the flux is proportional to the current and so the flux set up by each ampere is the same:

$$\Phi = \frac{\Phi}{I} \times I.$$

Multiplying both sides of this equation by N/t,

$$\frac{N\Phi}{t} = \frac{N\Phi}{I} \times \frac{I}{t}.$$

$N\Phi/t$ is the average rate of change of flux linkages and is equal to the average induced e.m.f.

$$\therefore E_{av} = \frac{N\Phi}{I} \times \frac{I}{t}.$$

$N\Phi/I$ is also a constant and it is known as the self-inductance (L) of the coil and it is expressed in henry(s).

$$\therefore E_{av} = L \times \frac{I}{t} \text{ volts,}$$

where I/t is the average rate of change of current.

Note that 1 henry (1 H) $= 10^3$ millihenrys (10^3 mH)
$= 10^6$ microhenrys (10^6 μH).

EXAMPLE. The current in a coil, whose (self)-inductance is 0·2 H, changes uniformly from 10 to 20 A in 0·05 sec. Calculate the average e.m.f. induced whilst the current is changing. If a current of 20 A is uniformly reversed in 0·1 sec, what is the value of the induced e.m.f.?

Change of current $= 20 - 10 = 10$ A.

$$\text{Induced e.m.f. } E_{av} = 0{\cdot}2 \times \frac{10}{0{\cdot}05} \text{ volts}$$

$$= \mathbf{40\ V}.$$

When the current is reversed, current change $= 20 \times 2 = 40$ A.

$$E_{av} = 0{\cdot}2 \times \frac{40}{0{\cdot}1} \text{ volts}$$

$$= \mathbf{80\ V.}$$

13.24. The effects of applying a voltage to a coil

When a voltage V is applied to a circuit, the current is initially zero. As the current begins to increase, a magnetic flux is set up and this links with the conductors carrying the current. This change of flux linkage induces an e.m.f. in the conductors which is in opposition to the applied voltage (Lenz's law). It is for a given circuit, therefore, the difference between these voltages which determines the current flowing at any instant. A full treatment of this is not possible here, but the effect is that the current takes a certain period of time before it reaches the value given by Ohm's law.

If the circuit is not surrounded by iron, little flux is set up and as the induced e.m.f. is small, the current reaches its final value very quickly. Such a circuit is said to be almost non-inductive.

If the circuit is associated with a magnetic material and has many turns, the induced e.m.f. results in the current taking several seconds before a steady current is reached. Such a circuit is said to be highly inductive. An example is the field coils on an electrical machine.

The formula for induced e.m.f. is often written

$$E_{av} = -L(I/t) \text{ volts.}$$

The minus sign implies that the induced e.m.f. E_{av} is in opposition to the applied voltage.

13.25. Mutual inductance

In Section 13.12 there is a brief description of the operation of a transformer. A changing current in one winding sets up a varying flux which links the second winding and induces an e.m.f. in it.

This may be said to be a mutually induced e.m.f. and its value is given by:

$$E_2 = M\frac{I_1}{t} \text{ volts,}$$

where E_2 (volts) is the e.m.f. induced in the secondary winding, I_1/t (amperes per second) the rate of change of primary current and M the mutual inductance between the windings in henrys.

Suppose that the current in the primary winding was changing at the rate of 1 A/s and as a result an e.m.f. of 1 V was induced in the secondary winding, then it would be said that there is a mutual inductance between the coils of 1 henry. If the mutual inductance is 1 H, then current changing at the rate of 1 A/s in either winding would induce an e.m.f. of 1 V in the other winding.

EXAMPLE. Two coils are wound on a common magnetic core and when a current of 20 A in one of them is reversed uniformly in 0·1 sec, the e.m.f. induced in the other winding is 800 V. What is the value of the mutual inductance between the coils?

If the current changes from 20 A in one direction to 20 A in the opposite direction, the total current change is 40 A. Hence

$$800 = M\frac{40}{0{\cdot}1}$$

$$M = \mathbf{2\ henrys\ (H).}$$

13.26. Force between parallel conductors carrying current

If a single conductor (1) carries a current as shown in Fig. 13.13 the magnetic field is concentric with the conductor. For the assumed direction of the current, the flux is in the direction shown. If a second conductor (2) carries a current parallel to the first, it may be regarded as lying in the field of the first conductor. By applying Fleming's left-hand rule, the force on conductor (2) is found to be directed towards conductor 1. As action and reaction are equal and opposite, the force between conductors carrying currents in

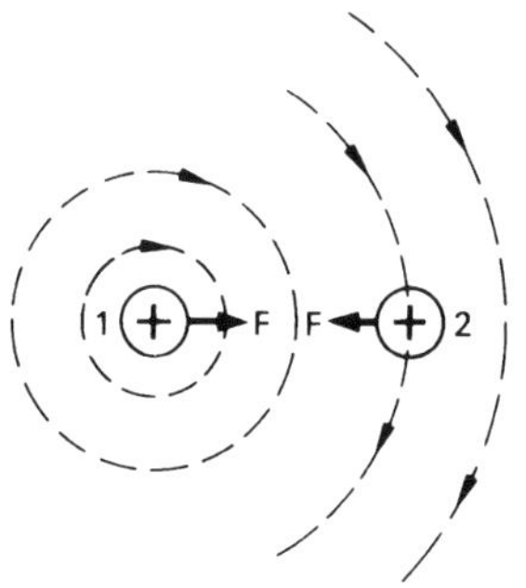

Fig. 13.13

the same direction is one of attraction. A force of repulsion exists when the currents are in opposite direction.

In all types of electrical machines there are conductors lying parallel to each other. When these carry current, considerable forces may be set up and windings may have to be braced to withstand them. This force is particularly great under fault conditions.

The force on the conductor (2) is $F = BlI$ newtons. In a vacuum the flux density B at a distance d metres from the centre of the conductor (1) is

$$B = \mu_0 H = \mu_0 \frac{NI_1}{2\pi d};$$

$2\pi d$ is the length of the flux line in the position of conductor (2). For single conductors, $N = 1$ turn. If the centre of conductor (2) is d metres from the centre of conductor (1), then the force between the conductors in a vacuum is

$$F = BlI = \frac{\mu_0 NI_1}{2\pi d} . \, l \, . \, I_2 \text{ newtons}$$

or $$F = \frac{2I_1 I_2 \times 10^{-7}}{d} \text{ newtons/metre length of conductor.}$$

(Note: $\mu_o = 4\pi \times 10^{-7}$.)

This expression is used for defining the ampere. If the conductors were 1 m apart and each carried a current of 1 A, the force between them per metre length of conductor would be 2×10^{-7}N.

For all practical applications this force would have the same magnitude if the conductors were separated in air.

13.27. Energy stored in a magnetic field

When current flows in a circuit, energy is stored in the magnetic field which is set up. This energy is considerable in an iron-cored coil, e.g. the field coil of a machine, and when the current in such a coil is switched off, an arc at the switch will be observed. This is caused by the magnetic field collapsing and most of the energy stored in the magnetic field is dissipated as heat in the arc.

It can be shown that

$$\text{Energy stored in the magnetic field} = \tfrac{1}{2}LI^2 \text{ joules,}$$

where L is the (self)-inductance of the circuit in henrys and I the current flowing in amperes.

Inductive circuits should never be broken by switches designed to operate only on an alternating current supply as the contacts do not have a wide separation and they will probably be welded together by the arc. The collapsing field cuts the windings and the induced e.m.f. tries, according to Lenz's law, to maintain the current, thus tending to prolong the arc.

Where an inductive circuit is to be broken, a special switch may be used which connects a resistor, called a field discharge resistor, across the field windings prior to disconnecting the supply. The stored energy is then dissipated as heat in the resistor instead of in an arc.

In some cases, arcing is suppressed by connecting a capacitor across the switch. This is one of the functions of the capacitor connected across the contact points in a car ignition circuit.

Problems

13.1. A conductor 50 cm long is moving at 200 m/s at right angles to a uniform magnetic field. If an e.m.f. of 40 V is induced in the conductor, what is the density of the magnetic field?

(0·4 Wb/m^2.)

13.2. A conductor 100 cm long is moving at 10,000 cm/sec at an angle of 30° to a uniform magnetic field of density 0·2 Wb/m². Determine the e.m.f. induced in the conductor.

(10 V.)

13.3. (a) A coil of 1000 turns, when carrying a current of 2 A, sets up a magnetic flux of 0·05 Wb. If the current is reversed in one-hundredth of a second, what is the average e.m.f. induced in the coil?

(b) An iron-cored coil of 2000 turns produces a flux of 0·008 Wb. When the switch in the circuit is opened, the flux falls to 0·001 Wb in 0·02 sec. What is the average e.m.f. induced?

((a) 10,000 V, (b) 700 V.) (U.L.C.I.)

13.4. A straight conductor, 100 cm long, is situated at right angles to a uniform magnetic field of density 0·6 Wb/m². Calculate the force on the conductor when carrying a current of 70 A.

(42 N.)

13.5. A uniform air-gap has a cross-sectional area of 5 cm². The total flux across the gap is 0·3 mWb. Calculate the flux density in the gap.

(0·6 Wb/m².)

13.6. To set up a certain magnetic flux in an air-gap of length 2 mm, a magnetising force of 200,000 AT/m is required. Calculate (a) the m.m.f. required for the air-gap, (b) the magnetic flux in the gap if its cross-sectional area is 40 cm².

((a) 400 AT, (b) $10{\cdot}04 \times 10^{-4}$ Wb.)

13.7. A circular ring has a radial saw-cut 6 mm wide. The cross-sectional area of the ring is 4 cm² and the mean length of the iron path is 75 cm. If the ring is uniformly wound with 300 turns of wire, calculate the current required to set up a flux of 600 μWb in the air-gap. The magnetisation curve for the material is:

B (Wb/m²)	0·63	1·22	1·47	1·54	1·58	1·61
H (AT/m)	400	800	1200	1600	2000	2400

(27·3 A.)

Answers are not given to the following questions:

13.8. An electromagnet is wound with a coil having 1500 turns. When the current is switched off, the magnetic flux falls to zero in 0·5 sec. If an e.m.f. of 100 V is induced in the coil due to the flux change, determine the flux in the magnet when the current is in the coil.

13.9. A straight conductor, 60 cm long, carries a current of 40 A in a direction at right angles to a uniform magnetic field of 0·8 Wb/m². Calculate (a) the force on the conductor, (b) the power in watts required to move the conductor against the above force at 15 m/sec, (c) the average e.m.f. induced in the conductor by this movement.

13.10. A round ferromagnetic bar of length 80 cm and cross-sectional area 0·8 cm^2 is formed into a closed ring and wound uniformly with a 500-turn coil of insulated wire. A current of 2 A is passed through the coil. Calculate the flux in milliwebers (mWb) in the iron. Under these conditions, the relative permeability of the iron may be taken as 800.
(Permeability of free space $= 4\pi \times 10^{-7}$.) (U.L.C.I.)

13.11. In a loud-speaker, the moving coil consists of 200 turns of wire wound on a circular former, so that the coil has an effective diameter of 3 cm. The coil is situated in the radial air-gap of the speaker magnet so that the whole of its length is perpendicular to the field, which has a flux density of 1·6 Wb/m^2. Calculate the force on the coil when it is carrying a current of 100 mA.

13.12. A coil wound with 1000 turns has a diameter of 50 cm and a length of 75 cm. It is revolving in a magnetic field having a uniform flux density of 0·75 Wb/m^2 at 1800 rev/min. Determine the e.m.f. induced in the coil when the plane of the coil is at 60° to the direction of the field.

13.13. A coil wound with 2000 turns has an inductance of 0·3 mH. Determine the flux set up in the coil by a current of 5 A.

13.14. A six-pole generator has a flux of 15 mWb per pole. If the armature speed is 900 rev/min, calculate the average value of the e.m.f. per conductor. (N.C.T.E.C.)

CHAPTER 14

Torque Production

14.1. General

It has been shown in Section 13.14 how a force is produced on a conductor carrying a current in a magnetic field. This force on the current would be present even if there was not a conductor carrying it. For example, in the case of an electric arc such as may occur if a switch is opened when carrying a direct current in certain circuits, a conveniently placed magnetic field will deflect the arc in such a direction as to increase its length and so help to extinguish the arc.

The operation of most instruments depends on the interaction of two magnetic fields. When an instrument pointer moves it is the result of electrical energy being converted into mechanical energy. This energy conversion occurs in all types of electrical machines. The following examples illustrate various ways in which a torque can be developed.

Suppose that the magnet N–S (Fig. 14.1) is fixed and the one

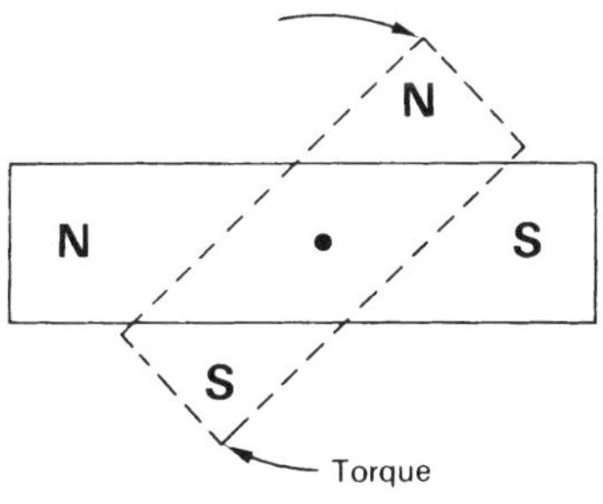

FIG. 14.1

N–S, drawn with dashed lines, is free to rotate in a horizontal plane. Considered separately, there are two magnetic fields but these fields must combine to form a resultant field. There is a tendency for the two fields to line up so that the flux distortion is at a minimum and for the magnetic lines of force to shorten themselves as much as possible. In this case, the magnet that is free to rotate will move to a stable position where the two magnets are in line. There is in this case a clockwise torque on the moveable magnet.

In Fig. 14.2(a), a cylindrical piece of initially unmagnetised soft iron is free to rotate between the poles of two magnets. As there is

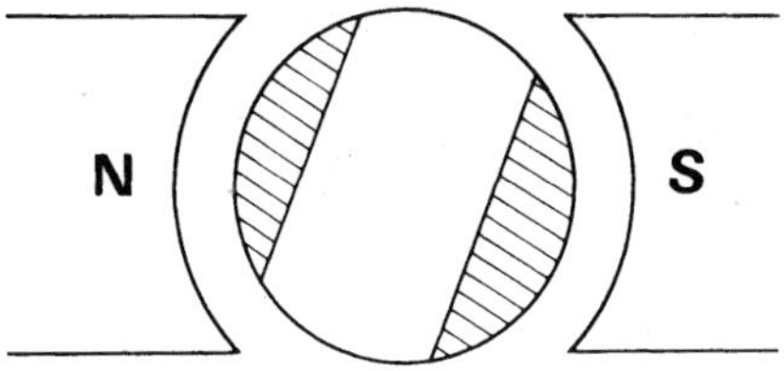

(a)

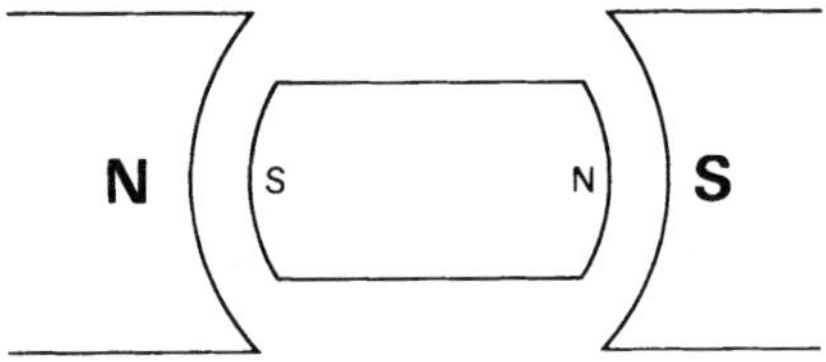

(b)

FIG. 14.2

complete symmetry, no torque is set up. If the shaded sections were removed, a torque would be set up in a clockwise direction so that the piece of soft iron moves to the position shown in Fig. 14.2(b), where the length of the flux path is as short as possible. In this position, the reluctance of the magnetic circuit is at its lowest value.

Sometimes the torque produced in this manner is referred to as a reluctance torque and it is utilised in some machines. It should be noted that poles are induced in the piece of soft iron, the polarity being as shown in Fig. 14.2(b).

A force of repulsion will also exist if two pieces of a magnetic material are magnetised with like polarities (Fig. 14.3).

If one piece is fixed and the other is so arranged that it is free to move, then if a current is passed through a coil surrounding both pieces of material, they will be magnetised with like polarities and so a force of repulsion exists. This principle is used in many moving iron instruments.

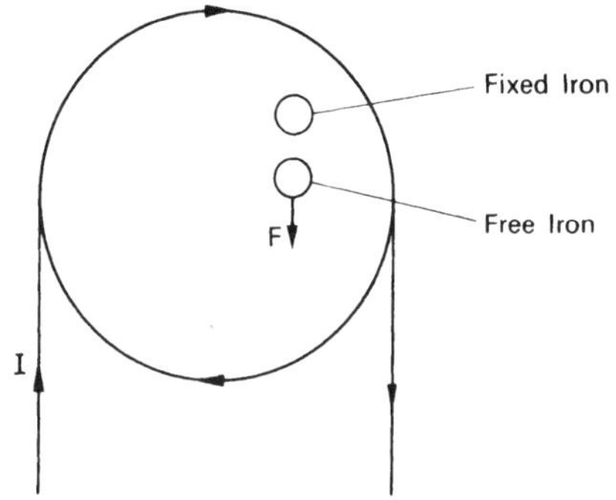

FIG. 14.3

A force of attraction between a fixed coil carrying current and a piece of initially unmagnetised material will exist if they are adjacent to each other as shown in Fig. 14.4. There is a force of attraction

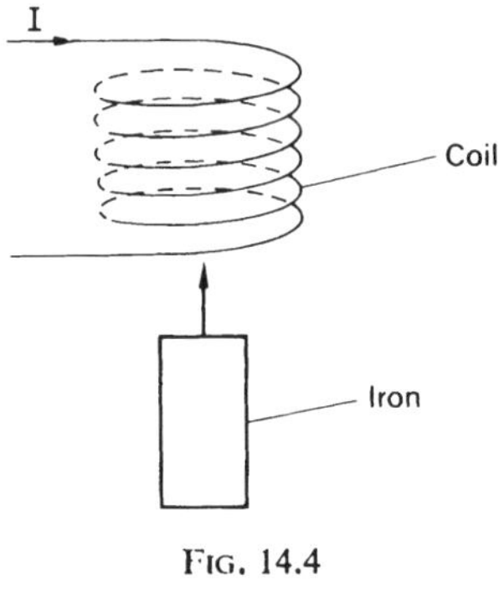

FIG. 14.4

in the direction shown and if the material were completely free to move, it would come to rest in the centre of the coil. Relays and some moving-iron instruments use this principle.

Some devices use fixed coils carrying one current to set up a magnetic flux Φ_1 and a moving coil carries a second current setting up a flux Φ_2 (Fig. 14.5). These two fluxes try to align themselves

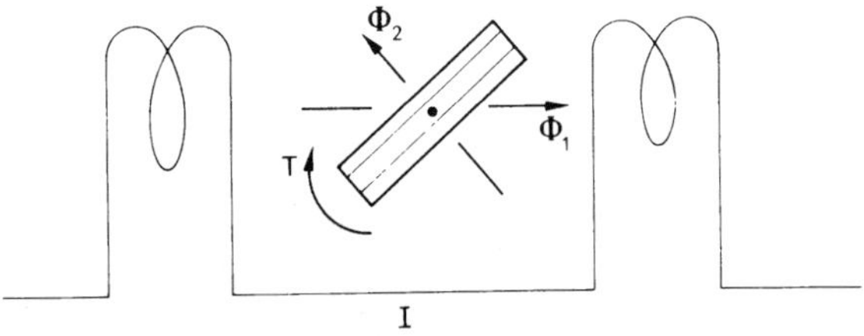

FIG. 14.5

and so a clockwise torque is produced as shown. Usually the rotation of the moving coil is restricted by a hair-spring. This principle of torque production is used in dynamometer type instruments.

If the fixed coils are replaced by permanent magnets, the principle of torque production is unchanged. This arrangement is utilised in the moving-coil instrument.

CHAPTER 15

Indicating Instruments

15.1. Moving-coil instruments

These instruments respond to the average value of the operating current and are thus normally used for indicating the values of direct currents and voltages. An instrument of this type cannot be used directly for measuring alternating quantities but can be so used if the alternating supply is first rectified. Some instruments have their zero at the centre of the scale and so readings can be taken for currents flowing in either direction through the instrument. They are referred to as centre-zero instruments.

Operating torque

A coil having a pointer attached rotates in the field of a permanent magnet. If a current-carrying conductor is lying in a magnetic field, a force is produced and if the direction of magnetic field is fixed and the conductor free to move, then the conductor will move in the magnetic field. The magnet assembly is so designed that the conductors always move in a uniform magnetic field and at right angles to it.

It is desirable that the strength of the magnetic field in which the coil is lying shall not change with time, otherwise the instrument will become inaccurate. In many instruments, horseshoe-shaped magnets of tungsten steel have been used. Their length of iron between poles is long compared with their cross-sectional area whereas when a material such as alnico is used, the magnet will be short and have a relatively large cross-sectional area.

More recently, anisotropic magnets have been used, these being made of an alloy of iron, nickel, cobalt, and aluminium. Whilst the magnet is being cast, the molten metal is subjected to a strong magnetic field and as it solidifies, a magnet is formed which possesses great magnetic strength in one direction only and this will be maintained over a very long period.

A sintered form of construction is used in some instruments. The powdered magnet alloy, together with some soft iron powder for the pole shoes, is preformed and compressed in a die. Under the influence of heat and pressure, the magnet and pole shoes become alloyed together.

Figure 15.1(a) shows the form of the magnetic field where the magnet system is of the coaxial pattern. The sintered radially magnetised cylindrical magnet is mounted inside the moving coil system and the magnetic circuit is completed by an annular outer yoke of soft iron, which offers a path of low reluctance for the flux (i.e. the flux is easily established in the yoke). The yoke also limits the amount of leakage flux (that is, flux not usefully employed) which may affect nearby apparatus and it also acts as a screen against external magnetic fields which may affect the instrument.

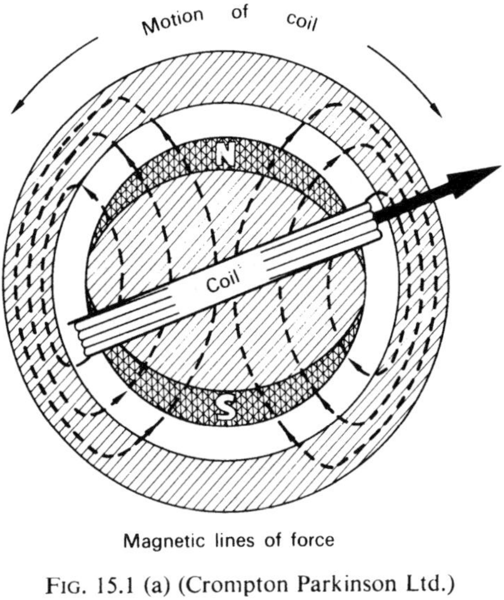

FIG. 15.1 (a) (Crompton Parkinson Ltd.)

The strength of the magnetic field set up by the magnets can vary from 0·14 to 2·2 Wb/m^2 depending on the type of instrument.

Figure 15.1(b) shows a moving-coil movement assembly. The coil will rotate when carrying a current such that its magnetic field tries to line up with the field of the permanent magnet.

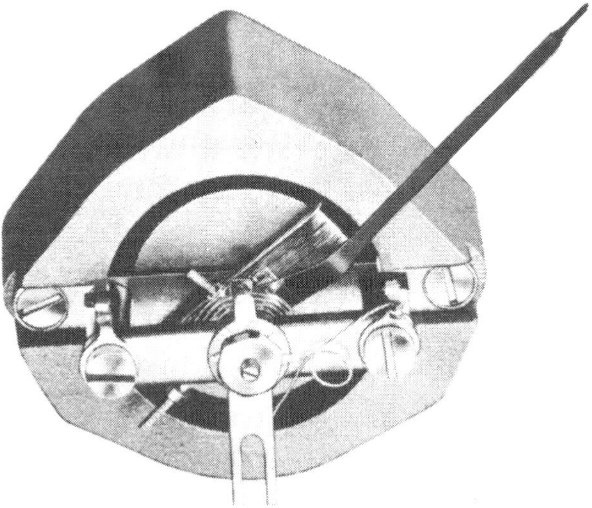

FIG. 15.1 (b) (Crompton Parkinson Ltd.)

The moving coil is wound on an aluminium former, this being very light and non-magnetic. The coil consists of a number of turns of fine enamelled-copper wire.

The current carried by the moving coil is, in most instruments, led into and out of it by two phosphor-bronze hair-springs. As currents above about 20 mA would cause a significant temperature rise in the springs and so alter their stiffness, larger currents are diverted from the coil by a shunt. These are dealt with in Section 15.2. Many instruments have a moving coil which gives a full-scale deflection (F.S.D.) when carrying a current of 15 mA. About

0·6 ampere-turns in the moving coil are required in these instruments. The force on a single conductor is given by

$$F = BlI \text{ newtons,}$$

where l is the effective length in metres of one conductor. The effective length is that length of conductor lying in the magnetic field.

Hence torque exerted by one conductor

$$T = Fr = BlIr \text{ newton-metres,}$$

where r is the radius of the coil in metres.

If the coil has N turns, there are $2N$ conductors and so the instrument operating torque

$$T_0 = 2BlIrN \text{ newton-metres.}$$

For a given instrument, only the current can be varied.

$$\therefore\ T_0 = kI, \text{ where the constant } k = 2BlrN.$$

EXAMPLE. A moving-coil instrument has a coil of effective length 2 cm and diameter 1·5 cm. The coil consists of 120 turns of wire moving in a uniform magnetic field of flux density 0·4 Wb/m^2. Calculate the deflecting (or operating) torque.

$$\begin{aligned} \text{Force on conductors} &= BlI \text{ newton} \\ &= 0{\cdot}4 \times (0{\cdot}02 \times 2 \times 120) \times 0{\cdot}015 \text{ N} \\ &= 0{\cdot}0288 \text{ N.} \end{aligned}$$

$$\begin{aligned} \text{Deflecting torque} &= 0{\cdot}0288 \times \frac{0{\cdot}015}{2} \\ &= \mathbf{0{\cdot}000216\ Nm.} \end{aligned}$$

CONTROLLING TORQUE

This is the torque which is equal and opposite to the operating torque whenever the instrument is indicating a steady value. In instruments using hair-springs to lead the current into and out of the moving coil, the springs also provide the controlling torque.

Thus if an instrument requires a torque of 0·04 g-cm for F.S.D., the springs will also be exerting the same value of torque. For a spring, the controlling torque T_c is proportional to the angle of twist (θ) of the spring:

$$T_c = k_1\theta.$$

It has already been shown that the operating torque,

$$T_0 = kI.$$

If $T_c = T_0$, then $k_1\theta = kI$.

Thus the angle of deflection indicated by the pointer is proportional to the current flowing and so the instrument has a uniform scale.

Damping torque

It is desirable when an indicating instrument shows a change of reading that the pointer comes quickly to rest without undue oscillation. Some method is necessary to damp out the oscillations.

When the pointer of a moving-coil instrument is moving, the aluminium former carrying the coil is moving in a magnetic field and so an e.m.f. is induced in it. The e.m.f. causes a current to flow in the former which is in effect a single-turn coil and this current, known as an eddy current, produces a torque which obeys Lenz's law and so is in opposition to the torque causing the coil to move. This torque, which is only present when the coil is moving, results in the coil, and hence the pointer, coming quickly to rest. A moving-coil instrument has a damping torque caused by eddy currents.

Recent developments

The moving coil in instruments having control springs usually turns on pivots in a spring-mounted jewel bearing. Latest developments include pivotless instruments, the movement being suspended by ribbons of heat treated beryllium copper. These also provide the controlling torque. Pivots and jewel bearings of any instrument introduce frictional errors and by this construction, these errors, and also the hair-springs, are eliminated.

15.2. The moving-coil instrument as ammeter or voltmeter

A given moving-coil movement can be used in an instrument designed to indicate either current or voltage. When the movement is used in an instrument calibrated to read current, most of the current flows through a resistor in parallel with the movement. This resistor, known as a shunt, is usually made of manganin. One reason for the use of manganin is that its resistance is very little affected by changes in temperature. The shunt is mounted inside the case of the instrument for ammeters having a current range up to perhaps 100 A, but with instruments calibrated to read very large currents, the shunt is mounted externally.

A shunt is manufactured with a high degree of accuracy so that its resistance is exactly that required for use with a given movement. If the shunt is external to the instrument, then the leads are also specially made in conjunction with the shunt and under no circumstances must the leads be changed.

If the movement is to be used so that the instrument indicates voltages, then an accurately made resistance must be connected in series with the movement in order to limit the current to that which will give full-scale deflection of the pointer when the maximum voltage to be measured is applied across the series combination. This, too, must be made from a material whose resistance is virtually independent of temperature.

EXAMPLE. The movement of a moving-coil instrument whose resistance is 5 Ω indicates full-scale deflection when a current of 15 mA flows through it. It is connected to a shunt by two leads each of resistance 0·05 Ω. Calculate the resistance of a shunt which will enable the movement to indicate full-scale deflection when a current of 20 A is taken from a supply.

What would the instrument read if the length of each lead were reduced to half the original length and the supply current reduced to 10 A?

Figure 15.2(a) shows the equivalent circuit of the instrument, its leads and the shunt.

For full-scale deflection $I_c = 15$ mA or 0·015 A.

$$\text{Voltage across the shunt } V_{AB} = 0{\cdot}015\,(5+2\times0{\cdot}05)\text{ V}$$
$$= 0{\cdot}0765\text{ V.}$$

$$\text{Current through the shunt } I_s = I - I_c$$
$$= 20 - 0{\cdot}015\text{ A}$$
$$= 19{\cdot}985\text{ A.}$$

$$\text{Resistance of shunt } R = \frac{V}{I_s}$$
$$= \frac{0{\cdot}0765}{19{\cdot}985}\ \Omega$$
$$= \mathbf{0{\cdot}00383\ \Omega.}$$

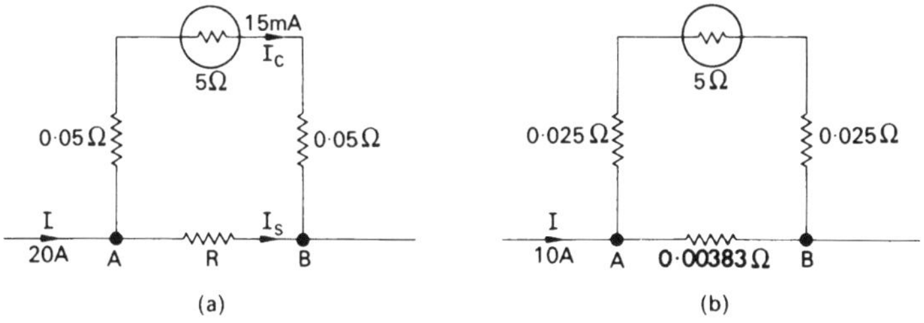

FIG. 15.2

If the instrument leads were cut to half their length, each would have a resistance of 0·025 Ω. Figure 15.2(b) shows the new condition.

$$\text{Total resistance} = \frac{5{\cdot}05\times0{\cdot}00383}{5{\cdot}05+0{\cdot}00383}\ \Omega.$$

$$\text{Voltage across shunt } V_{AB} = 10\times\frac{5{\cdot}05\times0{\cdot}00383}{5{\cdot}05383}\text{ V.}$$

$$\text{Current through the instrument} = \frac{V_{AB}}{5{\cdot}05}$$

$$= \frac{10 \times 0{\cdot}00383}{5{\cdot}0538}\ \text{A}$$

$$= \frac{0{\cdot}0383}{5{\cdot}0538}\ \text{A}$$

$$= \frac{38{\cdot}3}{5{\cdot}0538}\ \text{mA}$$

$$= 7{\cdot}578\ \text{mA.}$$

If the instrument originally indicated 20 A when 15 mA were flowing through the moving coil, then when 7·578 mA were passing through it, it would read

$$20 \times \frac{7{\cdot}578}{15} = \mathbf{10{\cdot}10\ A.}$$

EXAMPLE. The moving-coil assembly of the previous example is to be incorporated in an instrument calibrated to read 0–150 V. What value of series resistor must be used with the instrument?

As a current of 15 mA is required for F.S.D., the total circuit resistance is given by:

$$\frac{150 \times 10^3}{15}\ \Omega$$

$$= 10{,}000\ \Omega,$$

$$\therefore \quad R + 5 = 10{,}000\ \Omega,$$

$$R = \mathbf{9995\ \Omega.}$$

If a bad connection effectively increased the value of the series resistor by 5%, what would the instrument indicate when connected to a 150-V supply?

$$\text{Total circuit resistance} = 5 + \left(9995 \times \frac{105}{100}\right) \Omega$$
$$= (5 + 10{,}495) \ \Omega$$
$$= 10{,}500 \ \Omega.$$
$$\text{Current} = \frac{150 \times 10^3}{10{,}500} \text{ mA}$$
$$= 14 \cdot 285 \text{ mA}.$$
$$\text{Instrument reading} = 150 \times \frac{14 \cdot 285}{15} \text{ V}$$
$$= \mathbf{142 \cdot 85 \ V.}$$

15.3. Moving-iron instruments

These are by far the commonest type of switchboard instrument. They are used for measuring the effective (or r.m.s.) value of alternating currents and voltages (Section 19.10) as well as direct currents and voltages.

Operating torque

A moving vane fixed to the pivot of the instrument which carries the pointer is situated inside a circular electromagnet. Inside this coil is a fixed piece of iron shaped so as to obtain more uniform divisions on the scale of the instrument.

When current is passed through the coil, a magnetic field is set up and the fixed and moving irons are similarly magnetised so that the one which is free to move is repelled.

Depending on the size and angular movement of the pointer, the product of the current through the coil and the number of turns on the coil (the m.m.f.) varies from perhaps 100 to 250 AT for full-scale deflection. For example, an instrument may have 4000 turns and be designed to give full-scale deflection with a current of 50 mA, i.e. to have 200 AT. Likewise one manufactured to carry 5 A may have 40 turns.

With an m.m.f. of this magnitude, only a very weak magnetic

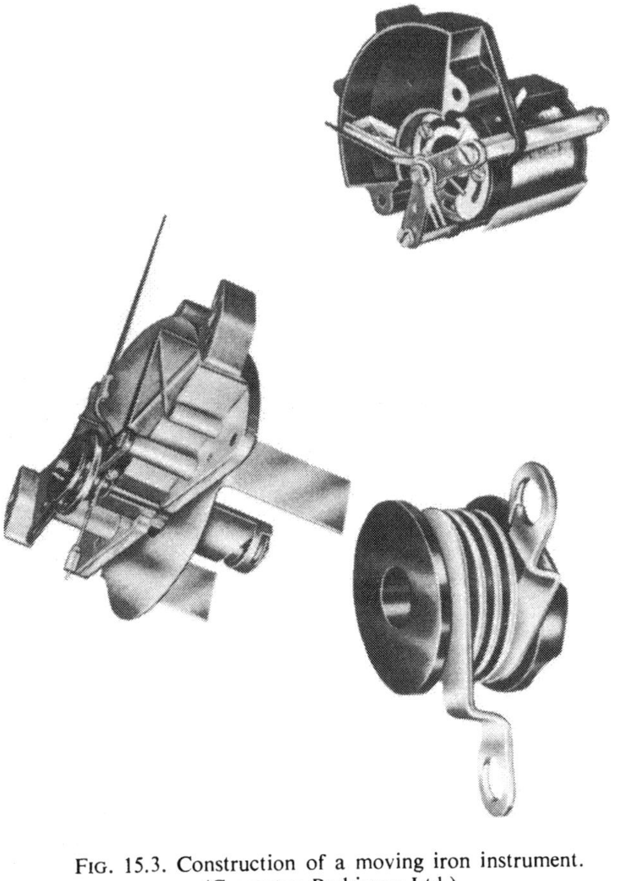

FIG. 15.3. Construction of a moving iron instrument.
(Crompton Parkinson Ltd.)

field is set up. A typical value may be 0·007 Wb/m^2. Unless such an instrument were adequately shielded against the effects of stray external magnetic fields, considerable errors in the readings obtained would almost certainly result. Nickel–iron screens around the field coil provide this shielding and in addition provide an easy (low reluctance) path for the flux set up by the core. This reduces the m.m.f. necessary to obtain full-scale deflection.

Moving-iron instruments are calibrated at the frequency at which they will be operating. If used at a different frequency errors will occur in the readings due to such factors as additional eddy currents, inductance of the coil, etc. Nevertheless, an instrument calibrated at 50 Hz will be sufficiently accurate for most uses over the range 25–100 Hz and for d.c. operation.

Controlling torque

In older type instruments, gravity control was frequently used. Present-day instruments use spring control as in moving-coil instruments but it should be noted that the spring has no other function. Pivotless instruments use the same type of controlling torque as moving-coil instruments.

Damping torque

In older type instruments, damping was obtained by a light vane attached to the spindle moving in a confined space (Fig. 15.3). This pneumatic damping is still used in certain instruments. Frequently eddy current damping is employed, a disc attached to the spindle rotating between the poles of a small permanent magnet. The principle involved is the same as was described for moving-coil instruments.

Scale shape

The operating torque developed is a function of the square of the operating current and is also dependent on the shape of the moving and fixed irons. Hence the scale is not linear. In general it tends to be cramped at the zero end and almost uniformly distributed in the centre of the scale and somewhat closed up at the top end of the scale.

15.4. Extension of current and voltage ranges for a.c. operation

Ammeters in power stations may have a scale of 0–500 A. The instrument actually will be carrying a maximum current of 5 A,

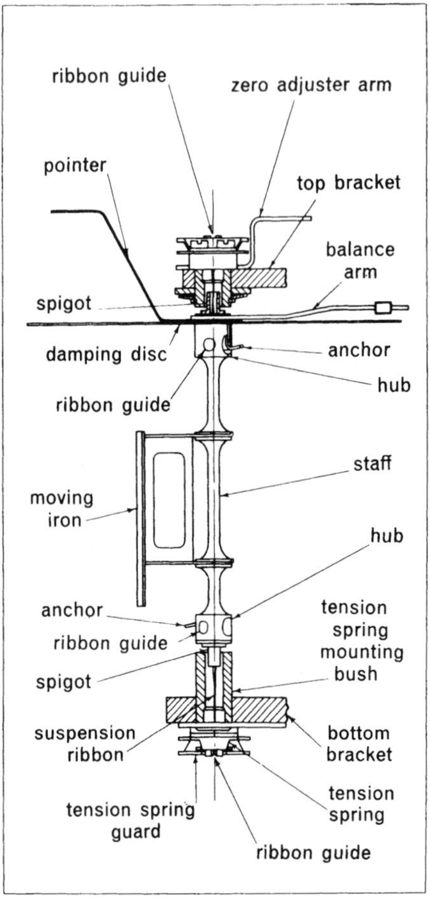

FIG. 15.4. Construction of the pivotless moving iron instrument. (Crompton Parkinson Ltd.)

the 500 A have been reduced in magnitude by a current transformer.

Similarly a voltmeter which is indicating 132,000 V may actually be operating at a voltage of 110 V, the voltage having been reduced to this value by a voltage (or potential) transformer.

These instruments are manufactured to a high degree of accuracy and when used in laboratories increase the range of values that can be measured with relatively few instruments.

15.5. Effect of frequency on the operation of moving-coil and moving-iron instruments

Suppose a centre-zero moving-coil and a moving-iron instrument were connected in series to a circuit in which a sinusoidally varying alternating current had a maximum value of 5 A (Fig. 15.5). Also assume that the frequency of this supply was 1 hertz (1 Hz) (see Section 19.4).

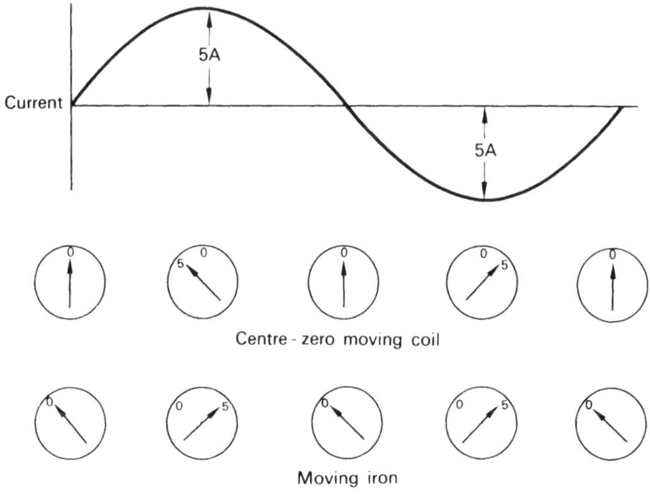

FIG. 15.5

During a period of 1 second, the pointer in each instrument would indicate as shown. If the frequency of the supply were gradually increased, it would be seen that the pointer of the moving coil instrument would start to oscillate about the zero and at frequencies above about 20 Hz, would remain at zero. The moving-iron instrument would soon take up a steady reading between zero and 5 A. Actually the value would be $5/\sqrt{2}=3{\cdot}53$ A. This value indicated by the moving-iron instrument is called the effective or r.m.s. value (Section 19.10).

15.6. Choice of voltmeter

As seen previously, the external resistance of a voltmeter to enable it to operate over a given voltage range depends on the characteristics of the instrument movement. For example, a 150-V voltmeter may have a total resistance of 150,000 Ω and it could be said that this instrument has a resistance of 1000 Ω/V. Another instrument may have a resistance of 100 Ω/V.

When a voltmeter is used to measure the voltage across part of a circuit, the accuracy of the reading depends on the resistance of the instrument relative to the resistances in the circuit. This may be illustrated by the following calculations.

EXAMPLE. Two 20-Ω resistors are connected in series to a 40-V supply. What is the voltage across each resistor? If a 20-V voltmeter having a resistance of 100 Ω/V were connected across one of the resistors, what would be the reading on the instrument?

Before the voltmeter was connected, the voltage across each resistor was 20 V as the resistances are equal.

The voltmeter has a resistance of

$$20\times100=2000\ \Omega$$

and so the circuit now consists of a 20-Ω resistor in series with a parallel circuit having resistors of 20 Ω and 2000 Ω.

$$\begin{aligned}\text{Total circuit resistance} &= 20+\frac{20\times2000}{20+2000}\ \text{ohms}\\ &=(20+19{\cdot}8)\ \Omega\\ &=39{\cdot}8\ \Omega.\end{aligned}$$

$$\text{Current flowing} = \frac{40}{39{\cdot}8} \text{ amperes}$$

$$= 1{\cdot}005 \text{ A.}$$

$$\text{Voltmeter reading} = 1{\cdot}005 \times 19{\cdot}8 \text{ V}$$

$$= \mathbf{19{\cdot}899 \text{ V.}}$$

Thus in this case, the voltmeter reading indicates the voltage as accurately as the instrument can be read.

EXAMPLE. Two equal resistors, each having a resistance of 10,000 Ω, are connected in series across a 200-V supply. What is the voltage across each resistor?

If a voltmeter is connected across one of the resistors, what will be its reading if the instrument has a resistance of (a) 10,000 Ω, (b) 100,000 Ω?

The voltage across each of the resistors without the voltmeter is 100 V as the resistances are equal.

(a) With the voltmeter connected

$$\text{Total circuit resistance} = 10{,}000 + \frac{10{,}000 \times 10{,}000}{10{,}000 + 10{,}000} \ \Omega$$

$$= 15{,}000 \ \Omega.$$

$$\text{Current from the supply} = \frac{200}{15{,}000} \text{ A}$$

$$= 0{\cdot}0133 \text{ A.}$$

$$\text{Voltmeter reading} = 0{\cdot}0133 \times 5000 \text{ V}$$

$$= \mathbf{66{\cdot}5 \text{ V.}}$$

(b) $$\text{Total circuit resistance} = 10{,}000 + \frac{10{,}000 \times 100{,}000}{10{,}000 + 100{,}000} \ \Omega$$

$$= (10{,}000 + 9090) \ \Omega$$

$$= 19{,}090 \ \Omega.$$

$$\text{Current from the supply} = \frac{200}{19{,}090} \text{ A}$$

$$= 0{\cdot}0105 \text{ A.}$$

$$\text{Voltmeter reading} = 0{\cdot}0105 \times 9090 \text{ volts}$$
$$= \mathbf{95{\cdot}4\ V.}$$

Note that using the voltmeter whose resistance is considerably greater than that of the circuit, the voltage indicated is nearer the value that existed before the voltmeter was connected. The voltmeter used must have a very much greater resistance than the circuit across which it is connected if accurate readings are to be obtained. This is important when measuring voltages in, for example, radio circuits.

15.7. The Wheatstone bridge

The Wheatstone bridge circuit is used in various forms for the accurate measurement of resistance from a fraction of an ohm to several megohms. The basic circuit is shown in Fig. 15.6.

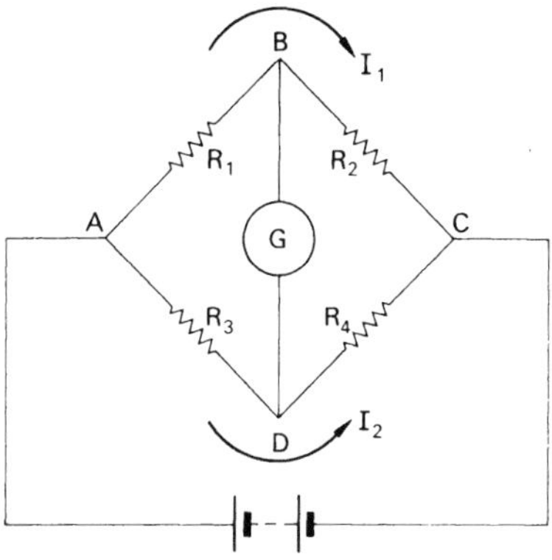

FIG. 15.6. The Wheatstone bridge.

The instrument connected to terminals *B* and *D* is a galvanometer which is a sensitive current-operated moving-coil instrument. Suppose that the values of the four resistors are such that no current passes through the galvanometer. Under this condition, the bridge is said to be balanced and the terminals *B* and *D* are at the same potential. Two currents will be flowing in the circuit: I_1 through R_1 and R_2 and I_2 through R_3 and R_4. As B and D are at the same potential, the voltage drop from A to B is the same as from A to D. Likewise the voltage drop from B to C is equal to that from D to C.

Hence $$V_{AB}=V_{AD}$$

and $$V_{BC}=V_{DC}.$$

By Ohm's law,

$$I_1R_1=I_2R_3$$

and $$I_1R_2=I_2R_4,$$

$$\therefore \quad \frac{I_1R_1}{I_1R_2}=\frac{I_2R_3}{I_2R_4}.$$

Hence $$\frac{R_1}{R_2}=\frac{R_3}{R_4}.$$

If under balance conditions any three of the resistors are of known value, the fourth may be found from the above formula. In commercial instruments, it is convenient to make the values of R_1 and R_2 to be variable in multiples of 10 Ω and for R_3 to be variable from 1 Ω to 10,000 Ω or more. The resistor R_4 then becomes that whose value is being determined.

Whilst the bridge is being balanced, the values of the resistors are varied in order to obtain zero reading on the galvanometer, and it is advisable to connect in parallel with the galvanometer a resistor of lower value than the galvanometer resistance (i.e. a shunt), so that during the initial adjustments some current is diverted from the instrument through the shunt. This reduces the sensitivity of the instrument during this period but protects it against damage caused by excessive current.

Methods of measurement which depend on obtaining circuit conditions in which zero p.d. exists between terminals of a bridge network to which a detector (in this case, the galvanometer) is connected, are referred to as null methods. An advantage of bridge methods is that variation in the magnitude of the bridge supply voltage does not affect the condition for balance.

15.8. A commercial Wheatstone bridge

The Post Office Box is a unit containing resistors corresponding to R_1, R_2, and R_3 and these were connected to a battery, a galvanometer and the resistance, to be measured, R_4. The appropriate value of resistance was obtained by removing a metal plug from a terminal strip. This device is still occasionally seen in some college laboratories.

The modern direct reading Wheatstone bridge may have an arrangement as shown in Fig. 15.7. The resistors R_1 and R_2, known as the ratio-arm resistors, are obtained by making connections to the appropriate terminals. The resistor R_3 is shown as being variable and is selected by use of the rotary switches. More expensive models will have the galvanometer included in the panel and a switch which has to be depressed in order to obtain maximum sensitivity of the instrument.

15.9. Effect of ratio arms on resistance measurement

Suppose that a resistor having a resistance of 5·82 Ω was being measured by a Wheatstone bridge circuit and the ratio arm resistors had values of $R_1 = 10\ \Omega$ and $R_2 = 10\ \Omega$. The nearest condition to balance would be obtained when $R_3 = 6\ \Omega$. Hence

$$\frac{10}{10} = \frac{6}{R_4}.$$

This gives an approximate value of $R_4 = 6\ \Omega$.

If now the ratio arms have values of $R_1 = 100\ \Omega$ and $R_2 = 10\ \Omega$, the nearest condition to balance would be obtained when $R_4 = 58\ \Omega$. Hence

$$\frac{100}{10} = \frac{58}{R_4}.$$

This gives a more accurate value of $R_4 = 5{\cdot}8\ \Omega$.

FIG. 15.7. A commercial Wheatstone bridge.
(Croydon Precision Instrument Co.)

When the ratio arms have values of $R_1 = 1000\ \Omega$ and $R_2 = 10\ \Omega$, exact balance would be obtained when $R_4 = 582\ \Omega$. Hence

$$\frac{1000}{10} = \frac{582}{R_4}.$$

This gives the accurate value of 5·82 Ω.

This illustrates the effect of the value of the ratio arm resistors on the accuracy with which measurements can be made. By making R_2 greater than R_1, resistances less than 1 Ω can be measured.

Problems

15.1. The resistance of the movement of a moving-coil instrument is 2 Ω and a current of 2 mA causes a deflection of one scale division. If the scale has 100 divisions, what value of resistance must be connected across the instrument in order that its full-scale reading corresponds to a supply current of 5 A?

(0·083 Ω.)

15.2. What value of series resistance must be used with the instrument movement described in problem 15.1. in order that it may have full-scale deflection as a voltmeter when connected to a 2-V supply?

(8 Ω.)

15.3. A moving-coil instrument has a resistance of 5 Ω and a full-scale deflection is given by a current of 15 mA. Determine the value of resistance required to enable this instrument to operate as (a) a 30-V voltmeter, (b) a 1·5-A ammeter.

((a) 1995 Ω, (b) 0·0505 Ω.)

15.4. A moving-coil instrument has a coil of resistance 5 Ω and gives full-scale deflection when a current of 15 mA is flowing. If a shunt of 0·005005 Ω is used to convert this instrument to an ammeter, what will be the current from the supply when full-scale deflection occurs?

(15 A.)

CHAPTER 16

Kirchhoff's Laws

16.1. General

There are two very important laws:

1. The sum of the currents flowing at any instant towards a junction in an electrical network is equal to the sum of the currents flowing away from it.
2. The algebraical sum of the e.m.f.s and p.d.s around any closed electrical circuit is zero.

When applying the second law a sign convention is necessary to decide which e.m.f.s or p.d.s are algebraically positive and which are algebraically negative. The one used in this book is given below.

16.2. Conventions

V_{AB} means the voltage or potential of A with respect to B and is positive if A is at a positive or higher potential than B.

Thus if $V_{AB} = +6$ V, then A is at a positive potential with respect to B by 6 V. It can be said that B is at a lower potential than A by 6 volts or

$$V_{BA} = -6 \text{ V}.$$

If

$$V_{AB} = +6 \text{ V},$$

then

$$-1 \times V_{AB} = -6 = V_{BA},$$

$$\therefore \quad -V_{AB} = V_{BA}.$$

It now remains to decide when a point is at a positive or negative potential with respect to another point.

In a resistor (or any form of electrical load in which there is no e.m.f. present) current conventionally flows from a high potential to a lower potential. Thus in Fig. 16.1 as current is flowing from A to B, A is at a higher potential than B by the amount IR volts.

Thus $V_{AB} = +IR$ or $V_{BA} = -IR$.

V_{AB} is also the voltage across the cell and is the e.m.f. of the cell ($=E$ volts) if its internal resistance is neglected.

$$V_{AB} = +E \text{ volts.}$$

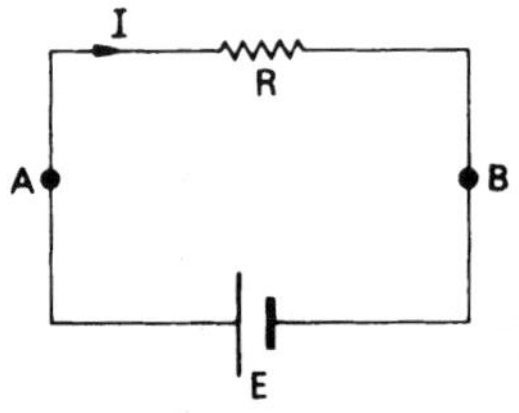

FIG. 16.1

Note that as far as a source of e.m.f. is concerned, if the first letter of subscript of V_{AB} (i.e. A) is the positive of the source of e.m.f., then V_{AB} is positive. If the first letter of the subscript had been the negative of the source of e.m.f., then the V_{AB} would have been negative. With currents, I_{AB} is understood to mean the current flowing from A to B. Figure 16.2 represents part of an electrical

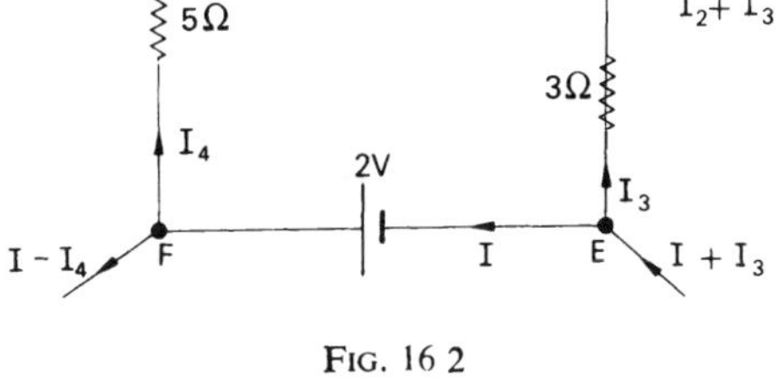

FIG. 16 2

circuit and path *ABCDEFA* is a closed circuit and so the algebraical sum of the e.m.f.s and p.d.s is zero.

PATH *ABCDEFA*

Taking the p.d.s and e.m.f.s in the letter order in which they appear in the path, their algebraical sum must be zero.

$$V_{AB}+V_{BC}+V_{CD}+V_{DE}+V_{EF}+V_{FA}=0.$$

V_{AB}—A is the **positive** of the battery $= +6$ volts,
V_{BC}—current flows **from** B **to** C $= +1I_1$ volts,
V_{CD}—C is the **positive** of the battery $= +4$ volts.

(N.B. The current I_2 flowing through a cell does not change its e.m.f., it only affects the terminal p.d. because the cell has internal resistance. As no value of internal resistance is given, it must be assumed to be zero.)

V_{DE}—current flows **towards** D $= -3I_3$ volts,
V_{EF}—E is the **negative** of the battery $= -2$ volts,
V_{FA}—current flows **from** F $= +5I_4$ volts.

Hence for path *ABCDEFA*,

$$+6+1I_1+4-3I_3-2+5I_4=0.$$

EXAMPLE. Figure 16.3 shows two batteries connected in a circuit with resistors of known value. A voltmeter connected across each battery in turn indicates e.m.f.s of 13 V and 6·4 V respectively.

The internal resistance of each accumulator is negligible and the reading on each ammeter is to be calculated.

The following steps should be taken to solve this problem.

1. Insert a letter (*A*, *B*, etc.) where a change of circuit component occurs. This has been done on Fig. 16.3.
2. Insert the least number possible of unknown currents (I_1, I_2, etc.) in the circuit until every component of the circuit has a current flowing through it. Indicate with an arrow the direction in which the current is assumed to be flowing. (If the assumed direction of any current is incorrect, the calculation will indicate this by having a negative value for that current.)

At every junction, the conditions of Kirchhoff's first law must be fulfilled. In this circuit only two unknown currents I_1 and I_2 are necessary.

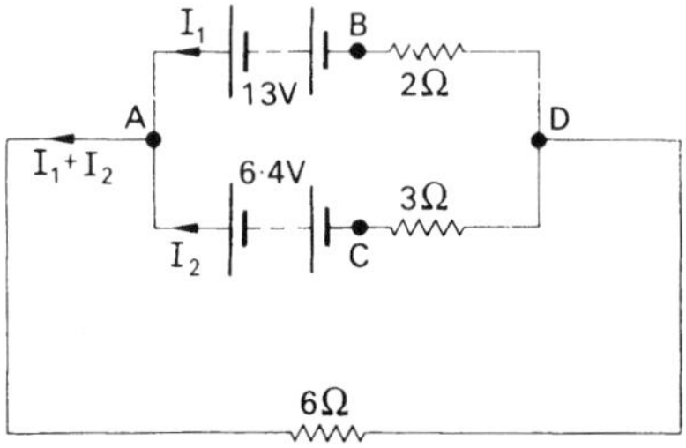

FIG. 16.3

As there are two unknown currents, two equations are necessary in order to solve the problem. It is possible to obtain three equations by using paths *ABDA*, *ACDA*, and *ABDCA* but any two of these will suffice. As the first two paths are shortest and so will probably involve fewer terms mathematically, these will be used. However, care must be taken to ensure that every part of the circuit is included at least once in the equations utilised.

PATH *ABDA*

$$V_{AB}+V_{BD}+V_{DA}=0,$$
$$+13-2I_1-6(I_1+I_2)=0,$$
$$8I_1+6I_2=13. \quad (1)$$

PATH *ACDA*

$$V_{AC}+V_{CD}+V_{DA}=0,$$
$$+6{\cdot}4-3I_2-6(I_1+I_2)=0,$$
$$6I_1+9I_2=6{\cdot}4. \quad (2)$$

Multiplying equation (1) by 3 and equation (2) by 2,

$$24I_1+18I_2=39, \quad (3)$$
$$12I_1+18I_2=12{\cdot}8. \quad (4)$$

Subtracting equation (4) from equation (3),

$$12I_1 = 26{\cdot}2,$$
$$I_1 = \mathbf{2{\cdot}18\ A.}$$

Substituting $I_1 = 2{\cdot}18$ in equation (1),

$$8 \times 2{\cdot}18 + 6I_2 = 13,$$
$$17{\cdot}44 + 6I_2 = 13,$$
$$6I_2 = -4{\cdot}44,$$
$$I_2 = \mathbf{-0{\cdot}74\ A.}$$

The current from the 13-V battery is 2·18 A in the assumed direction whilst a current of 0·74 A flows in the opposite direction to that assumed and is charging the 6·4-V battery.

The current in the 6-Ω resistor

$$(I_1 + I_2) = 2{\cdot}18 + (-0{\cdot}74)$$
$$= \mathbf{1{\cdot}44\ A} \text{ in the direction assumed.}$$

The following two examples are more difficult than those given in Chapter 11.

EXAMPLE. In the circuit shown (Fig. 16.4), calculate (a) the current flowing in each part of the circuit, (b) the potential of *B* with respect to *C*.

If the 1-Ω resistor is open-circuited, what now would be the potential of *B* with respect to *C*? (N.C.T.E.C.)

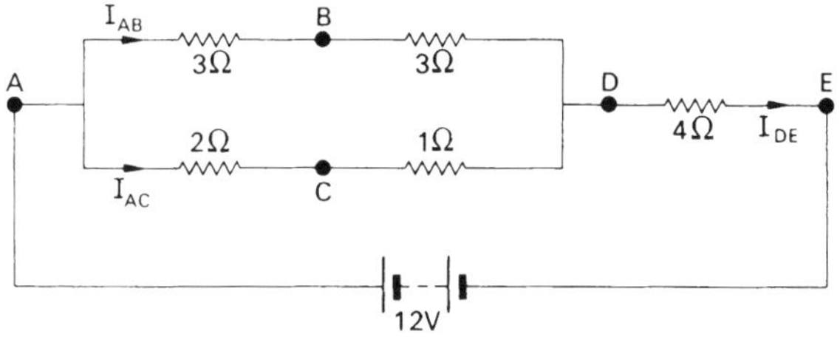

FIG. 16.4

$$\text{Total resistance} = \frac{6\times3}{6+3}+4=6\ \Omega,$$

$$I_{DE}=\frac{12}{6}\ \text{A}=\mathbf{2\ A}.$$

$V_{DE}=2\times4\ \text{V}=8\ \text{V}.$ $\qquad V_{AD}=12-8\ \text{V}=4\ \text{V}.$

$I_{AB}=\frac{4}{3+3}\ \text{A}=\mathbf{\tfrac{2}{3}\ A}.$ $\qquad I_{AC}=\frac{4}{2+1}\ \text{A}=\mathbf{1\tfrac{1}{3}\ A}.$

$$V_{BC}=V_{BA}+V_{AC}$$
$$=-3\times\tfrac{2}{3}+2\times1\tfrac{1}{3}=-2+2\tfrac{2}{3}=\mathbf{\tfrac{2}{3}\ V}.$$

With the 1-Ω resistor removed,

$$\text{Supply current}=\frac{12}{3+3+4}\ \text{A}=\frac{12}{10}\ \text{A}.$$

As there is no current through R_{AC}, A and C are at the same potential.

$$\therefore\quad V_{BC}=V_{BA}=-3\times\frac{12}{10}\ \text{V}=\mathbf{-3{\cdot}6\ V}.$$

EXAMPLE. In the circuit shown (Fig. 16.5), find the e.m.f. of the battery E. If the 15-Ω resistor were removed, what would be the p.d. between X and Y?

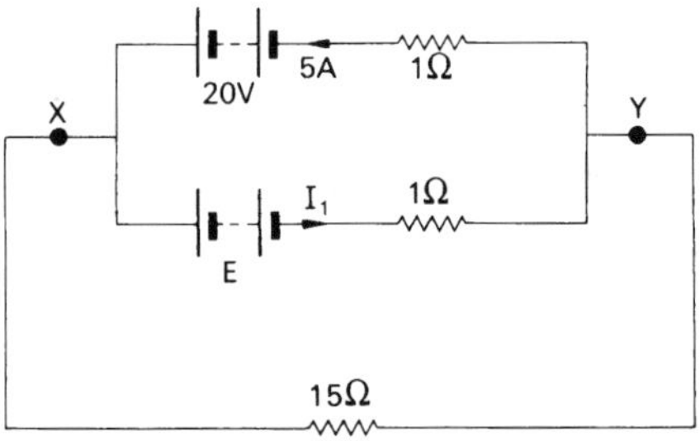

FIG. 16.5

$$V_{XY}=20-(5\times1)=15\ \text{V},$$

$$I_{XY}=\frac{15}{15}=1\ \text{A},$$

$$I_1 = 5 - 1 = 4 \text{ A},$$
$$E = 15 - (4 \times 1) = \mathbf{11\ V.}$$

With the 15-Ω resistor removed, a current I_C will flow in an anti-clockwise direction around the circuit.

$$I_C = \frac{20-11}{1+1} \text{ A} = 4{\cdot}5 \text{ A}. \qquad V_{XY} = 20 - (4{\cdot}5 \times 1) \text{ V} = \mathbf{15{\cdot}5\ V.}$$

Problems

16.1. Two batteries have e.m.f.s. of 12 V and 6 V and have internal resistances of 0·5 Ω and 1·0 Ω respectively. They are connected in parallel to feed a common load resistance of 1·33 Ω. Find the current in each part of the circuit.
(8 A (discharge) 12-V battery, 2 A (charge) 6-V battery, 6 A in load.)

16.2. Calculate the current in each part of the circuit shown (Fig. 16.6). What is the energy in joules consumed in the 0·5-Ω resistor in 1 hour?
(I_{BA} = 8 A, I_{CA} = 2 A, I_{AD} = 10 A; 115,200 J.) (N.C.T.E.C.)

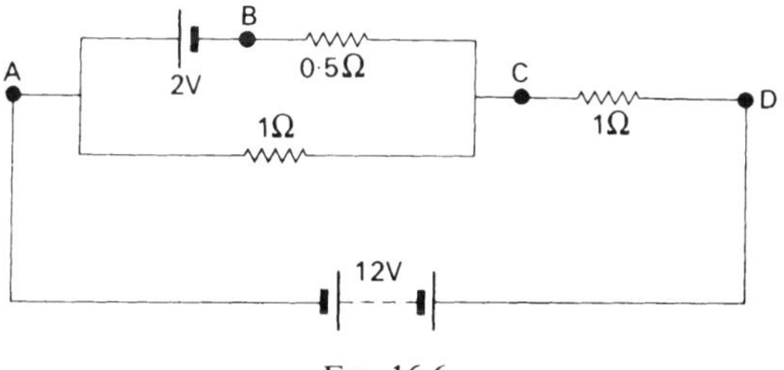

FIG. 16.6

16.3. Three batteries, *A*, *B*, and *C*, are connected in parallel to feed a common load resistance of 0·655 Ω. *A* has an e.m.f. of 20 V and an internal resistance of 0·1 Ω, *B* has an e.m.f. of 19·6 V and an internal resistance of 0·15 Ω and *C* has an e.m.f. of 22 V and an internal resistance of 0·2 Ω.

Calculate (a) the current in the load resistor, (b) the current from each battery.
(29 A, I_A = 10 A, I_B = 4 A, I_C = 15 A.)

16.4. In the circuit shown (Fig. 16.7), find the current flowing in each part of the circuit.
($I_{BA}=6$ A, $I_{CA}=4$ A, $I_{AD}=10$ A.) (N.C.T.E.C.)

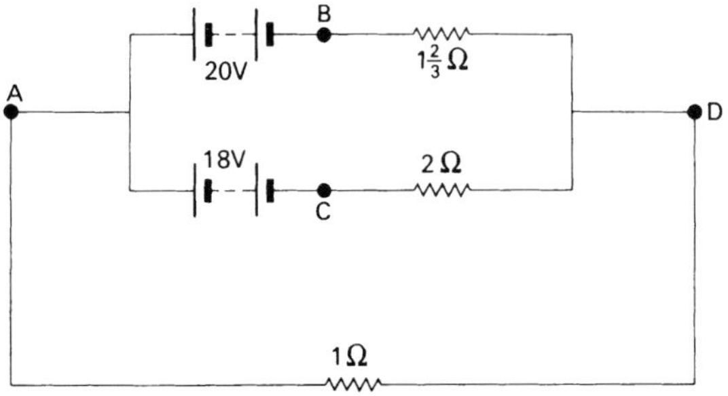

FIG. 16.7

16.5. Find the current in the branch *BD* (Fig. 16.8).
(1 A from *D* to *B*.)

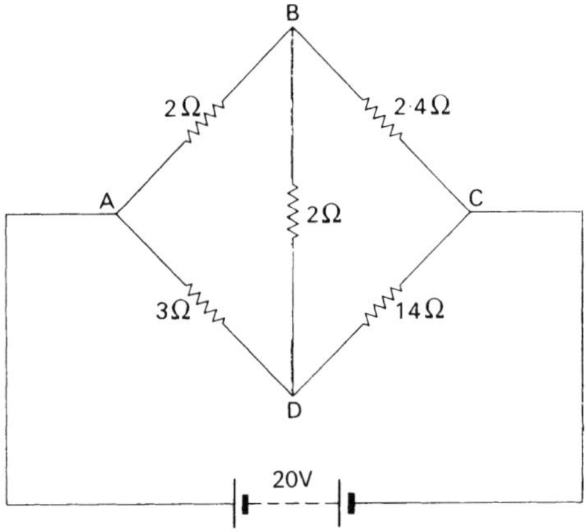

FIG. 16.8

Answers are not given to the following questions:

16.6. Two 12-V batteries are connected in parallel to an 18-V supply. If the internal resistance of the batteries are 0·11 Ω and 0·13 Ω, calculate the current taken by each battery. The internal resistance of the supply is negligible.

16.7. A battery on a car has an e.m.f. of 12 V and an internal resistance of 0·15 Ω. It is being charged by the car generator which has an e.m.f. of 15 V and an effective internal resistance of 0·1 Ω. The lights that are switched on can be represented by a resistor of 2 Ω.

Make a diagram of the circuit and insert symbols to represent the currents flowing and arrows to indicate their assumed directions.

Calculate the current through the battery and state whether the battery is being charged or discharged. (N.C.T.E.C.)

16.8. In the circuit shown (Fig. 16.9.), calculate the voltage V_{AB}.

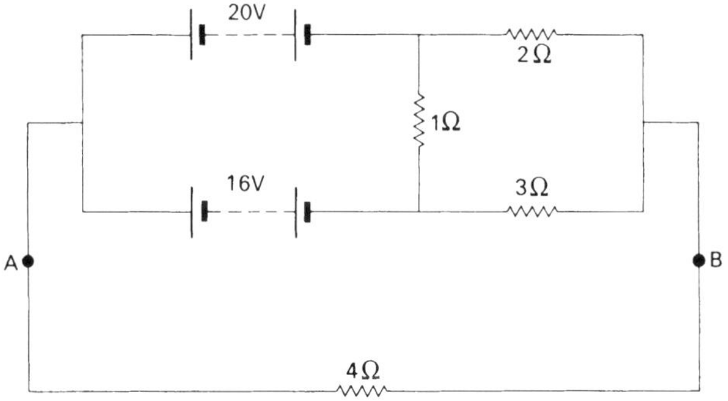

FIG. 16.9

16.9. Calculate the current distribution in the circuit shown (Fig. 16.10).

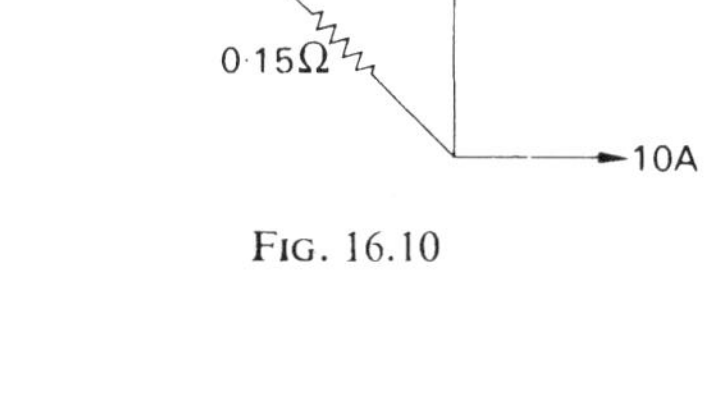

FIG. 16.10

CHAPTER 17

D.C. Two-wire Distributors

17.1. General

Electrical energy is usually supplied from the grid system to a main substation in a town or city. From here, it is fed to various smaller substations around the district through cables known as feeders. From these substations, cables are laid under the road and tappings are made to these cables which are known as distributors. These conduct the electrical energy into the houses, etc.

Whilst there is now relatively little direct current distribution in existence, the principles in the following calculations are applicable to alternating current distribution systems.

Consider the series resistance circuit shown in Fig. 17.1.

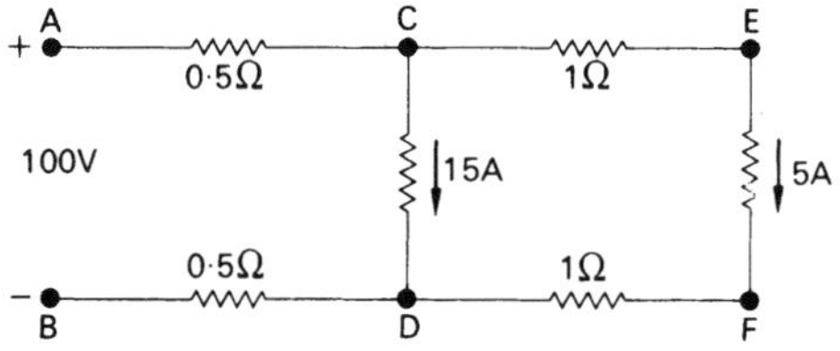

FIG. 17.1

It is assumed that the currents of 15 A and 5 A are flowing as shown. It is required to find (a) the current from the battery, (b) the voltages V_{CD} and V_{EF}, (c) the total power in the 0·5- and 1-Ω resistors, and (d) the power in the resistors R_{CD} and R_{EF}.

(a) To solve this, Kirchhoff's first law is applied.

Flowing from A is a current of $15+5=20$ A. The current $I_{AC}=\mathbf{20\ A.}$

Likewise the current flowing to B is $15+5=20$ A and so $I_{DB}=\mathbf{20\ A.}$

(b) The voltage across the resistor R_{CD} can be written down by inspection.

$$V_{CD}=100-(20\times 0{\cdot}5)-(20\times 0{\cdot}5)=\mathbf{80\ V.}$$

The voltage could have been found as follows:
Using path $BACDB$,

$$V_{BA}+V_{AC}+V_{CD}+V_{DB}=0,$$

$$-100+(20\times 0{\cdot}5)+V_{CD}+(20\times 0{\cdot}5)=0,$$

$$V_{CD}=\mathbf{80\ V.}$$

V_{EF} can be found in a similar manner.

$$V_{EF}=80-(5\times 1)-(5\times 1)=\mathbf{70\ V.}$$

Alternatively, using path $DCEFD$,

$$\mathrm{V}_{DC}+V_{CE}+V_{EF}+V_{FD}=0$$

as $$V_{CD}=+80\ \mathrm{V},\quad V_{DC}=-80\ \mathrm{V}.$$

$$-80+(5\times 1)+V_{EF}+(5\times 1)=0$$

$$V_{EF}=\mathbf{70\ V.}$$

(c) The power in the resistors can be found by calculating the I^2R loss for each resistor.

$$\text{Total power loss}=2(20^2\times 0{\cdot}5)+2(5^2\times 1)\ \mathrm{W}$$
$$=400+50\ \mathrm{W}$$
$$=\mathbf{450\ W.}$$

Alternatively, it could have been calculated from the knowledge of the voltages across the resistors.

The voltage drop across each 0·5-Ω resistor is

$$V_{AC}=V_{DB}=\frac{100-80}{2}\ \mathrm{V}=10\ \mathrm{V}.$$

Again alternatively,

$$V_{AC}=V_{DB}=20\times 0{\cdot}5\ \mathrm{V}=10\ \mathrm{V}.$$

Similarly, the voltage drop across each 1-Ω resistor is

$$V_{CE}=V_{FD}=\frac{80-70}{2}\text{ V}=5\text{ V}.$$

Power in the 0·5-Ω and 1-Ω resistors

$$=2(10\times 20)+2(5\times 5)\text{ W}$$
$$=400+50\text{ W}$$
$$=\mathbf{450\ W.}$$

(d) Power in resistor R_{CD}

$$=V_{CD}\times I_{CD}$$
$$=80\times 15\text{ W}$$
$$=\mathbf{1200\ W.}$$

Power in resistor R_{EF}

$$=V_{EF}\times I_{EF}$$
$$=70\times 5\text{ W}$$
$$=\mathbf{350\ W.}$$

The circuit could also have been conveniently drawn as showr in Fig. 17.2.

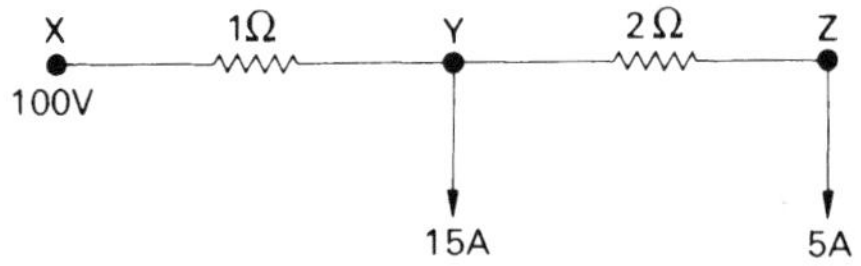

FIG. 17.2

Notice that R_{XY} is the total resistance of the two resistors R_{AC} and R_{BD}. Similarly for R_{YZ}.

Current flowing from X to Y, $I_{XY}=15+5\text{ A}=\mathbf{20\ A.}$

Voltage drop $V_{XY}=20\times 1\text{ V}=20\text{ V}.$

Voltage at $X=100-20\text{ V}=\mathbf{80\ V.}$

Voltage drop $V_{YZ}=5\times 2\text{ V}=10\text{ V}.$

Voltage at $Y=80-10\text{ V}=\mathbf{70\ V.}$

Power in the 1-Ω resistor$=20^2\times 1\text{ W}=400\text{ W}.$

Power in the 2-Ω resistor$=5^2\times 2\text{ W}=50\text{ W}.$

$$\text{Total power} = 400 + 50\ \text{W} = \mathbf{450\ W.}$$
$$\text{Power in the load at } Y = 80 \times 15\ \text{W} = \mathbf{1200\ W.}$$
$$\text{Power in the load at } Z = 70 \times 5\ \text{W} = \mathbf{350\ W.}$$

The circuit (Fig. 17.1) represents the electrical conditions that exist in a two-wire direct-current distributor supplying two loads. The two 0·5-Ω resistors can be considered to represent the resistance of the two cable conductors supplying the current for both the 15-A and the 5-A loads and the two 1-Ω resistors represent the resistance of the conductors supplying the 5-A load.

The power in the resistors in the example would be the copper losses in the cable and the power in the resistors R_{CD} and R_{EF} represents the power taken by the loads at these points.

EXAMPLE. A two-wire d.c. distributor fed at 250 V supplies loads of 70 A and 40 A at distances of 100 yd and 180 yd respectively from the supply end. Each conductor has an area of 0·1 in^2 for the first 100 yd from the supply point and 0·05 in^2 for the remainder of the distributor.

Calculate (a) the resistance of each section of the distributor, (b) the voltage at each load, (c) the power loss in the distributor, (d) the power taken by the loads.

The resistivity of the copper used in the conductors is $\frac{2}{3} \times 10^{-6}$ Ω-in.

Figure 17.3 represents this system.

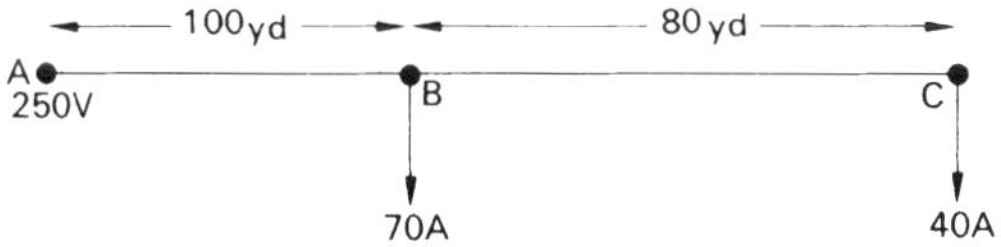

FIG. 17.3

$$\text{(a) Resistance } R_{AB} = \frac{\frac{2}{3} \times 10^{-6} \times 100 \times 2 \times 36}{0{\cdot}1}\ \Omega.$$
$$= \mathbf{0{\cdot}048\ \Omega.}$$

(Note that as there are TWO conductors, the effective length of conductor is $100\times 2=200$ yd.)

$$\text{Resistance } R_{BC}=\frac{\frac{2}{3}\times 10^{-6}\times 80\times 2\times 36}{0{\cdot}05}\ \Omega$$

$$=\mathbf{0{\cdot}077\ \Omega.}$$

(b) Supply current $I_{AB}=70+40=110$ A.

Voltage drop $V_{AB}=110\times 0{\cdot}048$ V $=5{\cdot}28$ V.

Voltage at $B=250-5{\cdot}28=\mathbf{244{\cdot}72\ V.}$

Voltage drop $V_{BC}=40\times 0{\cdot}077$ V $=3{\cdot}08$ V.

Voltage at $C=244{\cdot}72-3{\cdot}08=\mathbf{241{\cdot}64\ V.}$

(c) Power loss in the cable between A and B

$=110^2\times 0{\cdot}048$ W $=581$ W.

Power loss in the cable between B and C

$=40^2\times 0{\cdot}077$ W $=123$ W.

Total power loss $=581+123=$ **704 W.**

(d) Power in the load at $B=244{\cdot}72\times 70$ W $=$ **17,130 W.**

Power in the load at $C=241{\cdot}64\times 40$ W $=$ **9666 W.**

This example can be checked as follows:

Power in load at B =	17,130 W
Power in load at C =	9,666 W
Cable copper loss AB =	581 W
Cable copper loss BC =	123 W
Total power =	27,500 W

Power input to the network $=250\times 110$ W

$=$ **27,500 W.**

EXAMPLE. Part of a two-wire d.c. distributor system is shown in Fig. 17.4.

$R_{DA}=0{\cdot}1\ \Omega$, $R_{AB}=0{\cdot}2\ \Omega$, $R_{AC}=0{\cdot}25\ \Omega$ and $R_{BC}=0{\cdot}1\ \Omega$.

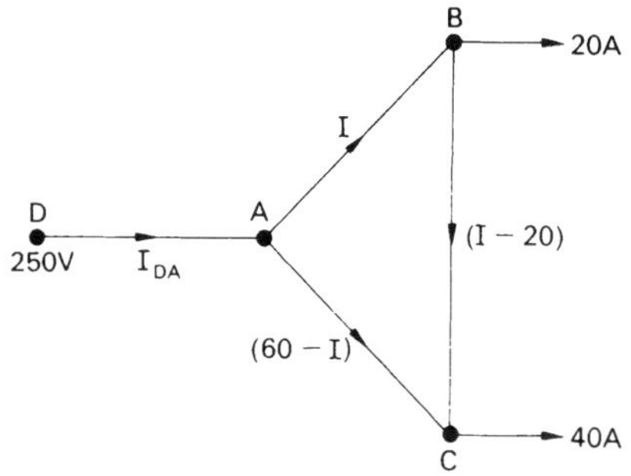

FIG. 17.4

The total resistance of both conductors is given and the voltage at D is 250 V. Calculate (a) the current in each part of the system, (b) the voltage at B and at C.

$$I_{DA} = 40 + 20 = \mathbf{60\ A.}$$

Let $I_{AB} = I$ amperes in the direction shown. Then $I_{AC} = (60 - I)$ and $I_{BC} = (I - 20)$.

Considering the path $ABCA$,

$$V_{AB} + V_{BC} + V_{CA} = 0$$

$$+0{\cdot}2I + 0{\cdot}1(I - 20) - 0{\cdot}25(60 - I) = 0,$$

$$0{\cdot}2I + 0{\cdot}1I - 2 - 15 + 0{\cdot}25I = 0,$$

$$0{\cdot}55I = 17,$$

$$I = \mathbf{30{\cdot}9\ A.}$$

Thus $I_{AB} = 30{\cdot}9$ A, $I_{BC} = (30{\cdot}9 - 20)$ A $= \mathbf{10{\cdot}9\ A}$ and $I_{AC} = (60 - 30{\cdot}9)$ A $= \mathbf{29{\cdot}1\ A.}$

Voltage at $A = 250 - 60 \times 0{\cdot}1 = \mathbf{244\ V.}$

Voltage at $B = 244 - 30{\cdot}9 \times 0{\cdot}2 = \mathbf{237{\cdot}8\ V.}$

Voltage at $C = 244 - 29{\cdot}1 \times 0{\cdot}25 = \mathbf{237{\cdot}2\ V.}$

EXAMPLE. A load of 240 A is supplied from a 460-V supply point by means of a two-core copper cable and a two-core aluminium cable connected in parallel. Each cable is 200 yd long and the cross-sectional area of each conductor is 0·06 in^2.

Calculate (a) the voltage at the load, (b) the current in each cable, (c) the total power loss in the cables.

The resistivities of copper and aluminium are respectively 0·7 and 1·1 microhm-inch.

(N.C.T.E.C.)

$$\text{Resistance of copper cable} = \frac{0{\cdot}7 \times 10^{-6} \times 400 \times 36}{0{\cdot}06}\ \Omega$$

$$= 0{\cdot}168\ \Omega.$$

$$\text{Resistance of aluminium cable} = 0{\cdot}168 \times \frac{1{\cdot}1}{0{\cdot}7}\ \Omega = 0{\cdot}264\ \Omega.$$

Total cable resistance as they are in parallel

$$= \frac{0{\cdot}168 \times 0{\cdot}264}{0{\cdot}168 + 0{\cdot}264}\ \Omega = 0{\cdot}102\ \Omega.$$

$$\text{Voltage drop} = 240 \times 0{\cdot}102\ \text{V} = 24{\cdot}6\ \text{V}.$$

(a) $$\text{Load voltage} = 460 - 24{\cdot}6 = \mathbf{435{\cdot}4\ V}.$$

(b) $$\text{Current in copper cable} = \frac{24{\cdot}6}{0{\cdot}168}\ \text{A} = \mathbf{147\ A}.$$

$$\text{Current in aluminium cable} = 240 - 147 = \mathbf{93\ A}.$$

(c) $$\text{Power loss in the cables} = 240 \times 24{\cdot}6\ \text{W} = \mathbf{5910\ W}.$$

Problems

17.1. A two-core cable supplies two loads from a 240-V supply. The first load is 100 yd from the supply end and takes a current of 30 A; the second load is 140 yd from the supply end and takes a current of 20 A. Draw a circuit diagram to illustrate the system.

Find the voltage across each load if each core of the cable has a cross-section of 0·04 in^2. (Assume a resistivity of $0{\cdot}7 \times 10^{-6}\ \Omega$-in.)

(233·7 V, 232·7 V.) (N.C.T.E.C.)

17.2. A current of 65 A is supplied to a direct-current motor through a two-core cable of length 240 ft and cross-sectional area 0·04 in^2.

(a) Calculate the voltage drop in the cable, (i) if the cable has copper cores, and (ii) if the wires are of aluminium.

(b) What size of aluminium cable would give the same voltage drop as copper cable?

(The resistivity of copper may be taken as 0·7 $\mu\Omega$-in. and that of aluminium as 1·1 $\mu\Omega$-in.)

((a) (i) 6·55 V, (ii) 10·30 V, (b) 0·063 in^2.) (C. & G.)

17.3. An electric motor connected to a 240-V d.c. supply develops 30 bhp. If its efficiency is 80% what current does it take from the supply?

If the motor is fed by a two-core cable 100 yd long and 0·04 in^2 conductor section, find (a) the cable resistance, (b) the power lost in the cable. (For copper, resistivity = $\frac{2}{3}$ $\mu\Omega$-in.)

(117 A, 0·126 Ω, 1724 W.)

17.4. A copper cable 100 yd long, whose twin-conductors are circular, delivers 1000 A at 240 V to a load. If the voltage drop in the cable is 5% of the voltage received, find (a) the power lost in the cable, (b) the diameter of each conductor. (For copper, resistivity = $\frac{2}{3}$ $\mu\Omega$-in.)

(12 kW, 0·71 in.) (U.L.C.I.)

17.5. A d.c. motor is supplied at 460 V by a two-core cable of length 100 yd and cross-sectional area 0·075 in^2. If the motor develops 50 bhp and its efficiency is 85%, find (a) the supply voltage to the cable, (b) power loss in the cable. (Resistivity of copper = $0{\cdot}67 \times 10^{-6}$ Ω-in.)

(466·1 V, 581 W.) (U.L.C.I.)

Answers are not given to the following questions:

17.6. A load of 230 A is supplied from a 460-V switchboard by means of a twin copper cable and a twin aluminium cable connected in parallel. The cables are 180 yd in length and each core of the cables has a cross-sectional area of 0·06 in^2. Calculate: (a) the voltage drop in the combined cables at full load, (b) the separate currents in each of the two cables. (The resistivities of copper and aluminium may be taken as 0·7 $\mu\Omega$-in. and 1·1 $\mu\Omega$-in. respectively.)

(C. & G.)

17.7. A two-core copper cable 100 yd long supplies two loads from a 230-V d.c. supply. The area of each conductor is 0·04 in^2. One load takes a current of 50 A at a point 60 yd from the supply end and the second takes 25 A at the far end of the cable. Find (a) the position and value of the minimum voltage, (b) the total power loss in the cable. (For copper, resistivity = $\frac{2}{3}$ $\mu\Omega$-in.)

(N.C.T.E.C.)

17.8. A sub-station supplies a load of 200 A at 11 kV to a factory situated 5000 yd away, through a two-core cable. The resistance of the cable is 0·01 Ω/core/100 yd. Find the power loss in transmission and express this as a percentage of the power supplied to the load. (U.L.C.I.)

17.9. Figure 17.5 represents a two-wire d.c. feeder system fed at A at 240 V. Cable AB has a resistance per conductor of 0·0125 Ω/100 yd, and cables BC and BD have resistances per conductor of 0·02 Ω/100 yd. AB=400 yd, BC=200 yd and BD=100 yd.

If the load at C is 15 A and the voltage at B is 236 V, calculate (a) the current at D, (b) the voltages at C and D. (U.L.C.I.)

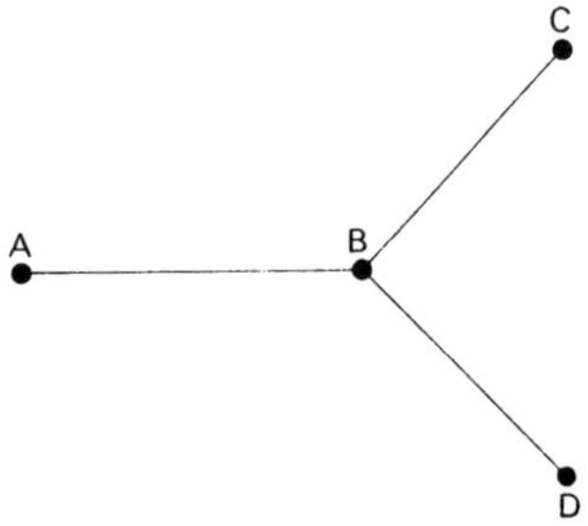

FIG. 17.5

CHAPTER 18

Capacitors

18.1. General

In 1745 it was discovered that electricity obtained from an electrostatic machine could be stored for a period of time and then released. The storage apparatus was known as a Leyden jar and it was the first capacitor.

CAPACITANCE exists wherever two conductors are separated by an insulating material which is called the DIELECTRIC. The term capacitor is generally reserved for components specifically designed so that the capacitance effect is predominant and of a desired value.

Commercially produced capacitors have many uses and are essential in the operation of computers, radio sets and television receivers. Their capacitance is usually very small, but larger units are made for uses such as the improvement of power factor. Some capacitors are designed to work in alternating current circuits whilst others should only be used in direct current applications. The latter are known as polarised electrolytic capacitors.

High-voltage capacitors and those used for power factor correction consist of a large number of small capacitor elements in a series–parallel arrangement. This type can be seen in Fig. 18.1. (See Section 18.10.)

Capacitance is associated with overhead lines and underground cables and its effect is important when considering the distribution of electrical energy.

A capacitor is able to store electrical energy for considerable periods. When a capacitor is connected to a d.c. supply, it will be found that if the supply is disconnected and the two capacitor terminals are shorted by a piece of wire, a spark results. The higher

FIG. 18.1. Three-phase 10-kVA, 415-V, 50-Hz capacitor fitted with discharge resistors. (B.I.C.C. Ltd.)

the voltage applied initially, the more intense the spark. Energy which was stored in the capacitor until its terminals were shorted is being dissipated as heat.

When a constant voltage V is applied to the capacitor, a quantity of electricity is stored in the form of electrical charges held on the surface of its plates.

By charging a capacitor to a known voltage and then discharging it by connecting the capacitor to an instrument known as a ballistic galvanometer, a deflection of the instrument pointer occurs which is proportional to the stored charge. By repeating this test for various capacitor voltages, it can be shown that the quantity of electricity stored is directly proportional to the applied voltage.

Thus $Q \propto V$

$\therefore \quad Q = CV$, where C is a constant.

If Q is in coulombs and V in volts, C is expressed in farads and is a measure of the capacitance of the capacitor.

A single capacitor having a capacitance of 1 farad (1 F) could not be manufactured because of its immense size and cost. Capacitors about $12 \times 4 \times 4$ in. may have a capacitance of 80 microfarads (80 μF) where 1 farad = 1,000,000 microfarads. In radio work, very much smaller capacitors are frequently used and these may have capacitances of a few picofarads (or micro-microfarads) where 1 microfarad = 10^6 picofarads (1 μF = 10^6 pF).

18.2. Capacitors in series

If a number of capacitors are connected in series to a supply (Fig. 18.2), the charge on each capacitor is the same, but if the capacitors have different capacitances, the voltage across each is

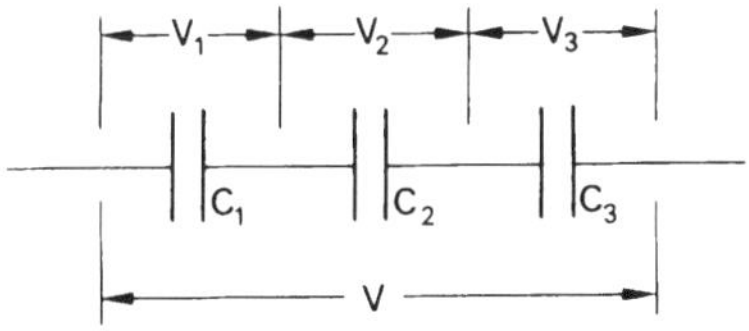

FIG. 18.2

different. However, the sum of the voltages across each capacitor must add up to the applied voltage. Thus

$$V = V_1 + V_2 + V_3 \text{ volts.}$$

As the charge on each is the same

$$Q = C_1V_1 = C_2V_2 = C_3V_3 \text{ coulombs.}$$

Suppose that these capacitors are replaced by a single capacitor C which is of equivalent capacity to the three capacitors. It will take the same charge from the same supply so that

$$Q = CV \text{ coulombs.}$$

Rearranging these expressions,

$$V_1 = \frac{Q}{C_1}, \quad V_2 = \frac{Q}{C_2}, \quad V_3 = \frac{Q}{C_3}, \quad V = \frac{Q}{C}.$$

As $$V = V_1 + V_2 + V_3,$$

then $$\frac{Q}{C} = \frac{Q}{C_1} + \frac{Q}{C_2} + \frac{Q}{C_3}.$$

Dividing through by Q,

$$\frac{1}{C} = \frac{1}{C_1} + \frac{1}{C_2} + \frac{1}{C_3}.$$

EXAMPLE. Find the capacitance of a single capacitor equivalent to three capacitors connected in series having capacitances of 20 μF, 30 μF, and 60 μF.

$$\frac{1}{C} = \frac{1}{20} + \frac{1}{30} + \frac{1}{60}$$

$$= \frac{3+2+1}{60} = \frac{6}{60},$$

$$\therefore \quad C = \frac{60}{6}\ \mu\text{F} = \mathbf{10\ \mu F}.$$

18.3. Capacitors in parallel

When capacitors are connected in parallel to a supply, as in Fig. 18.3, the voltage applied is the same across each capacitor, but if the capacitors have different capacitances, the charge held by each is different. Referring to Fig. 18.3,

Charge on $C_1 = Q_1 = C_1 V$ coulombs,
Charge on $C_2 = Q_2 = C_2 V$ coulombs,
Charge on $C_3 = Q_3 = C_3 V$ coulombs.

The total charge held by the capacitors is $Q = Q_1 + Q_2 + Q_3$ coulombs.

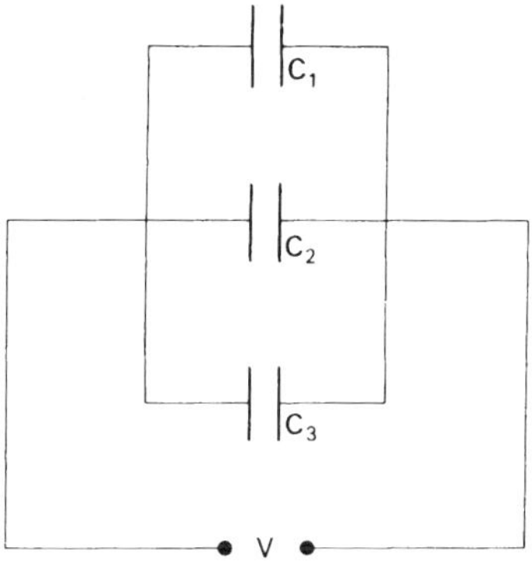

FIG. 18.3

Suppose a single capacitor of capacitance C held the same total charge when connected to the same voltage, then $Q=CV$. Hence

$$CV=C_1V+C_2V+C_3V.$$

Dividing through by V,

$$C=C_1+C_2+C_3.$$

EXAMPLE. If three capacitors having capacitances of 20 μF, 30 μF, and 60 μF are connected in parallel, what is the total capacitance?

$$\begin{aligned}\text{Total capacitance } C &= C_1+C_2+C_3\\ &= 20+30+60\ \mu\text{F}\\ &= \mathbf{110\ \mu F.}\end{aligned}$$

18.4. Electric field strength

If a voltage V volts is applied to two parallel plates an electric field will be set up between them. If they are separated by a distance of d metres (Fig. 18.4), then the electric field strength or electric force E existing between them is, neglecting fringe effects, given by

$$E=\frac{V}{d} \text{ volts per metre (V/m).}$$

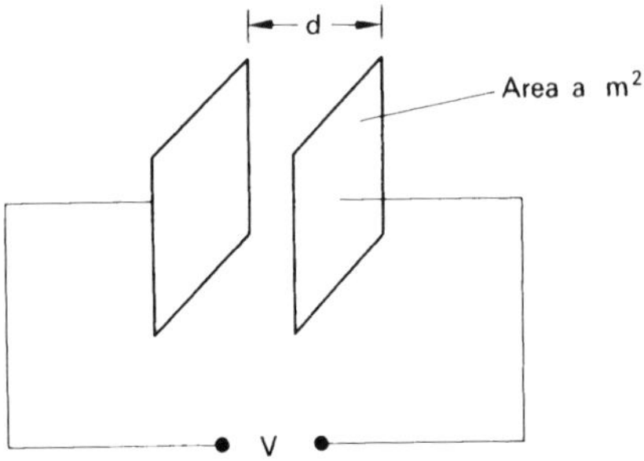

FIG. 18.4

18.5. Electric charge density

A capacitor stores a quantity Q coulombs of electricity and in the rationalised m.k.s. system of units, a charge of 1 coulomb (1 C) is considered to give rise to one line of electric flux. A line of electric flux indicates the direction in which a unit charge would move if free to do so when placed in the electric field. By convention, its direction is that taken by a positive charge.

The flux or charge density existing between the plates if each plate has an area of "a" square metres is (neglecting fringe effects)

$$D=\frac{Q}{a} \text{ coulombs per square metre (C/m}^2\text{)}$$

18.6. Permittivity of free space

For a vacuum, the ratio

$$\frac{\text{Electric flux density}}{\text{Electric field strength}}$$

is a constant which has a numerical value of $8{\cdot}85\times10^{-12}$ farad per metre (F/m). The constant is known as the permittivity of free space and is represented by the symbol ϵ_0.

Thus $$\epsilon_0=\frac{D}{E}=8{\cdot}85\times10^{-12} \text{ F/m.}$$

18.7. Capacitance of a parallel plate capacitor

With reference to Fig. 18.4,

Electric field strength $E = V/d$,
Electric flux density $D = Q/a$.

Assuming that the plates are separated in a vacuum,

$$\epsilon_0 = \frac{D}{E} = \frac{Q}{a} \times \frac{d}{V}.$$

For a capacitor

$$\frac{Q}{V} = C,$$

$$\therefore \quad \epsilon_0 = C \times d/a.$$

Hence the capacitance of a parallel plate capacitor is given by

$$C = \frac{\epsilon_0 a}{d} \text{ farads.}$$

18.8. Relative permittivity

If an insulating material is introduced between the plates, the capacitance of the parallel plate capacitor of Section 18.7 in general will be increased ϵ_r times and the capacitance becomes

$$C = \frac{\epsilon_0 \epsilon_r a}{d} \text{ F.}$$

ϵ_r is called the relative permittivity of the material used as the dielectric and typical values include

Vacuum	1
Air	1·00059
(i.e. for all practical purposes it is the same as for a vacuum)	
Glass	4–8
Mica	5·4
Mineral oil	2·2
Barium titanate	6000.

EXAMPLE. A parallel plate capacitor has the following dimensions.

Area of plates and of dielectric = 1600 cm^2.
Distance between the plates = 5 mm.

If the dielectric has a relative permittivity of 6, calculate the capacitance.

$$C = \frac{\epsilon_0 \epsilon_r a}{d} = \frac{(8{\cdot}85 \times 10^{-12}) \times 6 \times (1600 \times 10^{-4})}{5 \times 10^{-3}} \text{ F}$$

$$= \mathbf{1{\cdot}70 \times 10^{-9}\ F} \text{ (or } \mathbf{0{\cdot}0017\ \mu F}\text{).}$$

18.9. Multiplate capacitor

When a capacitor consists of n parallel plates with alternate plates connected together, then the multiplate capacitor has a capacitance of

$$C = \frac{\epsilon_0 \epsilon_r (n-1) a}{d} \text{ F.}$$

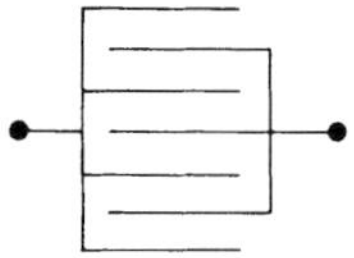

FIG. 18.5. Representation of multiplate capacitor.

Note that this applies to the parallel plate capacitor considered earlier as $(n-1) = 2-1 = 1$ dielectric between the two parallel plates. The multiplate capacitor is effectively $(n-1)$ similar capacitors connected in parallel.

18.10. Maximum working voltage of a capacitor

As stated in Section 9.4, if the electric force exceeds a certain value, electrical breakdown occurs. Although the capacitance of the parallel plate capacitor is increased as the distance between the plates is decreased, the permissible applied voltage for a given dielectric is decreased. Thus a maximum working voltage which is usually stated on the capacitor must not be exceeded.

When the maximum voltage is likely to be exceeded, series-parallel connections may be necessary.

EXAMPLE. A number of 5-μF, 200-V working capacitors are available. How many are required and how must they be connected to give a 2·5-μF capacitance, 800-V working combination?

Four of the capacitors connected in series will have 200 V across each when connected to a 800-V supply. The capacitance C of this group is given by:

$$\frac{1}{C}=\frac{1}{5}+\frac{1}{5}+\frac{1}{5}+\frac{1}{5},$$

i.e.

$$C=1{\cdot}25\ \mu\text{F}.$$

To make up a total capacitance of 2·5 μF, $\frac{2{\cdot}5}{1{\cdot}25}=2$ groups must be connected in parallel.

Thus there are two groups in parallel, each group having four 5-μF capacitors in series.

18.11. Energy stored in an electric field

As stated previously a capacitor is a device which will store electrical energy. When a capacitor is disconnected from a supply and the terminals are then shorted, this energy is given up as heat in the form of a spark. After use, the terminals of a capacitor should always be connected together (shorted) so that there is no danger of electric shock to anyone handling the capacitor.

When a p.d. is applied to a capacitor, there is a redistribution of the electric charge between the plates, the electrons tending to move towards the positive plate and the protons towards the negative plate. An electric strain occurs in the dielectric and an electric field exists between the plates. Energy is stored by the capacitor by virtue of the electric field.

It can be shown that the energy stored in a capacitor is given by:

$$\text{Stored energy} = \tfrac{1}{2}CV^2 \text{ joules,}$$

where C is the capacitance of the capacitor in farads and V is the potential difference across the capacitor in volts.

The expression for stored energy can be rewritten

$$\tfrac{1}{2}CV^2 = \tfrac{1}{2}\frac{Q}{V}\cdot V^2 = \tfrac{1}{2}QV \text{ joules}$$

$$= \tfrac{1}{2}\frac{Q}{a}\times\frac{V}{d}\times ad \text{ joules}$$

$$= \tfrac{1}{2}DE\times(\text{volume of the dielectric}) \text{ joules.}$$

Problems

18.1. A 200-F capacitor is charged to a voltage of 200 V. What quantity of electricity is stored by the capacitor?

(0·05 C.)

18.2. Two capacitors connected in series have a combined capacitance of 40 μF. If one capacitor has a capacitance of 80 μF, what is (a) the value of the other capacitor, (b) the voltage across each when the pair are connected to a 1000-V d.c. supply?

((a) 80 μF, (b) 500 V each.)

18.3. Capacitors whose capacitances are 20, 40 and 60 μF are connected in parallel to a 600-V d.c. supply. Calculate (a) the total capacitance, (b) the total quantity of electricity stored by the combination.

((a) 120 μF, (b) 0·122 C.)

18.4. When three capacitors are connected in parallel to a 1000-V d.c. supply, the total stored charge is 0·2 C. If two of the capacitors have values of 50 μF and 70 μF, calculate the capacitance of the third capacitor.

(80 μF.)

18.5. A 20-μF and a 60-μF capacitor are connected in parallel and the combination connected in series with a 40-μF capacitor to a 1000-V d.c. supply. Calculate the total charge stored by the capacitors.

($266{\cdot}7\times 10^{-4}$ C.)

18.6. A parallel plate capacitor has an area of dielectric between plates of 1600 cm^2, the distance between the plates being 5 mm. Calculate the capacitance of the capacitor (a) if a vacuum is maintained between the plates, (b) if the dielectric has a relative permittivity of 2.

((a) 283 pF, (b) 566 pF.)

18.7. A parallel plate capacitor with an air dielectric is charged from a 1000-V d.c. supply. It is then disconnected from the supply and connected to an electrostatic voltmeter of negligible capacitance. What will be the voltmeter reading (a) if the distance between the plates is doubled, (b) if, with the original spacing, a dielectric of relative permittivity 4 is inserted between, and compietely fills, the space between the plates?

((a) 2000 V, (b) 250 V.)

Answers are not given to the following questions:

18.8. Calculate the total capacitance when a 40-μF, a 60-μF and an 80-μF capacitor are connected (a) in series, (b) in parallel.

18.9. Two capacitors having capacitances of 10 μF and 20 μF are connected in parallel and the combination connected in series with an 8-μF capacitor to a 1000-V d.c. supply. What is (a) the total capacitance of the combination, (b) the total charge on the capacitors in coulombs?

18.10. A capacitor consists of two parallel metal plates, each 20×20 cm spaced 2 mm apart in air. Calculate the capacitance of this arrangement.

18.11. A 20-μF capacitor is charged from a 100-V d.c. supply. On being disconnected from the supply, it is immediately connected across an uncharged 5-μF capacitor. Calculate (a) the p.d. across the parallel capacitors, (b) the charge on each capacitor.

CHAPTER 19

Alternating Currents

19.1. Representation of pole systems

Consider a plane represented by the line *XY* drawn midway between the poles of the two magnets shown in Fig. 19.1(a). *XY* is referred to as either the Geometrically Neutral Plane (G.N.P.) as it lies geometrically midway between the poles or as the Magnetically Neutral Plane (M.N.P.) as it is the plane where on one side the north pole exerts more influence and on the other side, the south pole is more effective. Sometimes it is more convenient to draw the poles as shown in Fig. 19.1(b). The M.N.P. (and the G.N.P.) are midway between the poles.

Assuming that the poles are identical, the distance between the lines representing the M.N.P. is called a pole pitch. When poles are drawn in this manner, it is possible on a diagram to give an

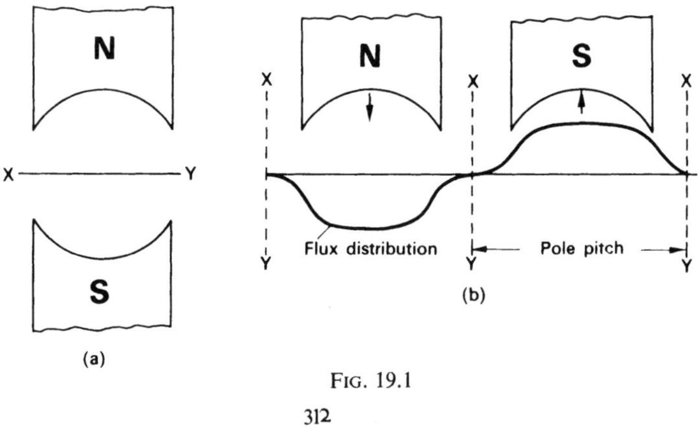

FIG. 19.1

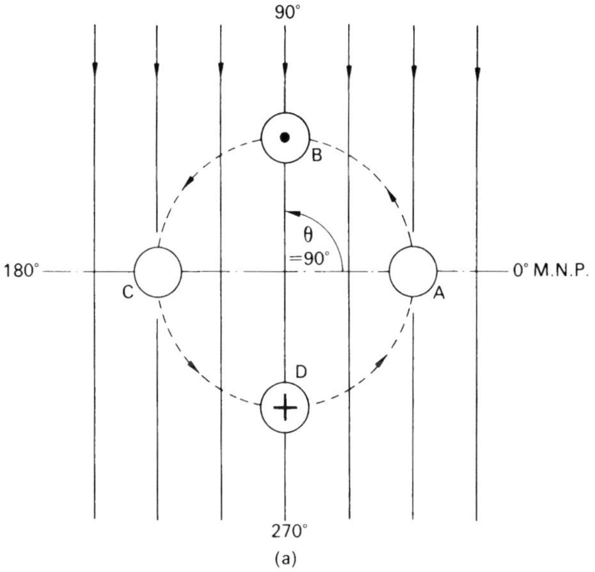
90°
B
θ
=90°
180°
0° M.N.P.
C
A
D
270°
(a)

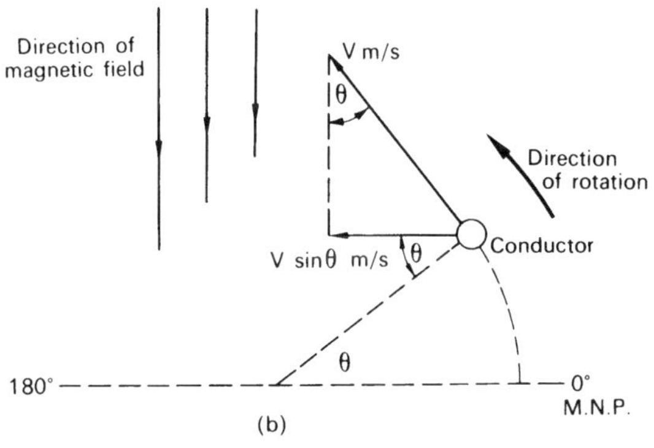
Direction of
magnetic field
V m/s
θ
Direction
of rotation
Conductor
V sinθ m/s
θ
θ
180°
0°
M.N.P.
(b)

FIG. 19.2

indication of the variation in the strength of the magnetic field over the pole pitch. The arrows adjacent to the poles indicate the direction of the magnetic flux.

It has been stated that the M.N.P. and the G.N.P. are the same. In a d.c. machine, this is only true under no-load conditions unless special windings are added to maintain the position of the M.N.P. Consideration of this must be left until later.

19.2. Generation of e.m.f.

Consider what happens to a single straight conductor as it is moved in a circular path in a uniform magnetic field (Fig. 19.2). At *A*, the conductor is moving parallel to the lines of magnetic force and so no e.m.f. is induced in it. At *B*, the conductor is moving at right angles to the magnetic field and the rate of cutting of the magnetic flux is at its greatest, so the e.m.f. induced in the conductor is at its maximum. This can be represented by E_M. Using Fleming's right-hand rule, the direction of the induced e.m.f. is vertically out of the book, shown ⊙. At *C*, the e.m.f. is again zero for the same reason as at *A*. When at *D*, the e.m.f. is again maximum but the direction of the e.m.f. is now into the book, shown ⊕. At all other points, the value of the e.m.f. at any instant is somewhere between zero and maximum.

19.3. E.M.F. waveform

The e.m.f. induced in the conductor at any instant may be as shown in Fig. 19.3. The waveform over 360° is called one cycle of e.m.f. and is obtained by plotting the value of the e.m.f. at any instant against the position of the conductor measured from the M.N.P., this being expressed by the angle θ. It is conventional to draw the first half of the graph above the axis. The graph implies that for the first 180° the e.m.f. is in one direction in the conductor and from 180° to 360° it is in the opposite direction. This e.m.f. is said to be alternating as it regularly changes its direction.

The graph shows values of e.m.f. at every instant. These are referred to as instantaneous values and the small letter e is used to represent them. The maximum value E_M is a particular value of the instantaneous values.

In Section 13.13 it was stated that the e.m.f. induced in a conductor was proportional to the component of the velocity of the conductor which was at right angles to the magnetic field. Figure 19.2(b) shows the position of the conductor when it has rotated by θ degrees from the M.N.P.

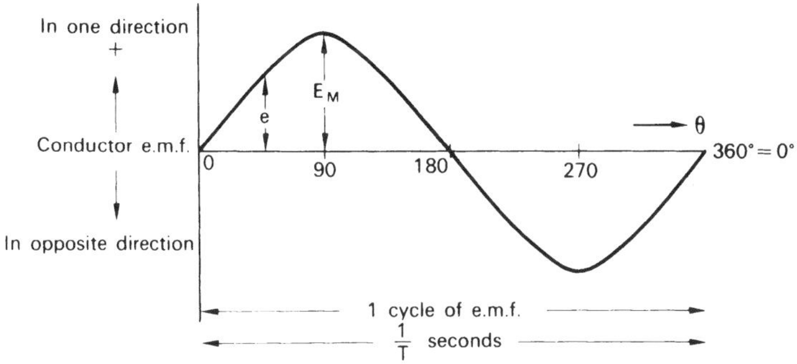

FIG. 19.3

The e.m.f. induced in the conductor is $e = BlV \sin \theta$ volts where V m/s is the tangential velocity of the conductor.

When $\theta = 90°$, the conductor is cutting the magnetic flux at its greatest rate ($\sin \theta = 1$) and so maximum e.m.f. is being generated.

$$\therefore \quad E_M = BlV$$

Hence $$e = E_M \sin \theta.$$

It follows that the waveform of the e.m.f. generated is sinusoidal and this is represented in Fig. 19.3.

19.4. Frequency

Every time a conductor passes a north and a south pole, the cycle of e.m.f. is repeated. The number of cycles of e.m.f. generated in 1 sec is called the frequency of the e.m.f. and it was expressed in cycles per second (c/s). Now the unit of frequency is the hertz (Hz). If the conductor in Fig. 19.2 were rotating at 3000 rev/min, the frequency would be 50 Hz. This is the frequency of the alternating current supply in Great Britain. In America it is 60 Hz.

Occasionally the term "periodic time" is used. This is the time in seconds of 1 cycle.

$$\text{Periodic time } T = \frac{1}{f} \text{ sec.}$$

Thus Fig. 19.3 can be plotted to a base of time as well as degrees.

19.5. Multipolar machine

Suppose a generator has six poles arranged as in Fig. 19.4. Every time the conductor passes a north and a south pole, 1 cycle of

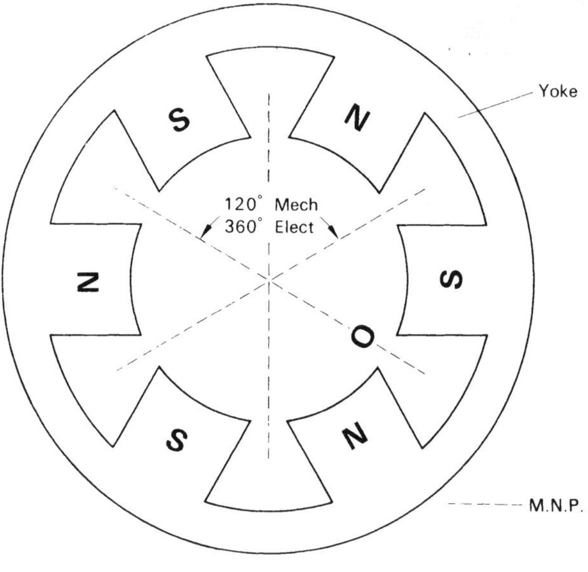

FIG. 19.4

e.m.f. is generated. Note that the poles are arranged alternately north and south and that there is an M.N.P. represented by the dashed lines between each pair of poles. In one revolution of the conductor, three cycles of e.m.f. are generated as there are three pairs of poles. In general, if a machine has p pairs of poles, p cycles of e.m.f. are generated. The number of cycles of e.m.f. generated per second is given by:

$$(\text{pairs of poles}) \times (\text{revolutions per second}),$$

$$\text{i.e. frequency } f = \frac{pN}{60} \text{ Hz},$$

where N is revolutions per minute.

19.6. Electrical and mechanical degrees

In mathematics, a complete sinusoidal waveform is obtained by plotting values of sin θ from 0° to 360°. For ease of calculation, each waveform of e.m.f. is assumed to be sinusoidal as this facilitates calculations in alternating current circuits. Thus each cycle of e.m.f. is considered as being spread over 360° electrical.

In the case of the six-pole machine, one revolution of 360° mechanical caused three cycles of e.m.f. to be generated, this being equivalent to $360^\circ \times 3 = 1080^\circ$ electrical. In general, if a machine has p pairs of poles, then θ degrees mechanical is equivalent to $p\theta$ degrees electrical. The circuit only reacts to a supply of a certain frequency and is not concerned with how it was generated. When studying alternating current theory, if an angle is given in degrees, it is understood to be degrees in the electrical sense. Mechanical degrees will be stated as such.

19.7. The simple alternator

Using a single conductor does not provide an easy way of using the e.m.f. generated, nor would an e.m.f. of a useful magnitude be obtained.

If a coil were rotated in a magnetic field, a larger e.m.f. could be generated, and if the ends of the coil were connected to two conducting rings mounted on, but insulated from, the shaft and from each other (these rings being known as slip-rings), then by having a carbon brush in contact with each ring, the e.m.f. could be more easily connected to an external circuit.

Consideration of the coil sides under each pole would show that the e.m.f. generated was alternating. If a resistor were connected to the two brushes, an alternating current would flow through it.

19.8. Generation of a sinusoidal e.m.f.

In power stations, the generators (called alternators) are designed to have an e.m.f. waveform as nearly sinusoidal as possible. This means that a graph plotted over 360° (electrical) of the instantaneous values of e.m.f. generated is almost a sine wave. In practice, the way in which an almost sinusoidal e.m.f. is generated is to

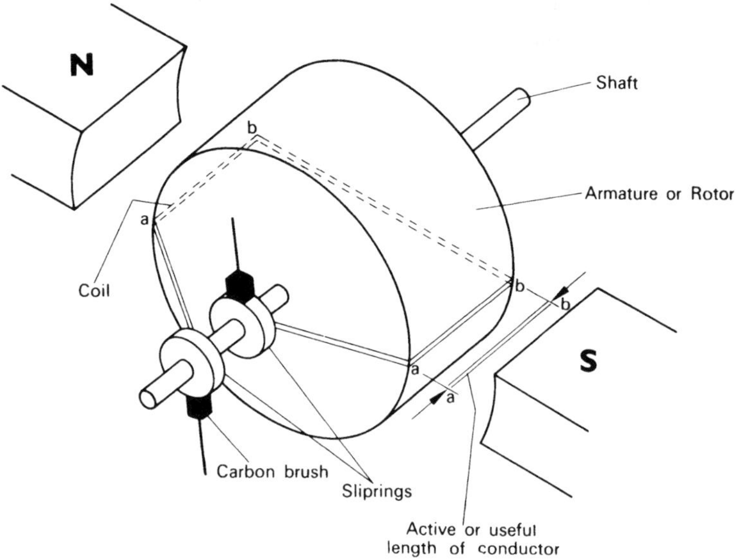

FIG. 19.5

arrange that the magnetic flux distribution over each pole pitch (180° electrical) is as nearly sinusoidal as possible and then making the conductors cut this flux at right angles.

It should be realised that this ideal sinusoidal flux distribution can never be obtained. Nevertheless, by suitable arrangement of the windings, the waveform of the e.m.f. generated is, for all practical purposes, sinusoidal.

19.9. Average value of a waveform

This value is used mainly in connection with rectifier calculations but it can be used for obtaining certain formulae and will be used for this purpose later in this chapter. For a symmetrical waveform, the average value is the average of the instantaneous values considered over half a cycle. In order to obtain the average value of a given waveform by a graphical method, the waveform should be drawn to scale. Figure 19.6 shows half a cycle of a sinusoidal waveform having a maximum value of 100 V.

The base should be divided into a number of equal parts representing equal periods of time of t sec: the greater the number of equal parts, the more accurate will be the answer. In Fig. 19.6 ten equal divisions have been taken and are shown by solid vertical

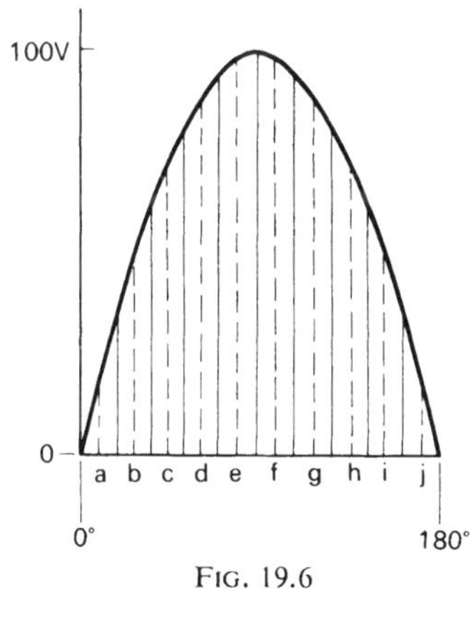

FIG. 19.6

lines. The mid-ordinate (shown dashed) for each interval should then be drawn and the length of each mid-ordinate (a, b, c, . . . , i, j) measured. The average value of the waveform is given by:

$$E_{av} = \frac{\text{sum of all mid-ordinates}}{\text{number of mid-ordinates}}$$

$$= \frac{a+b+c+d+e+f+g+h+i+j}{10}$$

In the case of the sine wave considered, the value for E_{av} is found to be about 64 V. For any sine wave, the average value is shown, in the following paragraphs, to be exactly

$$\frac{2}{\pi} \times (\text{the maximum value, } E_M),$$

i.e.

$$E_{av} = \frac{2}{\pi} \times E_M$$

$$\simeq 0{\cdot}637\ E_M.$$

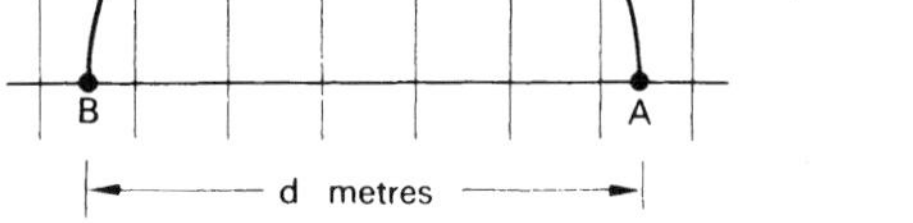

FIG. 19.7

Thus for $E_M = 100$ V, $E_{av} = 63{\cdot}7$ V $\simeq$ 64 V. The relationship can be shown as follows (see Fig. 19.7).

Consider a conductor of length l m on the rotor of a generator moving in a circular path at V m/s in a uniform magnetic field of density B Wb/m^2. Suppose that the diameter of the rotor is d m. As the conductor moves in a circular path AB, it cuts the same amount of flux as it would have done if it had gone directly from A to B.

$$\text{Total flux cut} = Bld \text{ Wb.}$$

The time taken for the conductor to travel the circular path from A to B is

$$t = \frac{\text{distance}}{\text{velocity}}$$

$$= \frac{\pi d/2}{V} \text{ sec} = \frac{\pi d}{2V} \text{ sec.}$$

Average rate of flux cutting (= average e.m.f. generated),

$$E_{av} = \frac{Bld}{\pi d} \times 2V = \frac{2}{\pi} BlV \text{ volts.}$$

But BlV is the maximum value of the e.m.f. generated,

$$\therefore \quad E_{av} = \frac{2}{\pi} \times E_M \text{ volts.}$$

19.10. Effective (or r.m.s.) value of a waveform

If a current which is continually varying in magnitude is passed through a resistor, the heat produced by it in a given time will be equivalent to the heat produced by some value of direct current flowing through the same value of resistance for the same length of time. The effective value of an alternating current is defined as being equal to that value of direct current which produces the same amount of heat when flowing through the same value of resistance for a time equal to one cycle of the alternating current.

Moving-iron instruments indicate the effective values of current and voltage and capital letters are always used to represent these values.

To obtain graphically the effective value of any waveform, the waveform should first be drawn and the construction used as was described for finding the average value of a waveform. Figure 19.8 represents a sinusoidal current whose maximum value is 10 A. Each half-cycle is divided into n equal intervals of t sec and the

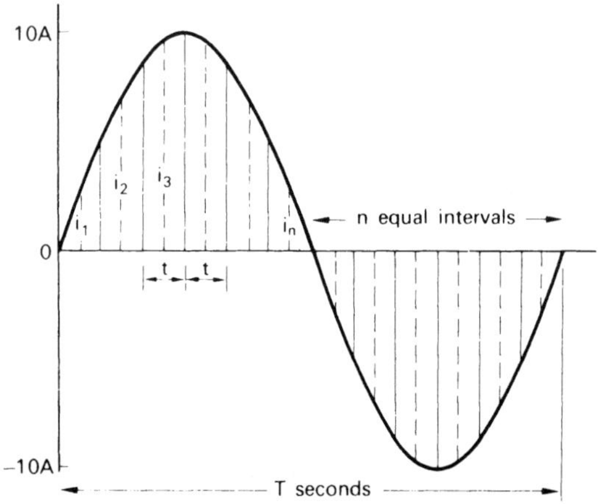

FIG. 19.8

mean values of current during these intervals are i_1, i_2, i_3, . . . , i_n. The heating effect during each half-cycle is the same so in a resistance of R ohms during a cycle the energy expended as heat is

$$2 \times (i_1^2 + i_2^2 + i_3^2 + i_4^2 + \ldots + i_n^2)\ Rt \text{ joules.}$$

As $t = (T/2n)$ sec, where T is the periodic time, this can be written as

$$\text{Heating effect} = 2 \times (i_1^2 + i_2^2 + i_3^2 + \ldots + i_n^2)\ R \times \frac{T}{2n}.$$

This must be the same as I^2RT joules where I is the effective value of the alternating current. Hence

$$I^2RT = 2 \times (i_1^2 + \ldots + i_n^2)\ R \times \frac{T}{2n},$$

$$\therefore \quad I^2 = \frac{(i_1^2 + \ldots + i_n^2)}{n},$$

$$I = \sqrt{\left(\frac{(i_1^2 + \ldots + i_n^2)}{n}\right)}$$

= the square root of the mean value of i^2 over one cycle.

Hence the alternative name of root-mean-square (r.m.s.) value.

If values of i^2 are plotted over a cycle as shown in Fig. 19.9, it can be shown that the line AB at a height of $I_m^2/2$ above the horizontal axis OX represents the mean value of i^2 over the cycle since the area $ABXO$ is equal to the (shaded) area below the curve

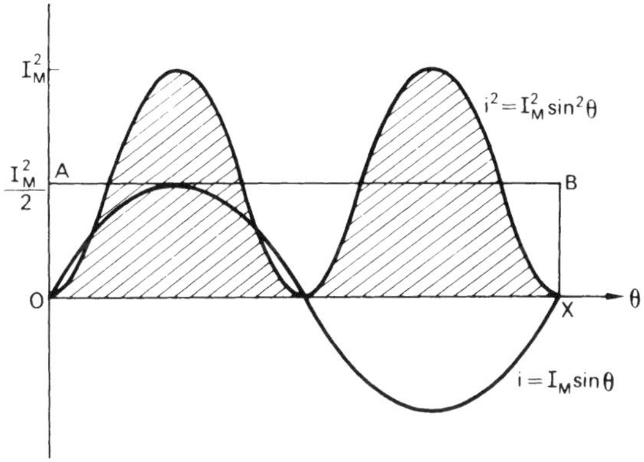

FIG. 19.9

of i^2. Thus for a sine wave, the average of the square of the instantaneous currents over one cycle is $I_m^2/2$,

$$\therefore \quad \text{effective or r.m.s. value} = \sqrt{\left(\frac{I_m^2}{2}\right)}$$

$$= \frac{I_m}{\sqrt{2}}$$

$$= 0{\cdot}707\ I_m.$$

Likewise, the effective value of a sinusoidal voltage is equal to 0·707 times the maximum value.

19.11. Form factor

The form factor of a wave may be used to find the r.m.s. value of a wave form if the average value is known. It is defined as:

$$\text{Form factor} = \frac{\text{r.m.s. value}}{\text{average value}}.$$

For a sine wave this becomes

$$\text{Form factor} = \frac{0{\cdot}707\ V_M}{0{\cdot}636\ V_M} = 1{\cdot}11.$$

The average value of a sine wave is taken over one half-cycle since it would be zero over a whole cycle.

19.12. Sinusoidal alternating voltage applied to a pure resistor

For a pure resistor, Ohm's law applies and so at every instant, $v = iR$ volts. Figure 19.10 shows waveforms of current and voltage over one cycle. As the two waveforms pass through maximum and zero at the same time, they are said to be in phase with each other.

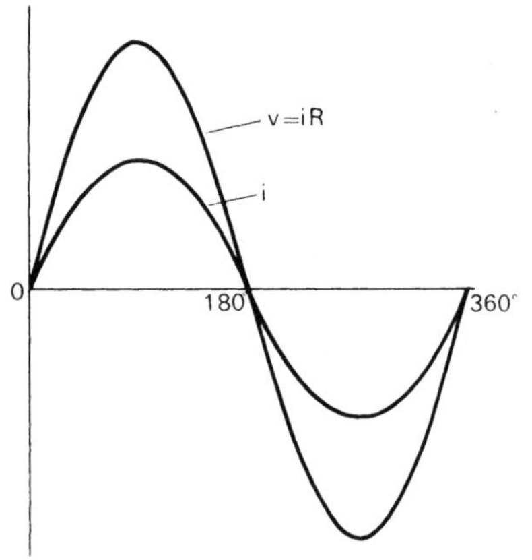

FIG. 19.10

19.13. Sinusoidal alternating voltage applied to a pure inductor

It is convenient to assume in this case that a sinusoidal current is flowing through a coil of negligible resistance. Also assuming that the coil has a non-magnetic core, the flux set up by a current will be proportional to it at every instant. Figure 19.11 illustrates this.

It has been stated that an e.m.f. is induced in a coil when the flux linking the coil is changing. When a rate of change of flux linkages is zero, the e.m.f. is zero and when the rate of flux change is at its maximum, the e.m.f. is also at its maximum.

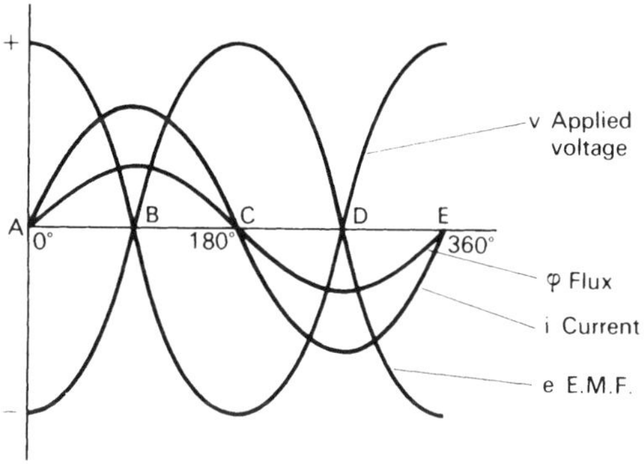

FIG. 19.11

At A, the flux is passing through zero and increasing in its arbitrary positive direction at its maximum rate. The e.m.f. is maximum at this instant and one point on the e.m.f. curve, shown below A, is found. It is drawn below A because at this instant, the current is increasing at its maximum rate in a positive sense so the direction of the induced e.m.f. must be so as to try to prevent the current from increasing (Lenz's law). It must, therefore, be in the opposite or negative direction.

At B, the rate of change of flux linkages is instantaneously zero as the flux has increased to a maximum value and is momentarily constant prior to commencing to decrease. Hence at this instant, the e.m.f. is zero.

The values of the e.m.f. at C, D, and E may be deduced in a similar manner and if the e.m.f. were determined at intermediate points, the resultant curve of e.m.f. would be as shown. As resistance has been considered to be negligible, the applied voltage necessary to cause the current flow will be numerically equal to but in the opposite direction to the e.m.f. at every instant and the applied voltage will be as shown in Fig. 19.11.

It will be observed that in this case, the current varies in the same way as the applied voltage but 90° after it. The current is said to lag the applied voltage by 90°.

19.14. E.M.F. equation for a pure inductor

Figure 19.12 shows the sinusoidal flux variation in a purely inductive coil over a period of $\frac{1}{4}$ cycle. If the frequency is f Hz, then the time of $\frac{1}{4}$ of a cycle (e.g. A to B) is $1/(4f)$ sec. During this period the flux changes by an amount $(\Phi_M - 0) = \Phi_M$ webers and if this flux change links a coil having N turns, the change in flux linkages is $\Phi_M N$ weber-turns.

The average rate of change of flux linkages is given by the change of flux linkages divided by the time of the change and this gives the average value of the e.m.f. induced in the coil. Thus,

$$E_{av} = \frac{\Phi_M N}{t} = \frac{\Phi_M N}{1/(4f)} = 4f\Phi_M N \text{ volts.}$$

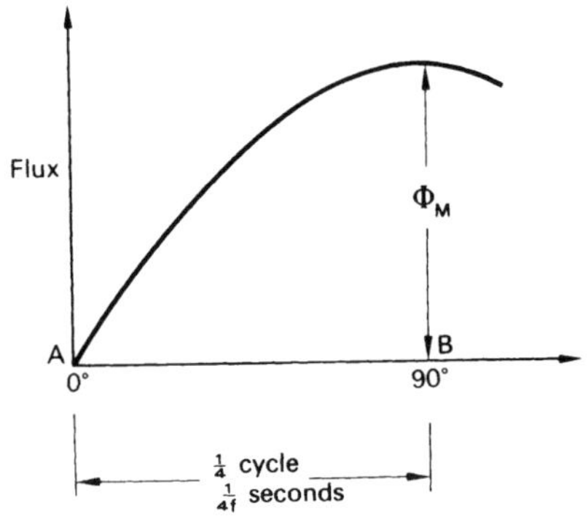

FIG. 19.12

This is the average value of a sinusoidal e.m.f. and the r.m.s. value of the voltage is obtained by multiplying the average value of the wave by its form factor. For a sine wave, the form factor is 1·11.

$$\therefore \quad \text{r.m.s. value of e.m.f. } E = 1{\cdot}11 \times 4f\Phi_M N$$
$$= 4{\cdot}44\Phi_M fN \text{ volts.}$$

Neglecting resistance, the applied voltage will be equal and opposite to the e.m.f. and so

$$\text{Applied voltage } V = 4{\cdot}44\Phi_M fN \text{ volts.}$$

It should be appreciated that the e.m.f. is in opposition to the applied voltage, and the negative sign for induced e.m.f. has been omitted in the preceding paragraphs. The above expressions are used for some calculations on transformers.

19.14. E.M.F. equation for a pure inductor

It has been shown that the inductance of a coil is given by

$$L = \frac{N\Phi}{I} \text{ henry,}$$

where L is the (self)-inductance of a coil of N turns when a current of I amperes sets up a flux of Φ webers. This can be rewritten

$$LI = N\Phi,$$
$$\therefore \quad LI_M = N\Phi_M$$

assuming that the flux set up is proportional to the current. From the previous section,

$$V = 4{\cdot}44\Phi_M fN \text{ volts.}$$

Substituting for $N\Phi_M$, $\quad V = 4{\cdot}44fLI_M$ volts.

But $\quad I_M = \sqrt{2}I$ for sinusoidal currents,

$$\therefore \quad V = 4{\cdot}44fL\sqrt{2}I \text{ volts}$$
$$= 6{\cdot}28fLI \text{ volts}$$
$$= 2\pi fLI \text{ volts.}$$

$2\pi f$ is called the angular velocity and is measured in radians per second. The Greek letter omega (ω) is used to represent it.

$$\therefore \quad V = \omega LI \text{ volts.}$$

The ratio V/I is called the inductive reactance of the coil and it is measured in ohms, being represented by the symbol X_L.

$$X_L = \frac{V}{I} = 2\pi fL = \omega L \text{ ohms.}$$

A variable voltage at a frequency of 50 Hz was applied to an air-cored coil and the corresponding values of current were:

Voltage across the coil	0	5	10	15	20	25	volts
Current through the coil	0	3·05	6·0	8·9	12·1	15·0	amperes

A graph of the current can be plotted against the voltage and the inductance of the coil can be calculated.

19.16. Angular velocity

If a voltage or a current varies sinusoidally, it can be represented mathematically by either $v = V_M \sin \theta$ or $i = I_M \sin \theta$ respectively. As a conductor cutting magnetic flux moves through 360° electrical when generating 1 cycle of e.m.f., it will move through ($360f$) degrees electrical in one second where f is the frequency of the supply. Its angular velocity could be expressed in degrees electrical per second. An angle of $360° = 2\pi$ radians and so the above angular velocity could also be expressed as $2\pi f$ radians per second.

The electrical angle θ turned through by a conductor in t sec is

$$\theta = 2\pi ft = \omega t \text{ radians.}$$

Thus $v = V_M \sin \omega t$ and $i = I_M \sin \omega t$, respectively.

EXAMPLE. Calculate the instantaneous value of current $\frac{1}{600}$ of a second after it has passed through zero value if the maximum value of the current is 10 A and its frequency is 50 Hz.

$$i = I_M \sin \theta$$
$$= I_M \sin 2\pi f t$$
$$= 10 \sin\left(2\pi \times 50 \times \frac{1}{600}\right) \text{A}$$
$$= 10 \sin \frac{\pi}{6} \text{ A.}$$

As 2π radians $= 360°$, 1 radian $= \frac{360°}{2\pi}$ and $\frac{\pi}{6}$ radians $= \frac{360°}{2\pi} \times \frac{\pi}{6} = 30°$.

$$\therefore \quad i = 10 \sin 30° \text{ A}$$
$$= 10 \times 0{\cdot}5 \text{ A}$$
$$= \mathbf{5\ A.}$$

19.17. Sinusoidal alternating voltage applied to a capacitor

For a capacitor,

$$Q = CV$$

or

$$It = CV \text{ coulombs,}$$

$$\therefore \quad I = C\frac{V}{t} \text{ amperes.}$$

Expressing this in words,

Current = capacitance × (rate of change of voltage).

Figure 19.13 shows the waveform of a sinusoidal voltage which is applied to a capacitor. The current associated with the movement of charge to or from the capacitor can be deduced from the rate of change of the voltage across the capacitor.

At 0°, 180°, and 360°, the rate of change of voltage and, therefore, the instantaneous value of current, is at its maximum whilst at 90° and 270° the rate of change of voltage and, therefore, the current is instantaneously zero. From 0° to 90° the current is in the same direction as the applied voltage whilst the capacitor is being charged, whereas from 90° to 180° the current is in the opposite direction to the direction of the applied voltage because whilst the voltage is decreasing, the capacitor is being discharged.

If other values of voltage were considered, it would be found that the variation of the current was sinusoidal but anticipating the voltage variations by 90°. The current in a capacitor is said to lead the applied voltage by 90°.

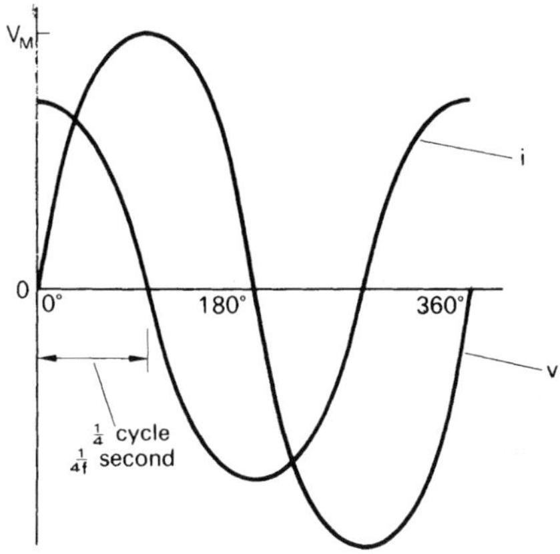

FIG. 19.13

19.18. Capacitive reactance

Figure 19.13 shows that the voltage across the capacitor changes from zero to V_M in $\frac{1}{4}$ of a cycle or in $1/(4f)$ sec. The change in voltage is $(V_M - 0) = V_M$ and therefore

$$\text{Average rate of change of voltage} = \frac{V_M}{\frac{1}{4}f} = 4fV_M \text{ volts per second,}$$

$$\therefore \text{ average current } I_{av} = (\text{average rate of change of voltage}) \times (\text{capacitance})$$

$$= 4fV_M C \text{ amperes.}$$

Changing to r.m.s. values of voltage and current,

$$\frac{I}{1{\cdot}11} = 4fC \times \sqrt{2}\ V \text{ amperes}$$

$$I = 4{\cdot}44 \times \sqrt{2} fCV \text{ amperes}$$
$$= 6{\cdot}28 fCV \text{ amperes}$$
$$= 2\pi fCV \text{ amperes.}$$

The ratio V/I for a capacitor is called the capacitive reactance and is represented by the symbol X_C.

$$X_C = \frac{V}{I} = \frac{1}{\omega C} \text{ ohms.}$$

Remember that the capacitance C in the above expressions is in farads. If, as is usual, the capacitance is given in microfarads, the factor 10^6 must be introduced.

EXAMPLE. Find the current through an 80-μF capacitor from a 200-V, 50-Hz supply. If the frequency were doubled, what now would be the current through the capacitor?

$$\text{Capacitive reactance } X_C = \frac{1}{2\pi fC} \text{ ohms}$$
$$= \frac{10^6}{2\pi \times 50 \times 80} \text{ ohms}$$
$$= 39{\cdot}8\ \Omega$$
$$\text{Current } I = \frac{200}{39{\cdot}8} \text{ amperes}$$
$$= \mathbf{5{\cdot}03\ A.}$$

$$\text{Current } I = 2\pi fCV$$
$$\therefore \quad I \propto f.$$

Hence if the frequency were doubled, the current would also be doubled.

$$I = \mathbf{10{\cdot}06\ A.}$$

19.19. Power in alternating current circuits

The power at any instant during a cycle is obtained by multiplying together corresponding instantaneous values of current and voltage.

Consider a battery. If the e.m.f. of the battery and the current through it are in the same direction, energy is being taken from the battery and is being used externally to it. If the current and the e.m.f. are acting in opposite directions, energy is being put into the battery from an external source.

In the case of the alternating circuit, if the current and the voltage at any instant are acting in the same direction as each other, the energy transfer is in one direction and if these quantities are instantaneously in opposite directions to each other, the energy transfer is in the opposite direction to previously. This direction of energy transfer is shown graphically by drawing the power curves either above or below the axis of the graph. In Fig. 19.14, the curve of instantaneous power has been drawn above the axis when the voltage and current are instantaneously in the same direction,

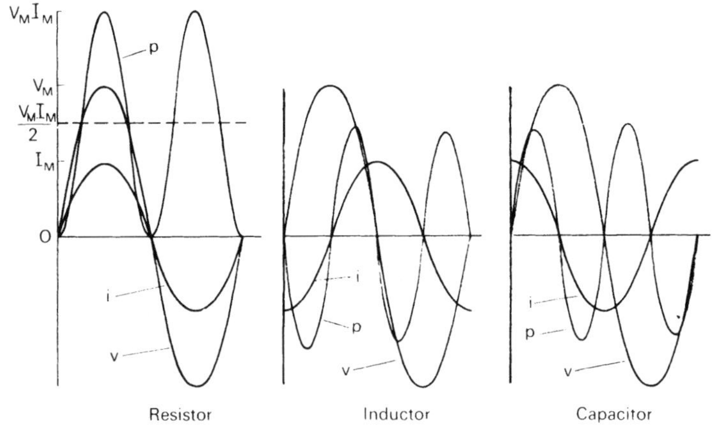

FIG. 19.14

and below the axis when they are instantaneously in opposite directions.

Figure 19.14 shows the voltage, current and power curves for resistors, inductors, and capacitors. The power at any instant, $d = vi$ watts. The product of the instantaneous values of quantities that are varying sinusoidally is another sinusoidally varying quantity.

Consider first the resistor. The power curve can be shown to be sinusoidally varying about the dotted line midway between zero power and the maximum value of the power curve, the maximum value being $V_M I_M$ watts. The average power, which is indicated on a wattmeter, is given by

$$P = \frac{V_M I_M}{2}$$

$$= \frac{V_M \cdot I_M}{\sqrt{2}\sqrt{2}}$$

$= VI$ watts (where V and I are r.m.s. values).

This is similar to the expression used for measuring power in direct current circuits. As the r.m.s. value was defined as being equivalent to the d.c. value, this was to be expected.

For both the inductor and the capacitor, the power wave is symmetrical about the zero axis and so the average power is zero. A wattmeter connected in a circuit containing only one or other of these circuit components would indicate zero watts. What is happening is that during $\frac{1}{4}$-cycle, energy is being taken from the supply and is being stored in the magnetic or electric field and this energy is returned to the supply during the next $\frac{1}{4}$-cycle.

Thus in any circuit containing resistors, inductors, and capacitors, energy is only expended in the resistors. The power in a resistor can also be calculated from the expressions I^2R and V^2/R watts, where I is the current through and V the potential difference across the resistor.

19.20. Volt-amperes, power, and power factor

When an alternating voltage is applied to a circuit containing any combination of resistors, inductors, and capacitors, the product of the applied voltage and the current, known as the apparent power and measured in volt-amperes, is normally greater than the true power in watts as indicated by a wattmeter. The ratio

$$\frac{\text{watts}}{\text{volt-amperes}}$$

is called the power factor of the circuit. Thus the real power (in watts) is equal to the apparent power (in volt-amperes) multiplied by the power factor. The power factor can be any value between unity, when the circuit is purely resistive, to zero when the circuit is purely inductive or capacitive. If the capacitive reactance is predominant, the power factor is said to be leading as the current will be leading the voltage. If the inductive reactance predominates, the power factor is said to be lagging as the current will be lagging the voltage.

Electrical equipment is usually rated in kilovolt-amperes (kVA) or megavolt-amperes (MVA)

$$1 \text{ MVA} = 10^3 \text{ kVA} = 10^6 \text{ VA}.$$

19.21. Vector (or phasor) representation of sinusoidal alternating quantities

Consider a vector OA rotating at constant velocity in a conventionally anticlockwise direction. If the lengths $AB = OA \sin \theta$ are plotted against θ for all values of θ from 0° to 360°, the resulting graph will be a sine wave. Should there be a second vector OC, rotating in the same direction at the same velocity but having a fixed angular displacement from the vector OA,—in Fig. 19.15 vector OC is lagging vector OA by α degrees—a second sine wave will result if its projections such as CD on to the line XOX are also plotted. This sine wave will lag the first by α degrees.

If the rotating vectors OA and OC represented, for example, the

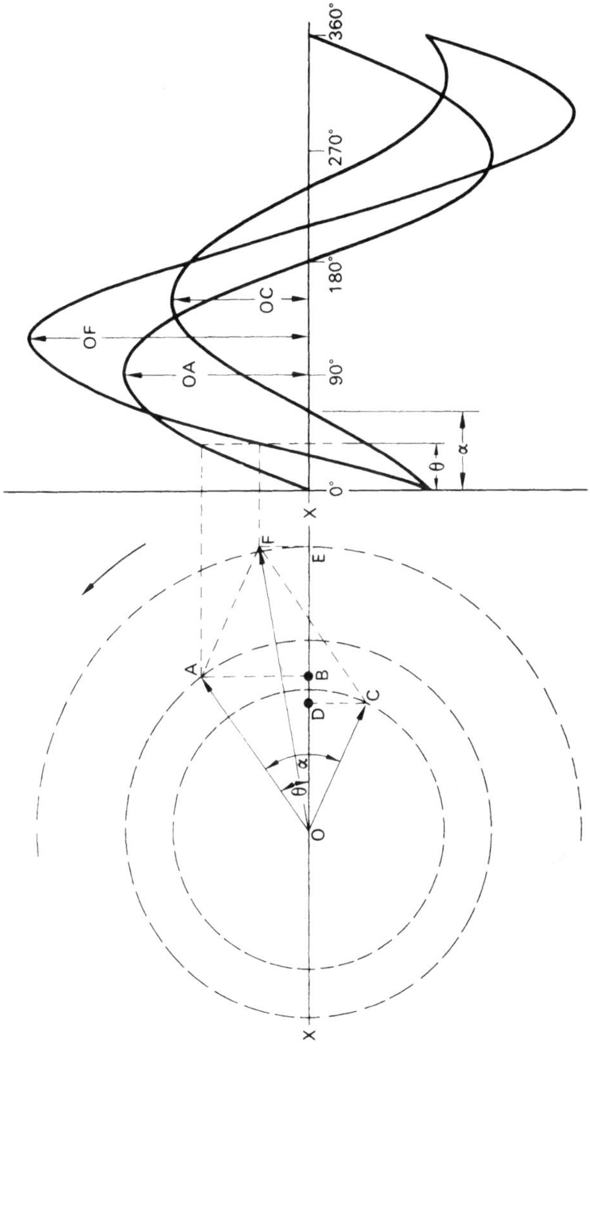

FIG. 19.15

maximum values of two sinusoidally varying voltages, the two waveforms would represent corresponding values of the instantaneous values of voltage for any given value of θ. As shown previously, $v = V_M \sin \theta$.

A series circuit supplied from an alternating voltage source will consist, or can be represented by, a combination of resistance, inductance, and capacitance. As the current is the same in all parts of the circuit, the voltages across the different types of circuit component will bear different phase relationships to each other and to the current. AT ANY INSTANT the value of the supply voltage is equal to the algebraical sum of the instantaneous voltages across the circuit components.

Referring again to Fig. 19.15, if the voltages were across two adjacent parts of a series circuit, the voltage across these parts of the circuit AT ANY INSTANT would be obtained by adding together the two instantaneous voltages and the resultant curve is also shown in the figure. This, too, is a sine wave. The same waveform would have been obtained if the two vectors *OA* and *OC* had been added by completing the parallelogram to give the vector *OF* and the projections of this vector such as *FE* had been plotted as described previously.

It is quicker and more convenient to find the resultant maximum value of the two alternating voltages by vector addition than by having to draw out the three sine waves. Thus if two voltages in a series circuit had maximum values of 30 V and 40 V and the former led the latter by 90°, the maximum value of the sum of the two voltages would be 50 V, this being found either by calculation or by means of vectors drawn to scale. If the three voltages were all divided by $\sqrt{2}$, the effective or r.m.s. values would be obtained. Thus r.m.s. values of voltages can be added vectorially. If a voltage were to be subtracted from another value, the method of vector subtraction is the same as that used for vectors in mechanical engineering, that is, reverse the direction of the vector and then add it to find the resultant.

It has been shown that $v = V_M \sin \omega t$ where ω is the angular velocity in radians per second. The rotating vector must have an angular velocity of ω rad/s when representing alternating voltages and currents.

Figure 19.16 shows vectorially the voltage and current phase relationships for a resistor, inductor, and capacitor. Vectors are now known as phasors when considering alternating current circuits.

In parallel circuits, phasors are used to find the resultant current when the current in each branch is known.

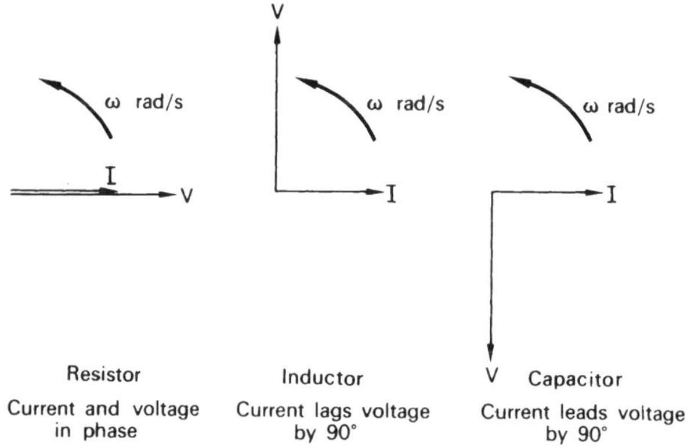

FIG. 19.16

19.22. Resistor and inductor in series

In a series circuit, the current is the same in all parts of it. When an alternating voltage is applied to such a circuit, the supply voltage

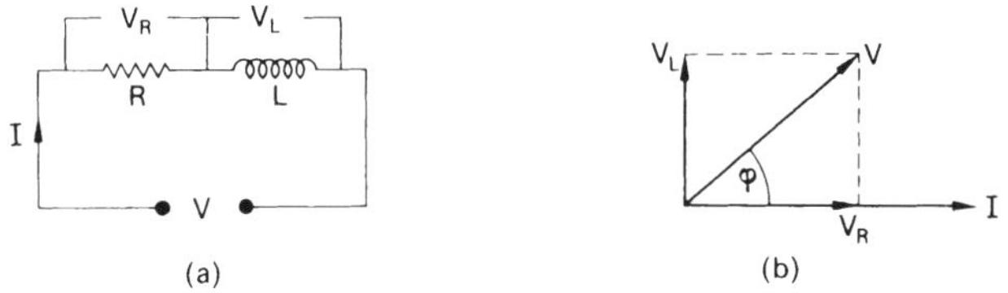

FIG. 19.17

V must equal the phasor sum of the voltages across the circuit components. In this case,

$V = V_R + V_L$ (phasor sum. This is NOT the arithmetical sum).

The voltages V_R and V_L are shown in the phasor diagram in Fig. 19.17(b). The phasor representing the voltage across the resistor V_R is in phase with the current and the phasor representing the voltage across the inductor is leading the current by 90°. The resultant voltage V is found by adding the phasors, that is, by completing the parallelogram. The phasor diagram shows that the supply current is lagging the supply voltage by less than 90°.

Considering the right-angled triangle in the phasor diagram,

$$V^2 = V_R^2 + V_L^2.$$

It is usual to put the term associated with resistance first in the equation so that when numbers are used, the first term is known to concern the resistive component.

From previous theory,

$$V_R = IR \quad \text{and} \quad V_L = I X_L,$$

$$\therefore \quad V^2 = (IR)^2 + (IX_L)^2$$

$$= I^2 (R^2 + X_L^2).$$

Hence $$V = I\sqrt{(R^2 + X_L^2)}.$$

The ratio V/I is called the impedance of the circuit.

$$\text{Circuit impedance } Z = \frac{V}{I} = \sqrt{(R^2 + X_L^2)} = \sqrt{(R^2 + (\omega L)^2)} \text{ ohms.}$$

Energy is only expended in the resistor and so the power taken by the circuit is

$$\text{power} = V_R I \quad \text{or} \quad I^2 R \text{ watts.}$$

The phasor shows that

$$V_R = V \cos \phi,$$

$$\therefore \quad \text{power} = VI \cos \phi \text{ watts.}$$

The power factor of the circuit is given by

$$\frac{\text{watts}}{\text{volt-amperes}},$$

$$= \frac{VI\cos\phi}{VI} = \cos\phi.$$

Whenever sinusoidal voltages and currents are present, the power factor is numerically equal to cos ϕ where ϕ is the phase angle between the supply voltage and current.

Again from the phasor diagram,

$$\cos\phi = \frac{V_R}{V} = \frac{IR}{IZ} = \frac{R}{Z},$$

$$\sin\phi = \frac{V_L}{V} = \frac{IX_L}{IZ} = \frac{X_L}{Z},$$

$$\tan\phi = \frac{V_L}{V_R} = \frac{IX_L}{IR} = \frac{X_L}{R}.$$

Figure 19.18 shows curves of instantaneous current, voltage and power when the current is lagging the voltage by ϕ degrees. The

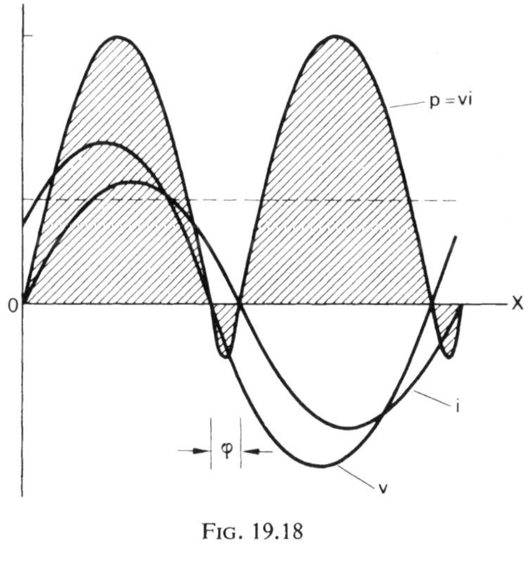

FIG. 19.18

large shaded areas above the axis OX represent the energy taken from the supply, part of which is converted into heat and the remainder stored in the magnetic field. The small shaded area below the axis represents the energy returned to the supply as the magnetic field collapses, this having previously been stored in the magnetic field.

The instantaneous power curve is symmetrical about the dotted line and its average value corresponds to the value of power given by the expression $VI \cos \phi$.

EXAMPLE. A circuit consists of a 20-Ω resistor and a 0·1-H inductor connected in series to a 200-V, 50-Hz supply. Calculate (a) the circuit impedance, (b) the supply current, (c) the voltage across each circuit component, (d) the power consumed, (e) the circuit power factor.

$$\text{Reactance of inductor } X_L = 2\pi f L = 2\pi \times 50 \times 0{\cdot}1 = 31{\cdot}4\ \Omega.$$

(a) $$\text{Circuit impedance } Z = \sqrt{(R^2 + X_L^2)} = \sqrt{(20^2 + 31{\cdot}4^2)} = \mathbf{37{\cdot}2\ \Omega}.$$

(b) $$\text{Current } I = \frac{V}{Z} = \frac{200}{37{\cdot}2}\ \text{A} = \mathbf{5{\cdot}38\ A}.$$

(c) $$\text{Voltage across resistor } V_R = IR = 5{\cdot}38 \times 20\ \text{V} = \mathbf{108\ V}.$$

Voltage across inductor $V_L = IX_L$

$= 5{\cdot}38 \times 31{\cdot}4$ V

$=$ **168 V.**

(With reference to the phasor diagram, note that $V^2 = V_R^2 + V_L^2$ or $200^2 = 108^2 + 168^2$.)

(d) Power consumed $P = I^2R$ or $V_R I$ watts

$= 5{\cdot}38^2 \times 20$ W or $108 \times 5{\cdot}38$ W.

$=$ **579 W.**

(e) $$\text{Power factor} = \frac{\text{watts}}{\text{volt-amperes}}$$

$$= \frac{579}{200 \times 5{\cdot}38}$$

$=$ **0·54** lagging (the current lags the supply voltage).

A power factor meter can only be read to two decimal places so there is no need to give answers more accurately than this.

An alternative method of solution is:

(a) Circuit impedance (as before) $Z =$ **37·2 Ω.**

(b) Supply current (as before) $I =$ **5·38 A.**

(e) $$\text{Power factor} = \cos\phi = \frac{R}{Z}$$

$$= \frac{20}{37{\cdot}2}$$

$=$ **0·54** lagging.

(c) Voltage across the resistor $V_R = V\cos\phi$

$= 200 \times 0{\cdot}54$ volts

$=$ **108 V.**

Voltage across the inductor $V_L = V\sin\phi$.

Using mathematical tables, cos $\phi = 0{\cdot}54$ and, therefore, sin $\phi = 0{\cdot}84$.

$$\therefore \quad V_L = 200 \times 0{\cdot}84 \text{ volts}$$
$$= \mathbf{168\ V.}$$

(d) Power consumed $P = VI \cos \phi$

$$= 200 \times 5{\cdot}38 \times 0{\cdot}54 \text{ watts}$$
$$= \mathbf{579\ W.}$$

19.23. Resistor and capacitor in series

The method of calculation for this series circuit is almost the same as for that described in the previous section. In this case, the voltage across the capacitor lags the current by 90°.

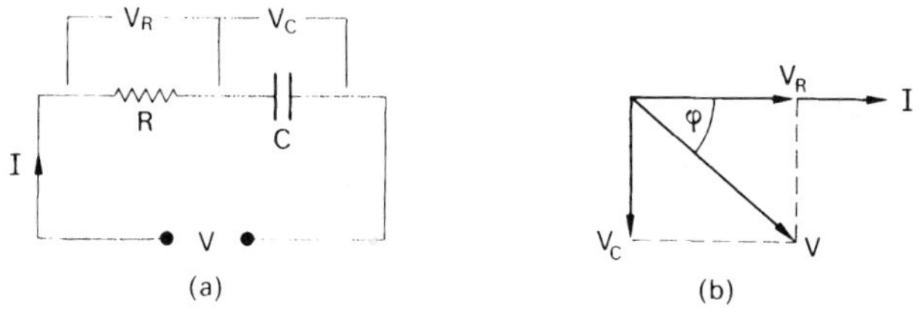

FIG. 19.19

Referring to the phasor diagram (Fig. 19.19),

$$V^2 = V_R^2 + V_C^2$$
$$= (IR)^2 + (IX_C)^2$$
$$= I^2 (R^2 + X_C^2)$$
$$\therefore \quad V = I\sqrt{(R^2 + X_C^2)}.$$

The circuit impedance

$$Z = \frac{V}{I} = \sqrt{(R^2 + X_C^2)} \text{ ohms}$$

$$= \sqrt{\left(R^2 + \left(\frac{1}{\omega C}\right)^2\right)} \text{ ohms.}$$

Energy is only expended in the resistor so, as in the previous section,

Power $P = V_R I$ or $I^2 R$ or $VI \cos\phi$,

and power factor $= \cos\phi$.

Also
$$\cos\phi = \frac{R}{Z},$$
$$\sin\phi = \frac{X_C}{Z},$$
$$\tan\phi = \frac{X_C}{R}.$$

A curve of the instantaneous power will be similar to that obtained for the *R-L* series circuit as shown in Fig. 19.18. In this case, however, energy is being stored in the electric field and then being returned to the system as the electric field collapses.

Calculations are carried out in a manner similar to that shown in the example on page 340.

19.24. A.C. and d.c. voltages applied to a coil

If equal a.c. and d.c. voltages are applied in turn to a coil, it will be found that the current with the applied a.c. voltage will be less than with an applied d,c. voltage.

This is because the alternating current sets up an alternating flux linking with the turns of the coil. This induces an e.m.f. in the coil which, by Lenz's law, must oppose the applied voltage. Thus part of the applied voltage is used in overcoming this back e.m.f. and a lower value is available for maintaining the current through the coil. The constant flux set up by the d.c. current does not cause any back e.m.f. to be set up in the coil.

EXAMPLE. An air-cored coil takes a current of 5 A from a 100-V d.c. supply, but when the same coil is connected to a 100-V a.c. supply the current taken is only 4 A. Explain why these two currents are different and calculate:

(i) the resistance of the coil,

(ii) the impedance of the coil,

(iii) the power input to the coil from the d.c. supply,

(iv) the power input to the coil from the a.c. supply,

(v) the power factor of the coil for the a.c. supply.

(U.L.C.I.)

(i) Coil resistance $R=\frac{100}{5}=\mathbf{20\ \Omega}$.

(ii) Coil impedance $Z=\frac{100}{4}=\mathbf{25\ \Omega}$.

(iii) Power input $=100\times5=\mathbf{500\ W}$ or $=5^2\times20=\mathbf{500\ W}$.

(iv) Power input $=4^2\times20=\mathbf{320\ W}$.

(v) Power factor $=\frac{320}{100\times4}=\mathbf{0{\cdot}8}$ lagging.

19.25. The dynamometer wattmeter

A wattmeter is an instrument which indicates the average power in a circuit. It has a moving coil which carries a current proportional to the supply voltage in a manner similar to that in a moving-coil voltmeter. The permanent magnet of the moving-coil instrument is replaced by two fixed air-cored coils which carry the current to the load and this current sets up a flux which is proportional to the current. The torque on the moving coil is dependent on the voltage and the current and the scale is calibrated to read the average power in the circuit. It is used on both a.c. and d.c. circuits.

Further consideration of the wattmeter is outside the scope of this book.

Problems

19.1. A coil of negligible resistance has an inductance of 0·2 H. Calculate the current it would take from a 240-V, 50-Hz supply.

(3·82 A.)

19.2. A capacitor takes a current of 2·5 A from a 240-V, 50-Hz supply. What is its capacitance in microfarads?

(33·2 μF.)

19.3. What is the difference between watts and volt-amperes in an a.c. circuit? When are these two quantities equal?

A 5-hp, single-phase, a.c. motor whose full-load efficiency is 84% takes a current on full-load of 22 A from a 240-V supply. Calculate: (a) the power input in kilowatts, (b) the power factor.

((a) 4·44 kW, (b) 0·84.) (N.C.T.E.C.)

19.4. A sinusoidal alternating voltage of 200 V is applied in turn to a resistor, an inductor and a capacitor and in each case the current is 5 A. Calculate for each case (a) the power, (b) the kVA taken from the supply, (c) the impedance.

((a) 1000 W, 0, 0; (b) 1 kVA in each case; (c) 40 Ω in each case.)

19.5. An a.c. generator has six poles and it is driven at 1000 rev/min. What is the frequency of the e.m.f. generated? If the e.m.f. generated has a value of 200 V and the current taken by a circuit connected to it takes a current of 10 A at a power factor of 0·4, calculate (a) the power taken, (b) the impedance of the circuit.

((a) 800 W, (b) 20 Ω.)

19.6. A 100-Ω resistor is connected in series with a coil. If the voltage across the resistor is 50 V and across the coil 100 V, calculate (a) the current in the circuit, (b) the reactance of the coil.

((a) 0·5 A, (b) 200 Ω.)

19.7. The instantaneous value of an alternating current is given by $i = 20 \sin 314t$. What is (a) the frequency, (b) the r.m.s. value of the current?

((a) 50 Hz, (b) 14·14 A.)

19.8. A 16-Ω resistor and a 0·0318-H inductor are connected in series to a 200-V, 60-Hz supply. Calculate (a) the current from the supply, (b) the voltage across each component, (c) the circuit power factor, (d) the power consumed.

((a) 10 A, (b) $V_R = 160$ V, $V_L = 120$ V, (c) 0·8 lagging, (d) 1600 W.)

19.9. In a circuit consisting of a resistor and inductor in series, the voltage across the resistor is 60 V and across the inductor 80 V. The value of the resistor is 30 Ω. Draw to scale the phasor diagram and find (a) the supply voltage, (b) the circuit power factor. Calculate the current from the supply and the power consumed.

((a) 100 V, (b) 0·6 lagging. 2 A, 120 W.)

19.10. An 80-μF capacitor is connected in series with a 30-Ω resistor across a 200-V, 50-Hz supply. Calculate (a) the current flowing, (b) the power factor, (c) the power consumed, (d) the voltage across each component.

((a) 4 A, (b) 0·6 leading, (c) 480 W, (d) $V_R = 120$ V, $V_C = 160$ V.)

19.11. A circuit consisting of a resistor and a capacitor in series takes a current of 6 A at a power factor of 0·6 leading from a 240-V, 50-Hz supply. Calculate (a) the power consumed, (b) the value of the capacitor in μF.

((a) 864 W, (b) 66 μF.)

19.12. A 40-Ω resistor and a capacitor are connected in series to a 200-V, 50-Hz supply. The voltage across the capacitor is 140 V and the circuit power factor is 0·72 leading. Determine the power consumed by the circuit.

(525 W.)

19.13. Three 110-V, 60-W lamps are connected in parallel and the combination connected in series with a capacitor to a 230-V, 50-Hz supply. Determine the capacitance of the capacitor so that the lamps are operating at normal voltage.

(25·7 μF.)

Answers are not given to the following questions:

19.14. An 0·15-H inductor takes a current of 5 A from a supply whose frequency is 50 Hz. What current would an 80-μF capacitor take from the same supply?

19.15. When an inductor of 10-Ω inductive reactance and negligible resistance is connected across a 50-Hz supply the current is 4 A. Draw a graph to scale to show the variation of inductive reactance with frequency from 25 Hz to 100 Hz in steps of 25 Hz. Also draw a graph to scale to show the variation of current for the same frequency range on a constant supply voltage. (Make the horizontal axis the frequency scale.) (U.L.C.I.)

19.16. Calculate the instantaneous value of a sinusoidally varying e.m.f. at a time 0·003 sec after it passes through zero if its maximum value is 200 V and its frequency 50 Hz.

19.17. What are the average and r.m.s. values of an isosceles triangular wave of maximum value 20 V?

19.18. A motor-driven bucket conveyor is used to raise vertically a distributed load of 2 tonf at a steady speed of 25 ft/min. The supply to the motor is 240 V single-phase a.c. and the motor works at a power factor of 0·7 lagging.

Assuming that the efficiency of the motor is 85% and that of the conveyor is 75%, calculate: (a) the output of the motor in horsepower, (b) the current taken by the motor from the supply. (C. & G.)

19.19. A sinusoidal alternating supply has an r.m.s. value of 100 V and a frequency of 50 Hz, and is connected to a series circuit having both resistance and inductance. The current taken from the supply has a r.m.s. value of 5 A and the power taken is 250 W. Calculate:

(a) the peak value of the current,
(b) the time taken for one complete cycle of the supply,
(c) the impedance of the circuit,
(d) the apparent power (volt-amperes) taken by the circuit,
(e) the power factor of the circuit,
(f) the resistance of the circuit. (C. & G.)

CHAPTER 20

Rectifiers

20.1. Introduction

For many years, rectifier equipment was almost exclusively of the mercury-arc and selenium types but they have been superseded in many applications by the high-power silicon and germanium diodes which were developed in the mid-1950's, and, more recently, by the thyristor.

All rectifiers operate on the principle that the resistance to current in one direction is very small and comparatively large currents can pass whereas current in the reverse direction is very small. Moreover, for these devices, current is not proportional to voltage. In other words, they have a non-linear voltage–current characteristic.

20.2. Copper oxide rectifier

First developed about 1920, this type consists of a semi-conducting layer of copper oxide formed on a thin copper sheet. The outer surface of the oxide has a graphite coating and a thin sheet of lead is pressed against this graphite. Each unit may be in the form of a washer and these are threaded onto an insulating tube along with some metal plates which have a bigger area than the units, these acting as cooling fins. These "stacks" are bolted together so as to ensure good electrical contact. Thus the rectifier units are connected in series, so enabling a higher voltage to be applied. The direction of current is from the oxide surface to the copper. These rectifiers have many uses in telecommunications.

20.3. Selenium rectifier

Developed a few years after the copper-oxide rectifier, this type has a layer of semiconducting selenium on a plate of steel or aluminium. Current direction is from the plate to the selenium. The rectifier units are built into stacks in a manner similar to that described for the copper-oxide rectifier. Selenium and copper-oxide rectifiers have traditionally been known as metal rectifiers.

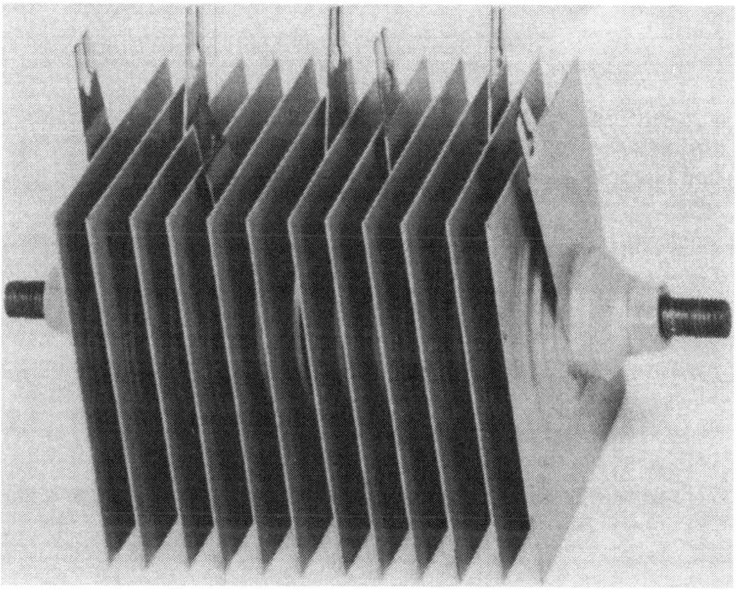

FIG. 20.1. Selenium rectifier unit.
(Westinghouse Brake and Signal Co. Ltd.)

20.4. Semiconductor point contact rectifiers

These are a more recent development. Typical voltage/current characteristics in the forward or conducting direction (+) and in the reverse or non-conducting direction (−) are as follows:

Silicon rectifier:

Voltage	+1·1	+1·0	+0·9	+0·8	+0·6	0	−1	−2	−3	−3·5	−4 V
Current	+5	+4	+3	+2	+1	0	−0·2	−0·3	−0·6	−1·5	−4 mA

Germanium rectifier:

Voltage	+1·0	+0·85	+0·7	+0·6	0	−40	−80	−100 V
Current	+4	+3	+2	+1	0	−0·2	−0·45	−0·8 mA

The reader should plot these points. In the case of the silicon rectifier, it will be seen that the current increases rapidly as the semiconductor is made positive with respect to the point electrode but when the polarity of the applied voltage is reversed, the current is small up to about 3 V, after which the reverse current increases rapidly.

When plotting the characteristic of the germanium rectifier, different scales should be used for the positive (forward) and negative (reverse) voltages. The reverse resistance is very much higher than that of the silicon rectifier. If the negative voltage exceeds about 100 V, the current increases with a decrease of voltage and if there is not a sufficiently high external resistance in the circuit, the rectifier may be destroyed by excessive current.

20.5. Junction diode rectifiers

The germanium or silicon crystals used in these rectifiers are quite small but the dissipation of heat produced in these rectifiers is a problem which can be overcome by mounting them in metal cans and fastening the can to large cooling fins. These rectifiers are capable of handling large currents. A 120-A silicon diode has a power dissipation of about 150 W when passing this current.

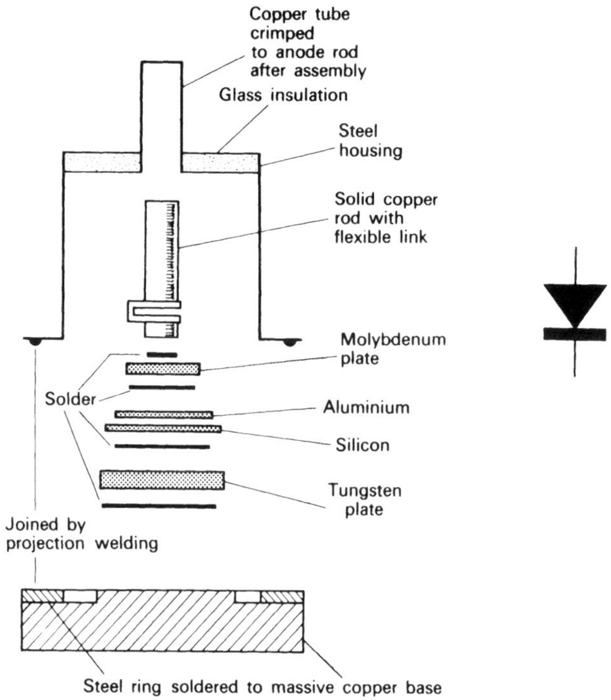

FIG. 20.2. Silicon junction rectifier.
(Westinghouse Brake and Signal Co. Ltd.)

20.6. Comparison of rectifiers

SELENIUM

They are economical for small power outputs but they have a limited life and ultimately have to be replaced. They can, however, recover from reasonable overloads.

Mercury arc

This valve rectifier is easy to instal and the bulb will last for many years. They are bulkier and more expensive than a semi-conductor rectifier of equivalent rating. In the U.S.A. production of this type of rectifier has virtually ceased.

Germanium

If the ambient temperature exceeds 45°C, they should not be used. Below voltages of about 220 V d.c., they are more efficient than the equivalent silicon rectifier. The forward voltage drop is about 0·5 V per diode.

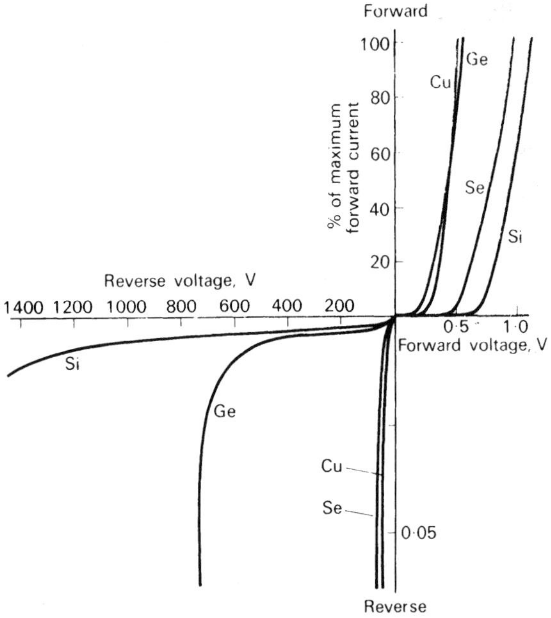

Cu—Copper oxide
Se —Selenium
Si —Silicon
Ge—Germanium

Forward and reverse characteristics of various rectifiers, based on d.c. measurements

Fig. 20.3. (Westinghouse Brake and Signal Co. Ltd.)

SILICON

They should always be considered if the ambient temperature is above 45°C. Their forward voltage drop is about 1·2 V per diode and this usually results in a lower efficiency than the germanium rectifier, particularly below about 440 V d.c.

20.7. Protection of rectifiers

All rectifiers must be protected against over-currents and excessive voltages. The mercury arc rectifier can be protected by the use of a.c. and d.c. circuit breakers, H.R.C. (high rupturing capacity) anode fuses and surge divertors. Selenium rectifiers will withstand heavy current over-loads of short duration. Since quite high over-voltages, again of short duration, do not normally cause damage, standard fuses and switchgear may be used.

Germanium and silicon rectifiers have a small thermal mass and so are liable to damage by overcurrents flowing for quite short periods. They may also be damaged by high voltages of a momentary nature and so the protection must include some means of limiting any surge voltages to a safe value. Special high-speed fuses have been produced for use with semiconductors and these can be used in conjunction with the more usual protective devices. Standard H.R.C. fuses must never be used to protect these rectifiers. Further discussion on rectifier protection is outside the scope of this book.

20.8. Current and voltage outputs

Semiconductor rectifiers can be connected in series to give the required voltage output and the combinations connected in parallel if larger currents are required. Rectifiers are normally used to obtain a direct current supply from an alternating current source which may be single- or poly-phase. The circuits used are dealt with in later studies. The symbol used to represent a rectifier bank is shown in Fig. 20.4 along with the conventional direction of current flow.

Selenium rectifiers can be manufactured for any voltage and

current. Mercury arc rectifiers are made for voltages above 50 V and currents greater than 5 A. Germanium and silicon rectifiers are normally manufactured to give an output of to 1000 V d.c. for any current but higher voltages can be obtained if required.

Rectifiers are used for direct current supplies, electrochemical applications, telecommunications, etc.

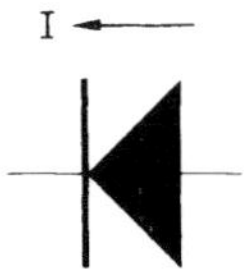

FIG. 20.4

20.9. Thyristors

A thyristor is a device which remains in a non-conducting state even with a suitable p.d. of the correct polarity across it until a suitable voltage is applied to what is called the gate. This switches it into a conducting state and its characteristic becomes similar to

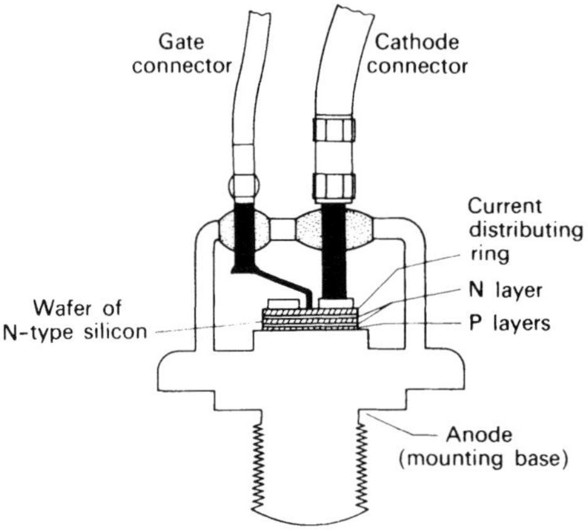

FIG. 20.5. Thyristor.
(Westinghouse Brake and Signal Co. Ltd.)

that of a junction diode. Neglecting inductance effects, the device ceases to conduct whenever the current becomes zero, until a further voltage is applied to the gate.

20.10. Rectifier circuits

Figure 20.6 shows typical half-wave and full-wave rectifier circuits. A half-wave circuit is one in which only half the full sine-

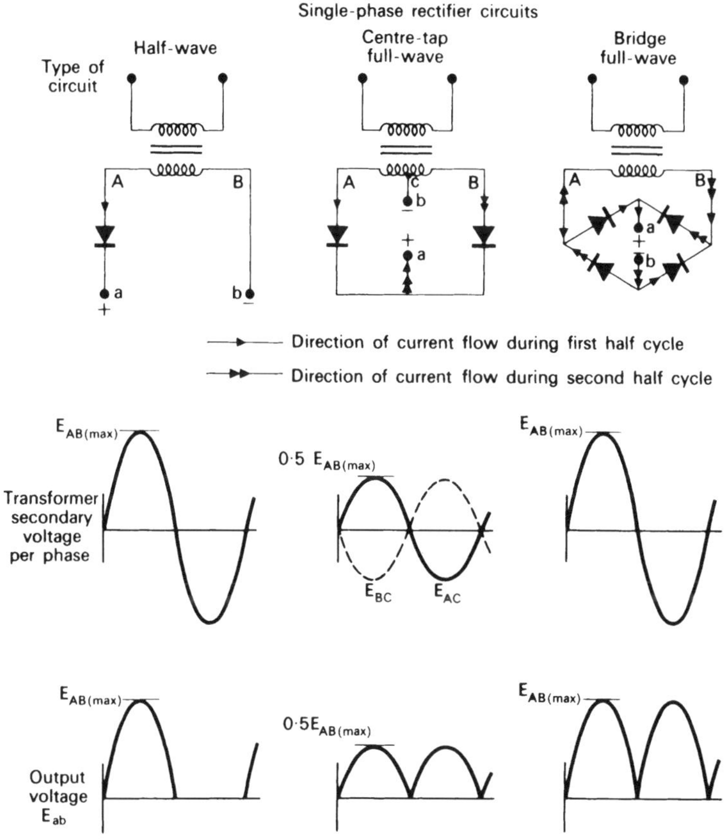

FIG. 20.6. Single-phase rectifier circuits.

wave of the voltage is rectified and, in a full-wave circuit, both halves of the alternating voltage waveform are rectified. Although being far from a smooth voltage output, a full-wave circuit gives a smoother output than does the half-wave rectifier. Poly-phase circuits give a much smoother output. Assuming a resistance load, the output current waveform is the same as that of the voltage E_{ab}.

If heavy currents are being rectified, it is usual to insert an iron-cored choke in series with the output terminals. It operates by inducing an e.m.f. in opposition to the ripple voltage of the rectified output. For small currents when the rectifier is supplying a high-resistance load such as may occur in electronic circuits, a capacitor is connected across the output terminals of the rectifier. This "reservoir" capacitor will charge up as the output voltage increases in magnitude and will discharge into the load during the periods when no energy is being transferred from the source of supply. This tends to smooth the output voltage as shown in Fig. 20.7(a).

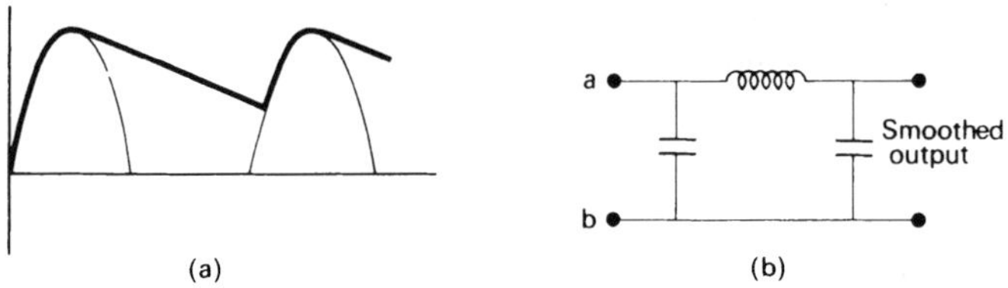

FIG. 20.7. (a) Smoothed output voltage. (b) Filter.

When no load is connected, the output voltage rises to the peak value of the alternating voltage. If an output is required having a negligible ripple, a combination of capacitor and a choke may be used such as is shown in Fig. 20.7(b), and this combination is called a FILTER.

20.11. Transistors

In Chapter 9 an explanation was given in terms of free electrons of the difference between conductors and insulators. Mention was made of a group of materials whose characteristics were similar to conductors and these were called resistors.

There is another class of materials whose electrical properties lie somewhere between conductors and insulators and these are called semiconductors. Germanium and silicon are typical semiconductor materials. By adding to germanium certain impurities in carefully-controlled quantities, it becomes a better conductor. There are two types of this impure germanium:

N-type. The particular impurity results in more free electrons than in pure germanium and if they are made to move in one direction, they constitute an electric current.

P-type. In this case, the impurity causes a deficiency of free electrons in the atom and the places where free electrons might have been are referred to as holes.

A hole can be filled by an electron from a neighbouring atom and the hole that has then been created will in its turn be filled by another electron from another nearby atom. Thus the hole "moves" in the opposite direction to the electrons and the "movement" of a hole in this manner is equivalent to the movement of a positive charge equal in magnitude to the negative charge carried by a single electron.

A typical *pnp* transistor (Fig. 20.8) consists of a thin slice of *n*-type germanium sandwiched between two layers of *p*-type germanium. The middle piece of *n*-type germanium is called the base and the other two pieces of *p*-type germanium are referred to as the collector and emitter respectively.

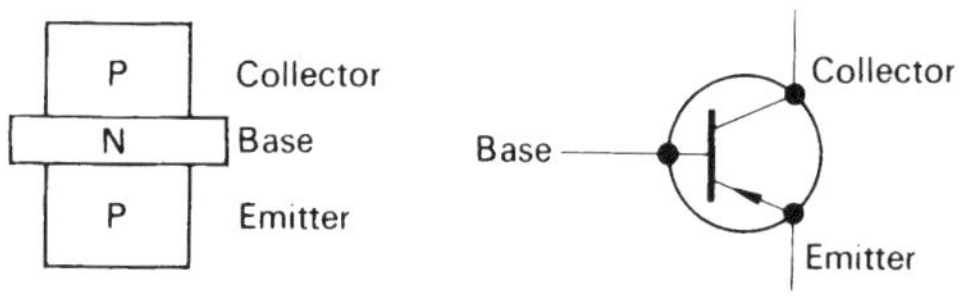

FIG. 20.8. Representation of transistor.

If the transistor is connected as in Fig. 20.9, which is a commonly used arrangement known as the common emitter connection, the action of the holes and the electrons across the two junctions under the influences of the two voltages is such as to cause the small base current to control the emitter–collector

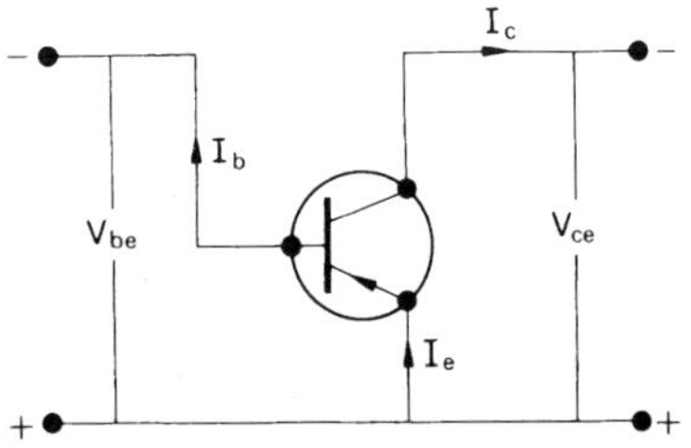

FIG. 20.9. Transistor-common emitter connection.

current. For instance, for a fixed value of V_{ce}, an increase in the base current I_b of only a few microamperes (which is obtained by increasing V_{be}) will result in a much larger increase in the collector current I_c in the order of milliamperes. As a small current is controlling a much larger current, the transistor is acting as a form of amplifier. The transistor is a current-controlled device whereas thermionic valves are voltage-controlled.

Compared with thermionic valves, transistors are very much smaller—overall dimensions may be less than $\frac{1}{8}$ in. As transistors have no heated cathode, they do not require a heater supply and they begin to operate as soon as the equipment is switched on. A valve may require over 100 V for its operation in a circuit whereas a transistor operates on a very much lower voltage.

Their applications include hearing aids, radios, televisions, computers, test equipment, etc.

CHAPTER 21

Electric Cables

21.1. General

Electric cables are normally installed on the assumption of a safe working life of at least 20 years. Changes in the insulating material take place with the passing of time and these changes, which may eventually result in an electrical breakdown, are accelerated at higher temperatures. Thus, if the working life is fixed, the limiting factor is the temperature at which the cable is required to operate.

During operation, the temperature at which the cable will operate depends upon the ambient temperature and the heating effects of the current produced due to the resistance of the cable conductors.

The heat dissipation of buried cables depends on the depth of laying, ground ambient temperature and its thermal resistivity, these being dependent on their geographical location and the season of the year. Nearby cables would also affect the ground temperature. Cables in air reach steady operating temperatures more quickly than similar cables underground and large cables take longer than small ones.

The heat may cause a change in the properties of an insulating material or in extreme cases, deformation may occur. It is important, therefore, to realise that there is "a cable for the job".

There is a very wide range of cables designed to operate at voltages up to 400 kV. It is not possible to discuss all these in this book, but the reader is referred to a publication, *Copper Cables*, published by the Copper Development Association.

The majority of cables have copper conductors and in a cable these may vary from a single conductor to stranded construction.

The number of wires contained in most common conductors is 3, 7, 19, 37, 61 or 91. Thus, 37/0·083 indicates that the conductor has 37 wires each having a diameter of 0·083 in.

21.2. Types of insulation (rubber)

RUBBER

Until about 1945 natural rubber was the only material used for insulating and sheathing electric cables used for the wiring of buildings. Rubber is one of a group of materials known as elastomers, these requiring vulcanising in order to make them sufficiently tough and elastic for use in cable manufacture. The unvulcanised material tends to remain in a stretched condition after a load has been applied.

Natural rubber is still extensively used in cable manufacture and is often used as a basis for comparison of other rubber-like insulating materials. Like other elastomers, it suffers from oxidation (generally referred to as ageing) at room and high temperatures but certain substances can be added to the rubber to reduce this "ageing".

If rubber cables are installed and operated in accordance with the Wiring Regulations of the Institution of Electrical Engineers, a maximum conductor temperature of about 50°C is reached if the cable is loaded in accordance with I.E.E. current ratings but a 20-year life could be reasonably expected with an operating temperature of 60°C using modern rubber compounds.

At temperatures as low as −40°C, rubber remains quite flexible. It does not, however, offer a good resistance to oil and solvents and should not be used if there is any likelihood of this combination occurring.

Natural rubber is combustible and will burn fairly easily once ignited. The carbonaceous residue is electrically conducting but as it is not expected that a cable will catch fire, this is not a serious disadvantage.

If a natural rubber cable is exposed to ozone or strong sunlight with its ultraviolet rays, the surface of the rubber cracks. This is more pronounced if there is any additional mechanical stress.

The insulation resistance of an installation using rubber cables is generally fairly high and will remain so for many years. In humid conditions, the electrical properties deteriorate to some extent.

Rubber compounds are often used for trailing cables operating at voltages up to 11 kV, these sometimes being used to supply electrical equipment in quarries.

Butadiene styrene (S.B.R.)

This used to be referred to as synthetic rubber and there is probably more S.B.R. than any other synthetic rubber.

Insulating compounds made from S.B.R. are not usually as good as those made from natural rubber. The tensile strength is usually considerably less and, whilst the electrical insulation resistance is a little less than natural rubber, it is quite suitable for the normal wiring of buildings. S.B.R. is similar to natural rubber in respect of its resistance to oil, solvents and fire and should be operated at about the same maximum temperature.

Cables with this insulation are somewhat cheaper than natural rubber-insulated cables, the economy being obtained at the expense of poorer physical and electrical properties.

Butyl rubber

Compared with natural rubber, butyl rubber has better ageing characteristics at the higher temperatures. It offers a greater resistance to the effects of ozone, and also to the effects of ultraviolet rays.

Butyl rubber is superior to natural rubber in moist conditions in respect of insulation resistance, but its tensile strength is very much less. In respect of oils, solvents, and fire, it is approximately equivalent to natural rubber.

The overall cost of producing cables with this insulation is such that they are normally only used where it is possible to utilise any of the better characteristics.

POLYCHLOROPRENE OR P.C.P.

This is much superior to natural rubber in its resistance to oils, weathering, and fire and for these reasons the cables are used underground in coal-mines. Should they catch fire, chlorine gas is produced which prevents oxygen from permitting the fire to continue.

Its maximum operating temperature is about 70°C. Unlike the materials previously mentioned, it is not suitable for very low temperatures as below about −20°C the compound stiffens significantly. Electrically, it is quite poor as the insulation resistance is about $\frac{1}{500}$ of that of natural rubber.

SILICONE RUBBER

This is one of the most expensive elastomers but it has a very high resistance to ageing at high temperatures and may be used continuously at 150°C and for short periods up to 200°C with a somewhat shorter life.

It is comparable to natural rubber in its resistance to oil and if it burns, the ash is quite a good insulator if it is not disturbed.

21.3. Types of insulation (thermoplastics)

These materials become softer as their temperature is increased. The material is usually extruded onto the conductor and cooled by passing through water. Examples are given below.

POLYVINYL-CHLORIDE (P.V.C.)

This is the most widely used alternative to natural rubber. There are many different compounds under this heading and the resultant properties also vary considerably.

For ordinary wiring purposes, P.V.C. is electrically inferior to natural rubber in respect of insulation resistance but electrical breakdown properties are not very much less than natural rubber. The usual P.V.C. cable can be worked up to 70°C but some types are suitable for use up to 85°C.

As the temperature falls, P.V.C. ultimately becomes brittle and special care is required in handling cables using this material at temperatures below 0°C.

P.V.C. has a very good resistance to fire and at normal temperatures it is oil and solvent resistant, and is not damaged by many acids. These qualities are poorer above 70°C, but there are some cables using P.V.C. compounds which have operated for years in hot transformer oil.

POLYTHENE

Polythene is used instead of P.V.C. wherever its excellent electrical characteristics are to be utilised. However, it has a sharp melting point at about 110°C so an installation must not be operated anywhere near this temperature. It is suitable for use at the lowest temperature met in normal installations.

It is resistant to most oils and solvents at room temperatures but this resistance decreases with increase in temperature. It burns easily, the molten material falling in flames off the conductor. Thus it must not be used where there is a fire risk.

Bright sunlight may cause deterioration of this material and cracking may take place through to the conductor. This effect can be greatly reduced by producing the polythene with a small percentage of carbon black uniformly dispersed in it.

Cables using this material are frequently used for the high-frequency currents associated with radio, television and radar.

POLYTETRAFLUORETHYLENE (P.T.F.E.)

This material can be used for long periods at temperatures of up to 250°C but above this value toxic fumes may be given off as it decomposes.

It is very costly and whilst having the good electrical properties of polythene, it is only used where the high cost is justified by the benefits obtained by using it at high temperatures.

Trade names of this material include Fluon and Teflon.

NYLON

A covering of nylon as thin as 0·004 inch has been satisfactorily used in field telephone cables and will withstand the handling that these cables normally receive.

It is not a particularly good insulator but has good ageing properties up to 100°C and is very resistant to oil and chemicals.

21.4. Other types of insulated cables

ASBESTOS-COVERED

This type of cable is used where there is considerable heat. For example, it should always be used for making the connections to an immersion heater.

GLASS FIBRE-INSULATED LIGHTING CABLES

Modern electric lamps have a much smaller glass bulb for a given rating than those manufactured a few years ago. This often results in a higher-wattage lamp being put into an existing fitting which may be supplied through a natural rubber or P.V.C.-covered cable. The use of a higher-powered lamp may result in the temperature of the lampholder terminals, and therefore the cable close to it, reaching values approaching 160°C. Fairly rapid deterioration of the cable occurs under these conditions of operation.

Some cables are now made using insulation of glass fibre and silicone varnish and these will operate satisfactorily at temperatures up to 180°C. It is good practice to utilise cables designed to withstand these temperatures when wiring lighting fittings.

MINERAL INSULATED COPPER-COVERED CABLES

M.I.C.C. cables are used where their mechanical strength and resistance to fire is of importance. They will operate continuously at quite high temperatures or under heavy current overload.

The outer covering is a copper tube and from one to seven solid copper conductors inside the tube are separated from each other by magnesium oxide. The original outer diameter of the tube is about 2 in. and the conductors are of such proportions that when the whole is passed through a considerable number of drawing passes, any size of conductor down to 0·0015 in^2 cross-sectional area may be obtained. Frequent annealing is necessary during manufacture as the drawing operation work-hardens the copper.

CHAPTER 22

Fuses

22.1. General

A fuse is a device which protects an electrical installation should a fault develop and cause an excessive current. If this excessive current were not disconnected the conductors carrying the current would heat up and the insulation would be damaged. In extreme cases, the heat generated in the cable conductors and/or apparatus may be large enough to cause a fire and many such fires occur each year. Also, a fuse protects human life from electrocution under faulty conditions.

Heat generated in any resistor (any conductor possesses some resistance) increases rapidly with increase in current magnitude. Heat energy in a resistor is given by the expression I^2Rt. Thus if a current is 10 times the normal permitted value, the heat developed in a given time will be $10^2 = 100$ times as great.

A fuse in a circuit is, in effect, the weak link in a chain. It depends for its operation on the production of heat to raise the temperature of the fuse element to melting point when the current is excessive.

The simplest type of fuse is the semi-enclosed rewirable porcelain type which is often found in private houses. Tinned copper is frequently used as the fuse element, the coating of tin providing a resistance to oxidation under normal operating conditions. The oxidation would in time reduce the cross-sectional area of the fuse and so reduce the value of the fusing current. These fuses are easily replaced and fuse replacement is cheap. However, they are likely to deteriorate with age and they may fail under normal operating conditions. They are also limited in use to fairly low currents and voltages.

During 1920–30 the electrical system rapidly grew so that the fault levels increased. Fault level can be expressed in a simple form as a measure of the current that would flow under fault conditions if the supply were not interrupted. If two 12-V batteries were connected in parallel, the current available if a short was placed on the batteries would be about twice as great as with only one battery.

Under these increased fault level conditions wire fuses and a later development, the cartridge fuse, often failed explosively and the uncontrolled arcing caused more damage than did the fault. Nearby equipment was, in some cases, also damaged, resulting in disconnection of healthy supplies. There was also the risk of fire and damage to personnel.

The H.R.C. fuse was introduced to overcome the disadvantages of the earlier fuses and Fig. 22.1 shows the construction of a modern fuse of this type.

There may be several elements in parallel and these are made from pure silver wire. Two sections of reduced current-carrying capacity are in series in the middle of the cartridge body. Heat produced under normal working conditions is quickly conducted away by the heavier element sections and by the granular filler which completely fills the remaining space in the cartridge body.

Under fault conditions, the reduced sections of the fuse becomes volatilised away. Most of the heat produced in the arc is absorbed in the thermo-chemical reaction between the silver vapour and the granular filler. The central operating elements in the fuse are quickly replaced by an insulating material having a high dielectric strength. As the fault current increases, the heat within the fuse increases and the reaction becomes more rapid.

Breaking capacity is a measure of the fault current with which the fuse is capable of dealing. For example, a 60-A fuse to British Standard 1361 (B.S. 1361) has a breaking capacity of 33,000 A at 0·3 power factor, i.e. it would interrupt the short-circuit current in a circuit where it could reach a maximum value of 33,000 A if not prevented from doing so by breaking the circuit. There is no emission of smoke or flame from these fuses when they operate on short-circuit.

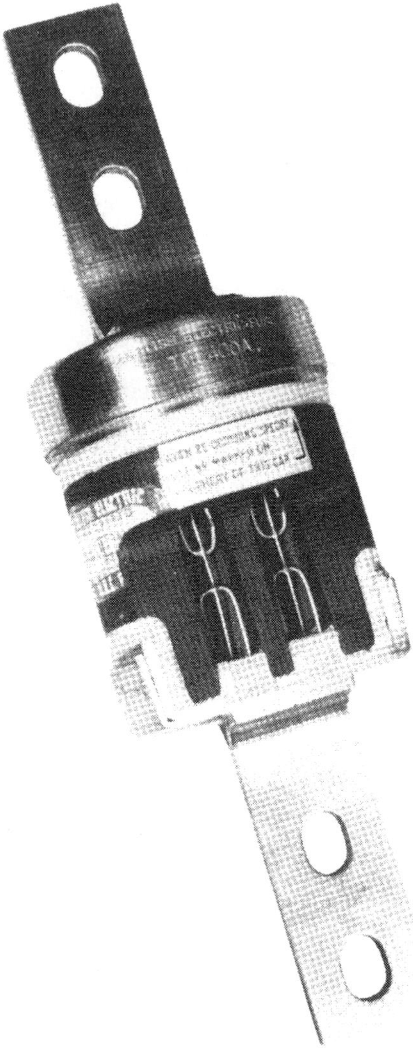

FIG. 22.1. H.R.C. fuse-link. (English Electric Co. Ltd.)

A fuse is designed so that the arrangement of the cartridge and filler is able to absorb the energy released by the fault. The energy required to melt a fuse is almost constant for very short periods of high current. It therefore follows that irrespective of the possible size of the final fault current, the fuse will, in many cases, limit the current to some predetermined value if a short-circuit takes place. This value is called the cut-off current.

A fuse must be able to withstand all the stresses set up when it is operating and so the containing tube and end caps must have the necessary mechanical strength. It must also not heat up when it is carrying its normal current nor must it operate with temporary over-currents (called transient currents) which may occur as, for example, when a motor is starting.

In most cases during operation, a voltage is set up across the fuse which tries to maintain the arc. This is an example of the application of Lenz's law. A well-designed fuse must in the later stages of arcing, build up its dielectric strength more rapidly than the build up of this recovery voltage.

Fuses in a system are, of course, only of value when a fault occurs and this is a condition which rarely happens. The fuse must, then, maintain its designed operating characteristics over a long period of time, remaining unaffected by all types of loading, atmospheric conditions, vibrations, etc. It has been verified that fuses having had nearly 30 years of continuous duty were still in a satisfactory condition.

Nevertheless, if a fuse is made to carry a current in excess of its normal rating for long continuous periods, then some deterioration may result. Most H.R.C. fuses have a fusing factor of either 1·25 or 1·6. This means that the minimum fusing current is 1·25 or 1·6 times the rated normal current depending on the type of fuse used.

Fuses must also possess the property of discrimination under fault conditions. Frequently in practice, a supply to a point may pass through two fuses in series. For example, suppose a 100-A fuse supplies a distribution board from which several supplies are taken through 30-A fuses. It is essential that if a fault occurred in one of the 30-A supplies that this fuse should blow and not the main 100-A fuse. A fuse can be designed to meet these requirements but external circuit factors can influence the discrimination. H.R.C. fuses have a discrimination comparable with other devices.

An important feature of cartridge fuses is that they have what is known as an "inverse-time characteristic". This means that the greater the value of the fault current in the circuit the more quickly will the fuse operate and break the circuit. Any damage caused by the excess current is therefore kept to a minimum.

A special application of H.R.C. fuses is the protection of semiconductor rectifiers. These rectifiers are small in size and have a relatively high output. Because of their small mass, they soon overheat due to excessive currents flowing for a short time, thus resulting in permanent damage. Also these rectifiers will not withstand very great over-voltages. Thus any protection device must react quickly to over-currents but must not generate an arc voltage during operation which would break down the reverse voltage of the rectifier.

The H.R.C. fuse is the only device that will offer this protection and special fuses have been designed which will fulfil the requirements stated above. Such fuses have various voltage ratings as well as different current ratings. They are much smaller than the normal industrial fuses. For example, an English Electric Co. Ltd. Industrial fuse rated up to 300 A weighs 470 g whereas a fuse rated up to 350 A for semiconductor rectifier protection weighs 152 g.

H.R.C. fuses are widely used throughout industry for the protection of distribution systems, cables, busbars, etc. They also have quite a wide use in marine work and are becoming increasingly popular in aircraft installations in view of the greater fault levels in modern aircraft.

CHAPTER 23

Thermionic Emission

23.1. Introduction

Thermionic emission is concerned with the production of charged particles at a heated surface. By far the most extensive application lies in the field of thermionic valves. Their use in radio and television receivers has in many cases been superseded by transistors.

In 1883 Thomas A. Edison found that an electric current would flow in the evacuated space between a heated metallic plate (cathode) and a nearly cool metallic plate (anode), when the latter was made positive with respect to the former. No noticeable effect occurred if the anode was negative with respect to the cathode. Near the end of the nineteenth century it was confirmed that the electrical charges given off from the cathode were the negatively charged particles which are now termed electrons.

Originally, the cathode was a filament made of a material such as tungsten and it was heated directly by passing a current through it. In modern valves, the heated filament is surrounded by a nickel cylinder, on the outside surface of which is sprayed a layer of a mixture of barium and strontium carbonates. This breaks down because of heat treatment given whilst the valve is being evacuated and it gives a surface on the cathode having a greater electron emission at a lower temperature than the original tungsten metal.

23.2. Space charge

Great care is taken in the construction of a valve to ensure that all gases have been removed so preventing contamination of the

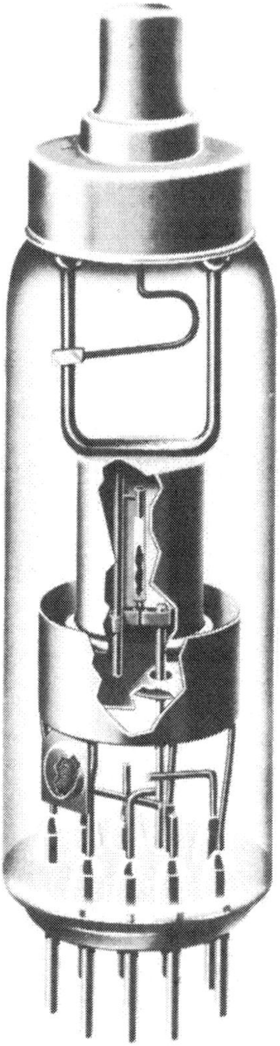

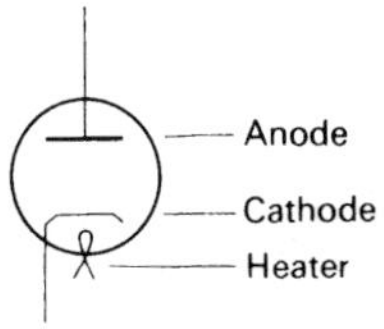

(b)

(a) DY 86 high-voltage half-wave rectifier. (Mullard Ltd.)

FIG. 23.1

cathode and also avoiding the possibility of gaseous discharge. The valve is said to have become soft if gas has entered and it may even "glow" during use.

When the cathode is heated, the heat energy results in the energy of certain electrons being increased and large numbers of them leave the surface of the cathode and enter the adjacent space. The loss by the atoms of some of their negative charges leaves the cathode with a resultant positive charge so the electrons are ultimately attracted back to the cathode. A SPACE CHARGE "cloud" of electrons forms round the cathode with a continuous intercharge of electrons between the "cloud" and cathode. The number of electrons in this space charge is a function of the temperature of the electrode.

Figure 23.1(a) shows the construction of a modern high-voltage diode valve whilst the sketch in Fig. 23.1(b) shows how this would be represented in a circuit diagram.

If the anode is made positive with respect to the cathode, the resultant electric field will induce electrons to leave the space charge and be attracted towards the anode. No electron emission normally takes place at the anode because the valve is designed to operate so that the anode does not reach a temperature sufficiently high to cause it to emit electrons.

The stronger the electric field existing between the anode and cathode, i.e. the more positive the anode is made with respect to the cathode, the more electrons there are leaving the space charge and crossing to the anode. Ultimately, the number of electrons being emitted by the cathode equals the number arriving at the anode. The current will then remain almost constant, even if the voltage between the anode and cathode is still further increased. Saturation is said to have occurred, but this does not normally happen if the value operates within the voltage limits specified by the manufacturer. Figure 23.2 illustrates the effect of cathode temperature on the anode current/anode voltage characteristic of a GRD 7 valve which is designed for use in laboratory experiments. (Note: Degrees Absolute $°K = °C + 273$.)

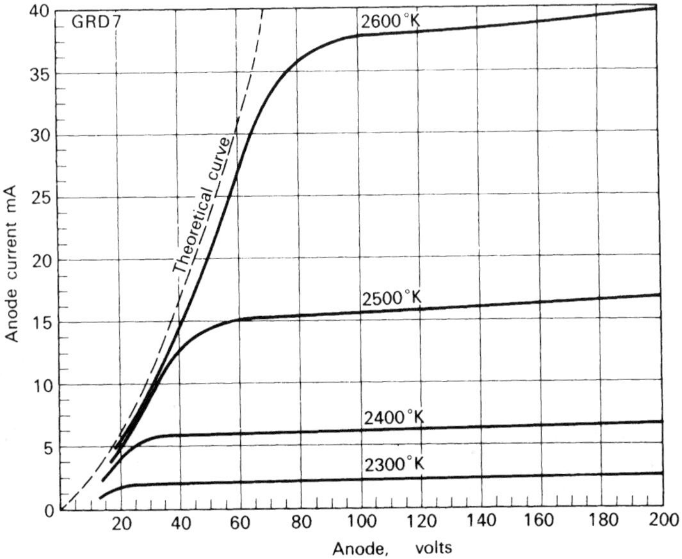

FIG. 23.2. Effect of cathode temperature on anode current/anode voltage characteristic. (Ferranti Ltd.)

23.3. Experimental determination of diode characteristics

The circuit diagram (Fig. 23.3) shows a circuit that could be used to obtain the results given in the worked example on page 375. The method of connecting the voltmeter is very important.

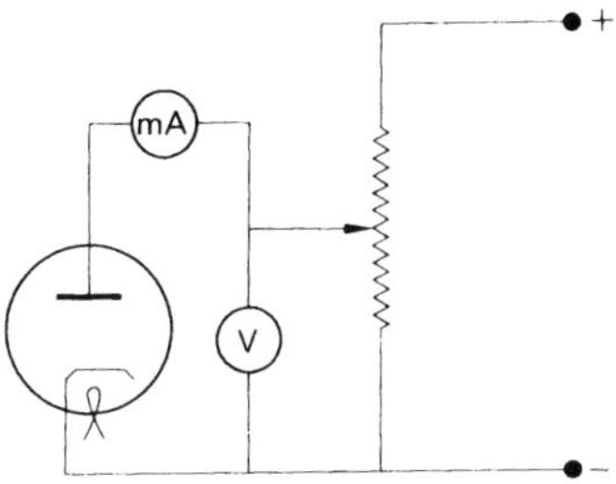

FIG. 23.3

A voltmeter takes a current of a few milliamperes and this current may be as great as that taken by the valve. Hence great care must be taken to ensure that the voltmeter is connected as shown, so that its current does not pass through the milliammeter.

23.4. Valve resistances

The resistance of a valve may be expressed under conditions of fluctuating voltages and currents or for a fixed operating voltage and current. In the former case, the term used is the anode slope resistance and in the latter case it is called the direct current (d.c.) resistance.

EXAMPLE. A diode valve has the following anode voltage/anode current characteristic:

Anode volts (V)	0	10	20	30	40	50
Anode current (mA)	0	3·2	9·6	22·0	44·7	73·0

Plot the above characteristic and use it to find the anode slope resistance and the d.c. resistance when the anode voltage is 35 V.

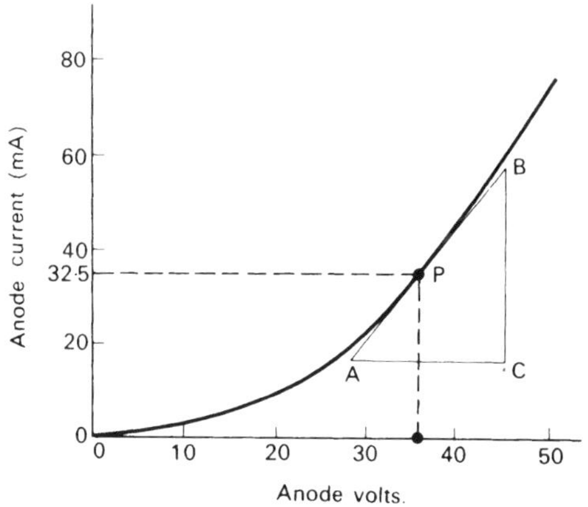

FIG. 23.4

Plot the curve with the anode current on the vertical axis (Fig. 23.4). Find the point on the curve corresponding to 35 V (point *P*). Draw a tangent *AB* to the curve at this point *P* and construct the right-angled triangle *ACB*.

$$\text{Anode slope resistance} = \frac{\text{Change in anode volts}}{\text{Corresponding change in anode current}}$$

$$= \frac{AC \times 10^3}{BC} \text{ ohms.}$$

(N.B. The 10^3 is necessary to convert the milliamperes to amperes.)

The reader should verify that the slope resistance is about 440 Ω.

The d.c. resistance at point *P*

$$= \frac{\text{Anode volts } (V_a) \text{ at point } P}{\text{Anode current } (I_a) \text{ at point } P}$$

$$= \frac{35}{32 \cdot 5 \times 10^{-3}}\ \Omega$$

$$= \mathbf{1080\ \Omega}.$$

Printed in Dunstable, United Kingdom

91338428R00221